Electrical Distribution Systems

Dale R. Patrick
Stephen W. Fardo

Published by
THE FAIRMONT PRESS, INC.
700 Indian Trail
Lilburn, GA 30047

Library of Congress Cataloging-in-Publication Data

Patrick, Dale R.
Electrical distrubution systems / Dale R. Patrick, Stephen W. Fardo.
p. cm.
Includes bibliographical references and index.
ISBN 0-88173-252-4
1. Electric power distrubution. 2. Electric power systems.
I. Fardo, Stephen W. II. Title.
TK3001.P37 1998 621.319--dc21 98-36065
CIP

Published by The Fairmont Press, Inc.
700 Indian Trail
Lilburn, GA 30047

Printed in the United States of America

10 9 8 7 6 5 4 3 2 1

ISBN 0-88173-252-4 FP

ISBN 0-13-011565-7 PH

Distributed by Prentice Hall PTR
Prentice-Hall, Inc.
A Simon & Schuster Company
Upper Saddle River, NJ 07458

Prentice-Hall International (UK) Limited, London
Prentice-Hall of Australia Pty. Limited, Sydney
Prentice-Hall Canada Inc., Toronto
Prentice-Hall Hispanoamericana, S.A., Mexico
Prentice-Hall of India Private Limited, New Delhi
Prentice-Hall of Japan, Inc., Tokyo
Simon & Schuster Asia Pte. Ltd., Singapore
Editora Prentice-Hall do Brasil, Ltda., Rio de Janeiro

Contents

Preface

Electrical Distribution Systems is intended as an introductory guidebook for self study or for use as a textbook in technical programs in electrical technology at vocational-technical schools, industrial training programs or college technical programs. The book uses a "systems" format to teach electrical distribution and associated power system concepts. Key concepts are presented by stressing applications-oriented theory. Through this approach, the student is not burdened with an abundance of information needed only for engineering design. "Real world" applications and operations are stressed throughout the book Mathematical problems are solved by basic algebraic and trigonometric applications.

There are few texts on the market dealing with the topic of electrical distribution systems which are applications-oriented. Some texts are available which cover engineering design of systems; however, the two-year vocational-technical and college market has been neglected. There seems to be a demand for a text dealing with electrical distribution and associated power system operation from a user's or technician's point of view.

Concepts are presented in this book through an "electrical power systems" model which includes power distribution as a key element. The other subsystems of this model are important associated systems for a comprehensive understanding of electrical distribution systems. The five subsystems of the "electrical power systems" model include: Electrical Power Production, Electrical Power Distribution, Electrical Power Control, Electrical Power Conversion, and Electrical Power Measurement.

A limited understanding of basic electrical terms is assumed in the organization of this book. However Appendix A—Important Terms—is included to provide assistance in defining basic electrical terms which may be used when dealing with electrical power systems.

Through this comprehensive "systems" approach, the reader will gain a more complete understanding of electrical distribution systems. The authors have used this instructional method in teaching classes dealing with electrical power systems for over 25 years in a large university technical program.

Dale R. Patrick
Stephen W. Fardo
Eastern Kentucky University
Richmond, KY 40475

Preface

Chapter 1

Power System Fundamentals

One of the most important areas of electrical knowledge is the study of electrical power systems. Complex transmission and distribution systems supply the vast need of our country for electrical power. Due to this tremendous power requirement, we must constantly be concerned with the efficient operation of our power distribution and associated systems.

BRIEF HISTORY OF ELECTRICAL POWER SYSTEMS

Electrical power systems have been in existence for many years. The applications of power systems have expanded rapidly since their development. At the present time, applications continue to increase, placing additional requirements on power production, distribution systems and associated systems.

Thomas Edison is given credit for developing the concept of widespread generation and distribution of electrical power. He performed developmental work on direct-current (DC) generators which were driven by steam engines. Edison's work with electrical lights and power production led the way to development of electric motors, distribution systems and associated control equipment.

Most early discoveries related to electrical power dealt with direct-current (DC) systems. Alternating-current power generation became widespread a short time later. The primary reason for converting to AC power production and distribution was that *transformers* could be used to increase AC voltage levels for long-distance distribution of electrical power. Thus the discovery of transformers allowed the conversion of electrical power from DC to AC systems. Presently, almost all electrical power systems produce and distribute *three-phase* alternating current. Transformers allow the voltage produced by AC generators to be increased while decreasing current level by a corresponding amount. This allows long-distance electrical power distribution at a reduced current level, reduces power losses, and increases overall power system efficiency. The increased use of electrical motors for home appliances and industrial and commercial equipment has increased the need for electrical power to be distributed to various locations.

In the early days of electrical power, the distribution systems were only an extension of the power generating plant. There was little planning for the efficient transfer of energy from the generating plant to the limited number of consumers. The expansion of electrical energy use has placed greater demands on the distribution system. Not only are more customers served, but today's equipment requires closer attention to voltage variation and little toleration of service interruption.

The design and operation of electrical power distribution systems has become a very important science. Well-engineered power systems of today are connected together in such a way that if a problem occurs in one system, it can be supplemented by another system. Electrical loads can be transferred easily from one system to another. The United States has a very reliable "grid" system which maintains electrical power to customers at the proper voltage level without interruption. It is extremely rare for "blackouts" or "brownouts" to occur. These conditions are avoided by proper planning for situations of extremely high demand. A blackout is a complete interruption of electrical power, while a brownout is a reduction of voltage level to the consumer. A brownout could be purposely done in order to deliver available power at a reduced voltage to avoid a blackout during a problem of extremely high demand. High demand usually occurs during abnormally hot or cold temperatures over an extended period of time.

Early power distribution systems supplied direct current (DC) at low voltage levels over relatively short distances. The invention of the transformer and the problems associated with delivering power over long distances brought about a change to the use of alternating current (AC) power systems. Today, greater electrical power demand can be supplied with long-distance, high voltage transmission. Voltage levels may be easily increased and reduced by transformers in order to supply electrical energy.

Not only has the efficiency of the electrical power distribution system been improved, but also the materials, equipment, and associated control systems have been continually updated. Examples of such improvement include the quality of steel towers, wood poles of long-lasting design, better conductors and insulators, and more reliable computer systems for monitoring and controlling the electrical distribution system.

THE SYSTEM CONCEPT

For a number of years, people have worked with jigsaw puzzles as a source of recreation. A jigsaw puzzle contains a number of discrete parts that must be placed together properly to produce a picture. Each part then plays a specific role in the finished product. When a puzzle is first started, it is difficult to imagine the finished product without seeing a representative picture.

Studying a complex field such as electrical power systems by discrete parts poses a problem that is somewhat similar to the jigsaw puzzle. In this case, it is difficult to determine the role that a discrete part plays in the operation of a complex system. A picture of the system divided into its essential parts therefore becomes an extremely important aid in understanding its operation.

The system concept will serve as the "big picture" in the study of electrical power systems. In this approach, a system will first be divided into a number of essential blocks. The role played by each block then becomes more meaningful in the operation of the overall system. After the location of each block has been established, discrete component operation related to each block then becomes more relevant. Through this approach, the way in which some of the "pieces" of electrical systems fit together should be more apparent.

BASIC SYSTEM FUNCTIONS

The word *system* is commonly defined as an organization of parts that are connected together to form a complete unit. There are a wide variety of electrical systems used today. Each system has a number of unique features, or characteristics, that distinguish it from other systems. More importantly, however, there is a common set of parts found in each system. These parts play the same basic role in all systems. The terms energy *source,* transmission *path, control, load,* and *indicator* are used to describe the various system parts. A block diagram of these basic parts of the system is shown in Figure 1-1.

Each block of a basic system has a specific role to play in the overall operation of the system. This role becomes extremely important when a detailed analysis of the system is to take place. Hundreds and even thousands of discrete components are sometimes needed to achieve a specific block function. Regardless of the complexity of the system, each block must still achieve its function in order for the system to be operational. Being familiar with these functions and being able to locate them within a complete system is a big step toward understanding the operation of the system.

The *energy source* of a system converts energy of one form into something more useful. Heat, light, sound, chemical, nuclear, and mechanical energy are considered as primary sources of energy. A primary energy source usually goes through an energy change before it can be used in an operating system.

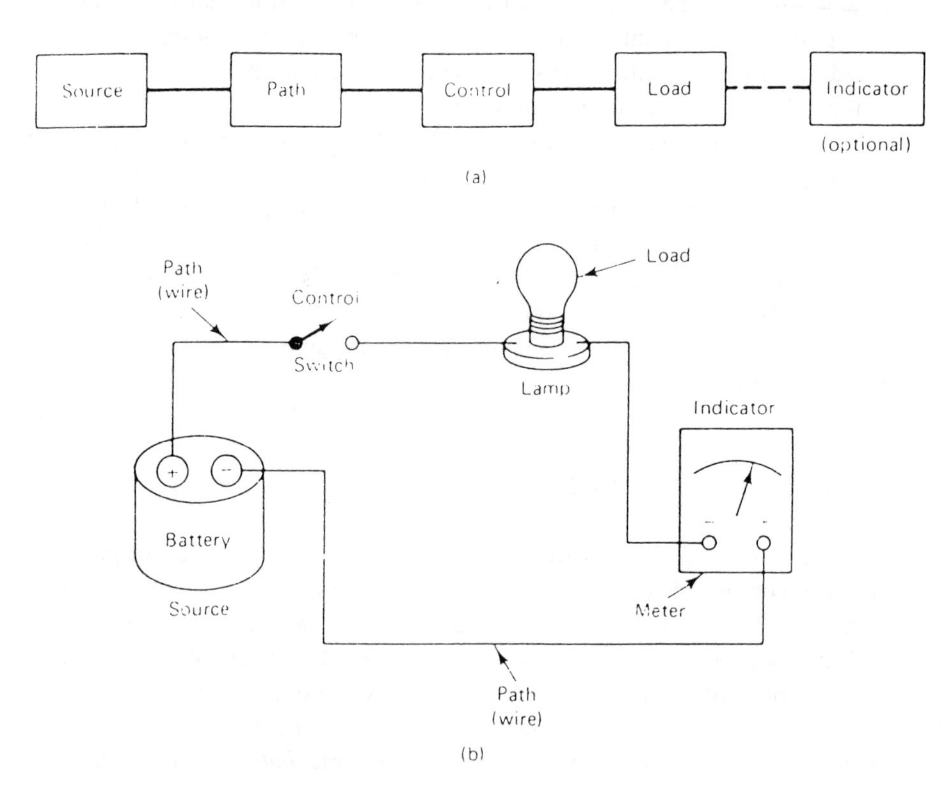

Figure 1-1. Electrical system: (a) block diagram; (b) pictorial diagram.

The *transmission path* of a system is somewhat simplified when compared with other system functions. This part of the system simply provides a path for the transfer of energy. It starts with the energy source and continues through the system to the load. In some cases, this path may be a single electrical conductor, light beam, or other medium between the source and the load. In other systems, there may be a supply line between the source and the load and a return line from the load to the source. There may also be a number of alternate or auxiliary paths within a complete system. These paths may be *series* connected to a number of small load devices or *parallel* connected to many independent devices.

The *control* section of a system is by far the most complex part of the entire system. In its simplest form, control is achieved when a system is turned on or off. Control of this type can take place anywhere between the source and the load device. The term *full control* is commonly used to describe this operation. In addition to this type of control, a system may also employ some type of *partial control*. Partial control usually causes some type of an operational change in the system other than an on or off condition. Changes in electric current or light intensity are examples of alterations achieved by partial control.

The *load* of a system refers to a specific part or number of parts designed to produce some form of work. Work, in this case, occurs when energy goes through a transformation or change. Heat, light, chemical action, sound, and mechanical motion are some of the common forms of work produced by a load device. As a general rule, a very large portion of all energy produced by the source is consumed by the load device during its operation. The load is typically the most prevalent part of the entire system because of its obvious work function.

The *indicator* of a system is primarily designed to display certain operating conditions at various points throughout the system. In some systems the indicator is an optional part, whereas in others it is an essential part in the operation of the system. In the latter case, system operations and adjustments are usually critical and are dependent upon specific indicator readings. The term *operational indicator* is used to describe this application. *Test indicators* are also needed to determine different operating values. In this role, the indicator is only temporarily attached to the system to make measurements. Test lights, meters, oscilloscopes, chart recorders, and digital display instruments are some of the common indicators used.

A SIMPLE ELECTRICAL SYSTEM EXAMPLE

A flashlight is a device designed to serve as a light source in an emergency or as a portable light source. In a strict sense, flashlights can be classified as portable electrical systems. They contain the four essential parts needed to make this classification. Figure 1-2 is a cutaway drawing of a flashlight, with each component part shown associated with its appropriate system block.

The battery of a flashlight serves as the primary *energy source* of the system. Chemical energy of the battery must be changed into electrical energy before the system becomes operational. The flashlight is a synthesized system because it utilizes two distinct forms of energy in its operation. The energy source of a flashlight is an expendable item. It must be replaced periodically when it loses its ability to produce electrical energy.

The *transmission path* of a flashlight is commonly achieved via a metal case or through a conductor strip. Copper, brass, and plated steel are frequently used to achieve this function.

The *control* of electrical energy in a flashlight is achieved by a slide switch or a push-button switch. This type of control simply interrupts the transmission path between the source and the load device Flashlights are primarily designed to have full control capabilities. This type of control is achieved manually by the person operating the system.

The *load* of a flashlight is a small incandescent lamp. When electrical energy from the source is forced to pass through the filament of the lamp, the lamp produces a bright glow. Electrical energy is first changed into heat and then into light energy. A certain amount of work is achieved by the lamp when this energy change takes place.

The energy transformation process of a flashlight is irreversible. It starts at the battery when chemical energy is changed into electrical energy. Electrical energy is then changed into heat and eventually into light energy by the load device. This flow of energy is in a single direction. When light is eventually, produced, it consumes a large portion of the electrical energy coming from the source. When this energy is exhausted, the system becomes inoperative. The battery cells of a flashlight require periodic replacement in order to maintain a satisfactory operating condition.

Flashlights do not ordinarily employ a specific *indicator* as part of the system. Operation is indicated when the lamp produces light. In a

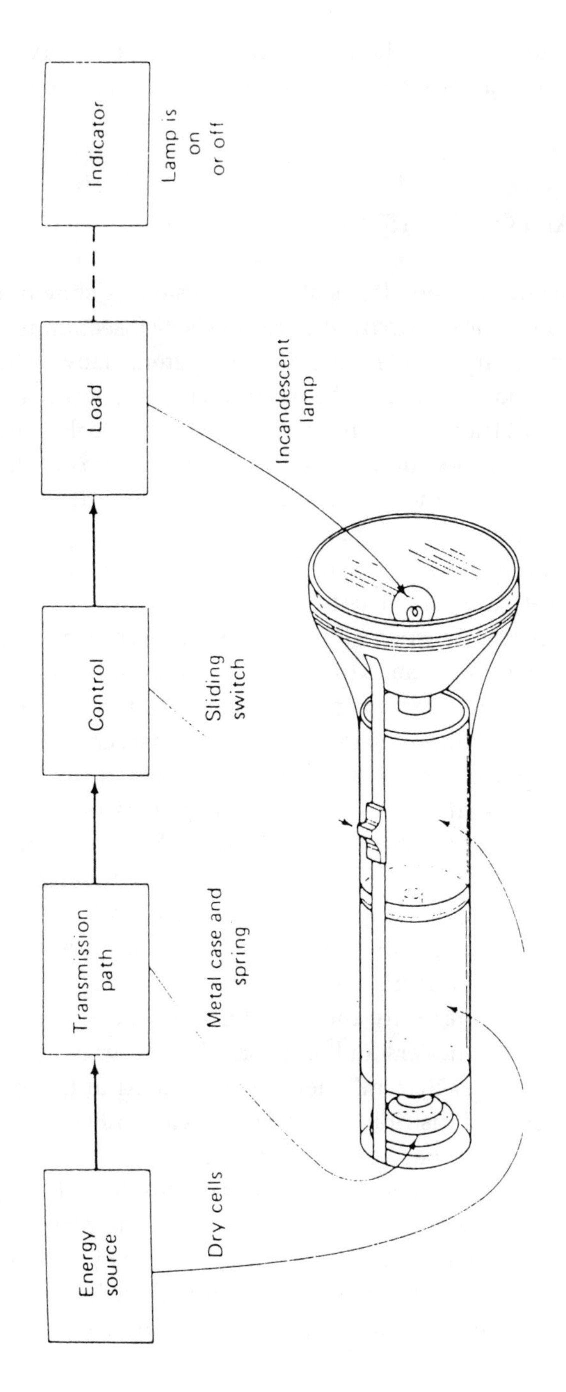

Figure 1-2. Cutaway drawing of a flashlight.

strict sense, we could say that the load of this system also serves as an indicator. In some electrical systems the indicator is an optional system part.

THE ELECTRICAL POWER SYSTEM

The block diagram of an electrical power system is shown in Figure 1-3. The first block or the *electrical power production* section is an important part of the complete electrical power system. However, once electrical power is produced, it must be distributed to the location where it will be used, so electrical *power distribution* systems (block 2) transfer electrical power from one location to another. Electrical *power control* systems (block 3) are probably the most complex of all the parts of the electrical power system as there are unlimited types of devices and equipment used to control electrical power. Then, the electrical *power conversion* systems (block 4), also called *loads,* convert the electrical power into some other form of energy, such as light, heat, or mechanical energy. Thus, conversion systems are an extremely important part of the electrical power system. Another part of the electrical power system is *power measurement* (block 5). Without electrical power measurement systems, control of electrical power would be almost impossible.

Each of the blocks shown in Figure 1-3 represents one important part of the electrical power system. Thus, we should be concerned with each part of the electrical power system rather than only with isolated parts. In this way, we can develop a more complete understanding of how electrical power systems operate. This type of understanding is needed to help us solve our energy problems that are related to electrical power. We cannot consider only the distribution aspect of electrical power systems. We must understand and consider each pan of the system. The "Electrical Power System" model will be used in this book to help understand electrical distribution systems. Refer to Figure 1-3 as a reference as you study the chapters of this book.

Figure 1-4 shows the generation and transmission of electrical power as an example. Power is produced at a generating plant (*source*). *Distribution* occurs between the plant and the consumer by power lines. Transformers are used to *control* the voltage and current levels. *Conversion* of electrical power to another form (light, heat, mechanical) occurs at the home.

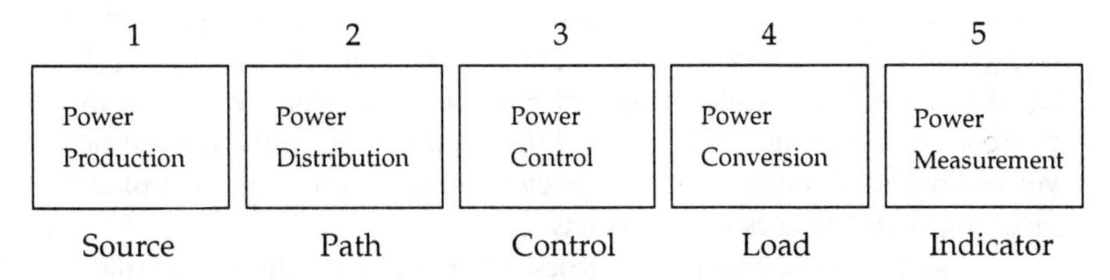

Figure 1-3. The Electrical power system model.

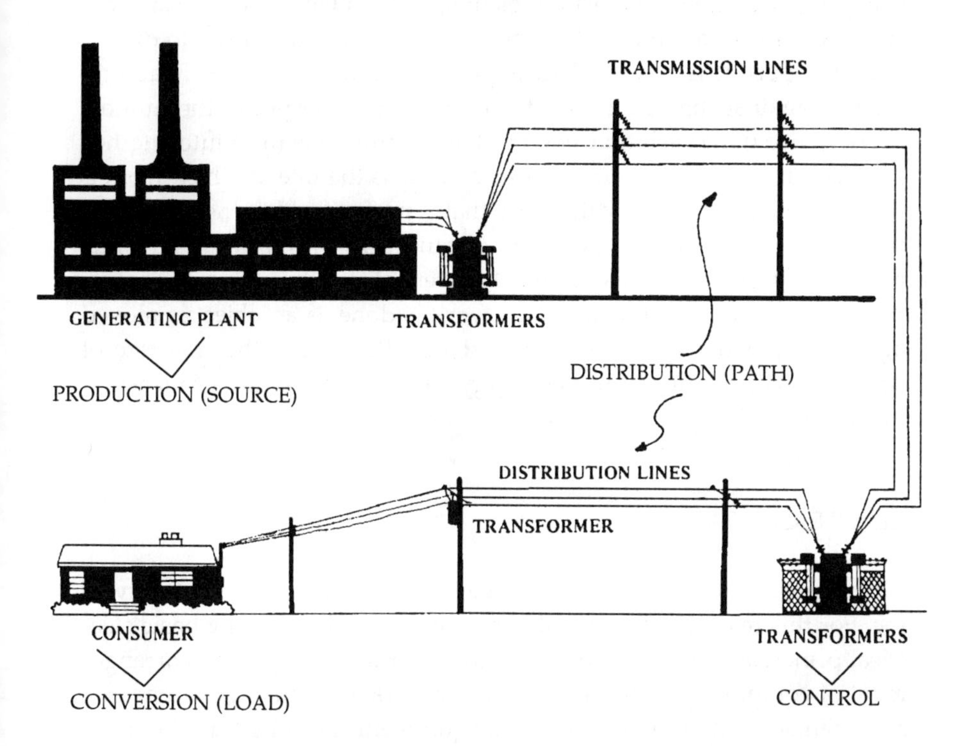

Figure 1-4. The generation and transmission of electrical power.

ENERGY, WORK, AND POWER

An understanding of the terms energy, work, and power is necessary in the study of electrical power systems. The first term, *energy,* means the capacity to do work. For example, the capacity to light a light bulb, to heat a home, or to move something requires energy. Energy exists in many forms, such as electrical, mechanical, chemical, and heat. If energy exists because of the movement of some item, such as a ball rolling down a hill, it is called *kinetic energy*. If energy exists because of the position of something, such as a ball that is at the top of the hill but not yet rolling, it is called *potential energy.* Energy has become one of the most important factors in our society.

A second important term is work. *Work* is the transferring or transforming of energy. Work is done when a force is exerted to move something over a distance against opposition, such as when a chair is moved from one side of a room to the other. An electrical motor used to drive a machine performs work. Work is performed when motion is accomplished against the action of a force that tends to oppose the motion. Work is also done each time energy changes from one form into another.

A third important term is power. *Power* is the rate at which work is done. It considers not only the work that is performed but the amount of time in which the work is done. For instance, electrical power is the rate at which work is done as electrical current flows through a wire. Mechanical power is the rate at which work is done as an object is moved against opposition over a certain distance. Power is either the rate of production or the rate of use of energy. The *watt* is the unit of measurement of electrical power.

ELECTRICAL SAFETY

Electrical safety is very important. Many dangers are not easy to see. For this reason, safety should be based on understanding basic electrical principles. Common sense is also important. The physical arrangement of equipment in the electrical lab or work area should be done in a safe manner. Well designed electrical equipment should always be used. For economic reasons, electrical equipment is often improvised. It is important that all equipment be made as safe as possible. This is especially true for equipment and circuits that are designed and built in the lab or shop.

Work surfaces in the shop or lab should be covered with a material that is nonconducting and the floor of the lab or shop should also be nonconducting. Concrete floors should be covered with rubber tile or linoleum. A fire extinguisher that has a nonconducting agent should be placed in a convenient location. Extinguishers should be used with caution. Their use should be explained by the teacher.

Electrical circuits and equipment in the lab or shop should be plainly marked. Voltages at outlets require special plugs for each voltage. Several voltage values are ordinarily used with electrical lab work. Storage facilities for electrical supplies and equipment should be neatly kept. Neatness encourages safety and helps keep equipment in good condition. Tools and small equipment should be maintained in good condition and stored in a tool panel or marked storage area. Tools that have insulated handles should be used. Tools and equipment plugged into convenience outlets should be wired with three-wire cords and plugs. The purpose of the third wire is to prevent electrical shocks by grounding all metal parts connected to the outlet.

Soldering irons are often used in the electrical shop or lab. They can be a fire hazard. They should have a metal storage rack. Irons should be unplugged while not in use. Soldering irons can also cause burns if not used properly. Rosin-core solder should always be used in the electrical lab or shop.

Adequate laboratory space is needed to reduce the possibility of accidents. Proper ventilation, heat, and light also provide a safe working environment. Wiring in the electrical lab or shop should conform to specifications of the National Electrical Code (NEC). The NEC governs all electrical wiring in buildings.

Lab or Shop Practices

All activities should be done with low voltages whenever possible. Instructions should be written, with clear directions, for performing lab activities. All lab or shop work should emphasize safety. Experimental circuits should always be checked before they are plugged into a power source. Electrical lab projects should be constructed to provide maximum safety when used.

Electrical equipment should be disconnected from the source of power before working on it. When testing electronic equipment, such as TV sets or other 120-V devices, an isolation transformer should be used. This isolates the chassis ground from the ground of the equipment and eliminates the shock hazard when working with 120-V equipment.

Electrical Hazards

A good first-aid kit should be in every electrical shop or lab. The phone number of an ambulance service or other medical services available should be in the lab or work area in case of emergency. Any accident should be reported immediately to the proper school officials. Teachers should be proficient in the treatment of minor cuts and bruises. They should also be able to apply artificial respiration. In case of electrical shock, when breathing stops, artificial respiration must be immediately started. Extreme care should be used in moving a shock victim from the circuit that caused the shock. An insulated material should be used so that someone else does not come in contact with the same voltage. It is not likely that a high-voltage shock will occur. However, students should know what to do in case of emergency.

Normally, the human body is not a good conductor of electricity. When wet skin comes in contact with an electrical conductor, the body is a better conductor. A slight shock from an electrical circuit should be a warning that something is wrong. Equipment that causes a shock should be immediately checked and repaired or replaced. Proper grounding is important in preventing shock.

Safety devices called ground-fault circuit interrupters (GFCIs) are now used for bathroom and outdoor power receptacles. They have the potential of saving many lives by preventing shock. GFCIs immediately cut off power if a shock occurs. The National Electric Code specifies where GFCIs should be used.

Electricity causes many fires each year. Electrical wiring with too many appliances connected to a circuit overheats wires. Overheating may set fire to nearby combustible materials. Defective and worn equipment can allow electrical conductors to touch one another to cause a short circuit, which causes a blown fuse. It could also cause a spark or arc which might ignite insulation or other combustible materials or burn electrical wires.

Fuses and Circuit Breakers

Fuses and circuit breakers are important safety devices. When a fuse "blows," it means that something is wrong in the circuit. Causes of blown fuses could be:

1. A short circuit caused by two wires touching
2. Too much equipment on the same circuit
3. Worn insulation allowing bare wires to touch grounded metal objects such as heat radiators or water pipes

After correcting the problem a new fuse of proper size should be installed. Power should be turned off to replace a fuse. Never use a makeshift device in place of a new fuse of the correct size. This destroys the purpose of the fuse. Fuses are used to cut off the power and prevent overheated wires.

Circuit breakers are now very common. Circuit breakers operate on spring tension. They can be turned on or off like wall switches. If a circuit breaker opens, something is wrong in the circuit. Locate and correct the cause and then reset the breaker.

Always remember to use common sense whenever working with electrical equipment or circuits. Safe practices should be followed in the electrical lab or shop as well as in the home. Detailed safety information is available from the National Safety Council and other organizations. *It is always wise to be safe.*

Chapter 2

Basics of Electrical Circuits

There are several basic fundamentals of electrical power systems. Therefore, the basics must be understood before attempting an in-depth study of electrical power systems. The types of circuits associated with electrical systems are either resistive, inductive, or capacitive. Most systems have some combination of each of these three circuit types. These circuit elements are also called *loads*. A load is a part of a circuit that converts one type of energy into another type. A resistive load converts electrical energy into heat energy.

In our discussions of electrical circuits, we will primarily consider alternating-current (AC) systems at this time as the vast majority of the electrical power which is produced is alternating current. A review of AC circuits is also included here. Before reviewing electrical circuits, you may want to look at Appendix A—Important Terms—and Appendix B—Electrical Symbols.

REVIEW OF DIRECT CURRENT (DC) ELECTRICAL CIRCUITS

To understand electrical power systems, it is necessary to know how to apply basic electrical theory. Electrical power is a somewhat mathematical subject. The mathematics is easy to understand, since it has practical applications that are easy to see. The basic theory used is called *Ohm's law*. Ohm's law should be learned, as it applies to the basic theory of electrical circuits. The examples which follow are DC circuits. AC circuits are more complex.

Ohm's law is the most basic and most used of all electrical theories. Ohm's law explains the relationship of voltage (the force that causes current to flow), current (the movement of electrons), and resistance (the opposition to current flow). Ohm's law is stated as;, follows: An increase in voltage increases current if resistance remains the same. Ohm's law stated in another way is: An increase in resistance causes a decrease in current if voltage remains the same. The electrical values used with Ohm's law are usually represented with capital letters. For example, voltage is represented by the letter *V* current, by the letter *I*, and resistance, by the letter *R*. The mathematical relationship of the three electrical units is shown in the following formulas. These should be memorized. The Ohm's law circle of Figure 2-1 is helpful in remembering the formulas.

$$V = I \times R$$

$$I = \frac{V}{R}$$

$$R = \frac{V}{I}$$

Voltage (*V*) is measured in *volts*. Current (*I*) is measured in *amperes*. Resistance (*R*) is measured in ohms.

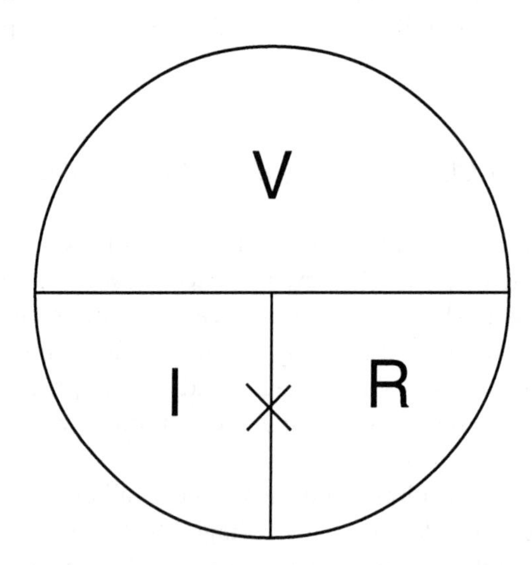

Figure 2-1. Ohm's law circle; V, *voltage;* I, *current;* R, *resistance. To use the circle, cover the value you want to find and read the other values as they appear in the formula:* V = I × R, I = V/R, R = V/I.

Series DC Circuits.

There are several important characteristics of series DC circuits:

1. The same current flows through each part of a series circuit.
2. The total resistance of a series circuit is equal to the sum of the individual resistances.
3. The voltage applied to a series circuit is equal to the sum of the individual voltage drops.
4. The voltage drop across a resistor in a series circuit is directly proportional to the size of the resistor.
5. If the circuit is broken at any point, no current will flow.

Sample series circuits problems are shown in Figure 2-2.

Parallel DC Circuits

There are several important basic rules for parallel circuits:

1. There are two or more paths for current flow.
2. Voltage is the same across each component of the circuit.
3. The sum of the currents through each path is equal to the total current that flows from the source.

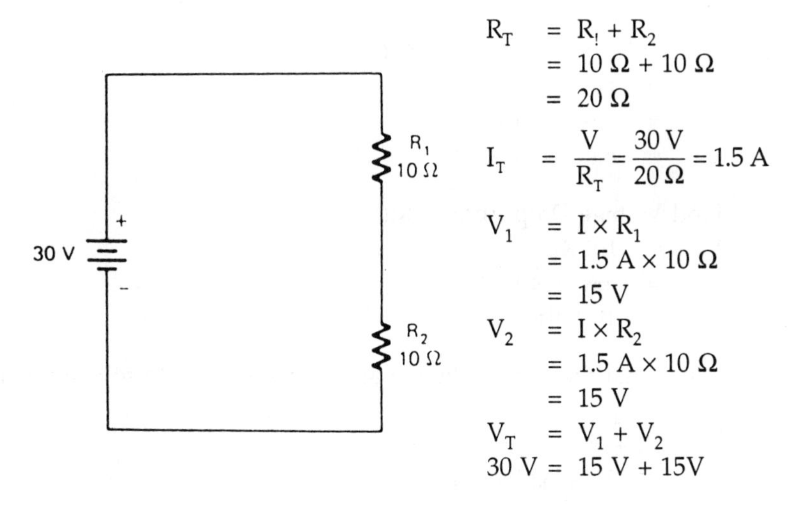

Figure 2-2a. Series-circuit example.

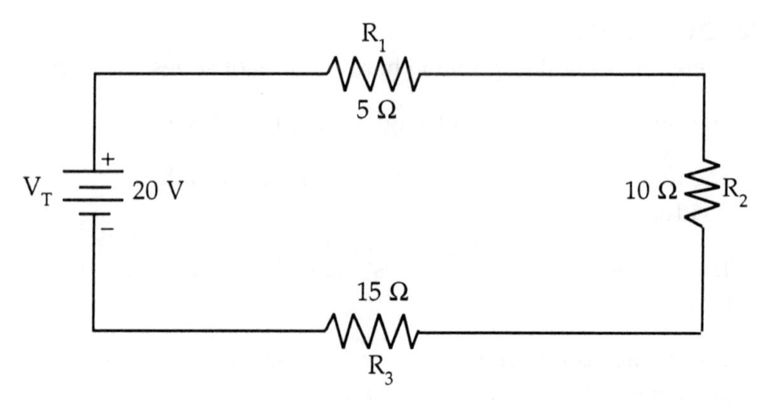

(1) Find Total Resistance (R_T)

$R_T = R_1 + R_2 + R_3$

$= 5\Omega + 10\Omega + 15\Omega$

$R_t = 30\Omega$

(2) Find Total Current (I_T)

$$I_T = \frac{V_T}{R_T} = \frac{20\text{ V}}{30\ \Omega} = .667 \text{ Amperes}$$

(3) Find Voltage Drop Across R_1 (V_1)

$V_1 = I \times R_1$

$= .667\text{A} \times 5\Omega$

(4) Find Voltage Drop Across R_2 (V_2)

$V_2 = I \times R_2$

$= .667\text{A} \times 10\Omega$

$V_2 = 6.67$ Volts

(5) Find Voltage Drop Across R_3 (V_3)

$V_3 = I \times R_3$

$= .667\text{A} \times 15\Omega$

$V_3 = 10$ Volts

(6) Verify that the sum of the voltage drops is equal to source voltage (V_T) –

$V_T = V_1 + V_2 + V_3$

$20\text{V} = 3.33\text{V} + 6.67\text{V} + 10\text{V}.$

Figure 2-2b. A series DC circuit problem-solving procedure.

4. Total resistance is found by using the formula

$$\frac{1}{R_T} = \frac{1}{R_1} + \frac{1}{R_2} + \frac{1}{R_3} + \ldots + \frac{1}{R_n}$$

5. If one of the parallel paths is broken, current will continue to flow in all the other paths.

Sample parallel circuit problems are shown in Figure 2-3.

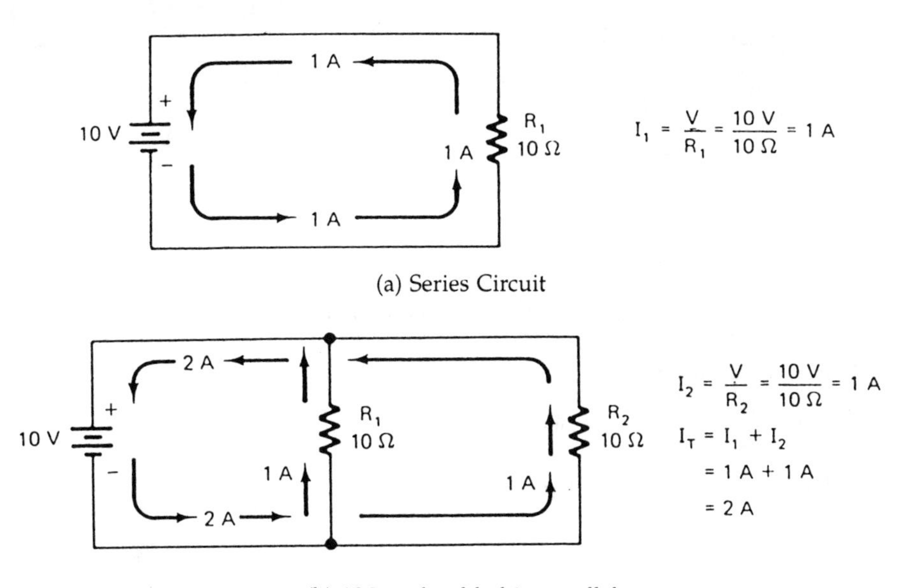

(b) 10Ω path added in parallel

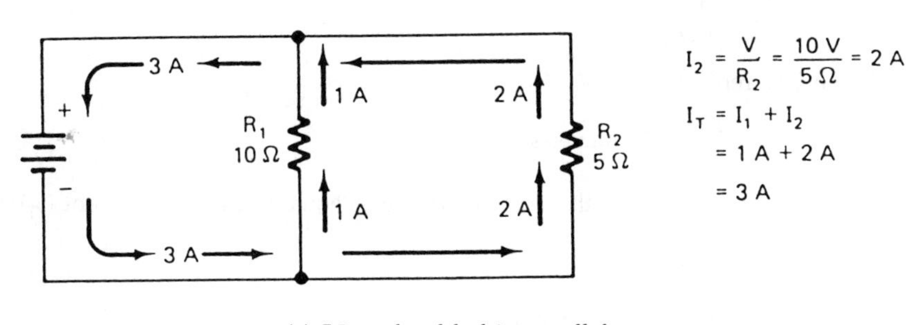

(c) 5Ω path added in parallel

Figure 2-3. Current flow in a parallel circuit: (a) one path; (b) two paths; (c) R_2 *changed to 5Ω.*

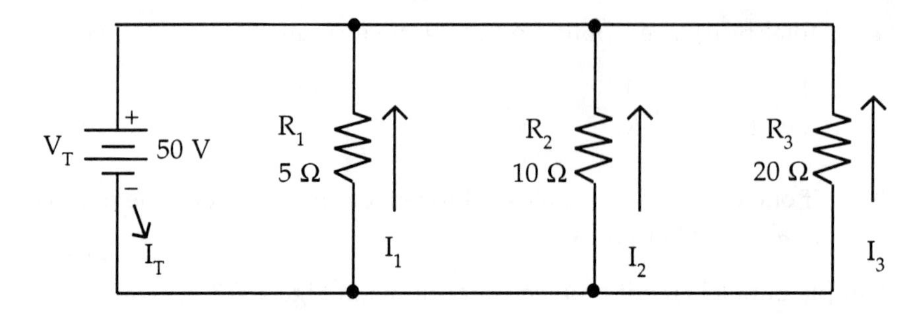

(1) Find total resistance (R_T) –

$$\frac{1}{R_T} = \frac{1}{R_1} = \frac{1}{R_2} + \frac{1}{R_3}$$

$$\frac{1}{R_T} = \frac{1}{5} = \frac{1}{10} + \frac{1}{20}$$

$$\frac{1}{R_T} = .2 = .1 + .05$$

$$\frac{1}{R_T} = .35$$

$$\frac{1}{R_T} = \frac{1}{.35} = 2.857\ \Omega \text{ (this is less than any branch resistance)}$$

(2) Find total current (I_T) –

$$I_T = \frac{V_T}{R_T} = \frac{50\text{ V}}{2.85\ \Omega} = 17.5 \text{ Amperes}$$

(3) Find current through R_1 (I_1) –

$$I_1 = \frac{V_1}{R_1} = \frac{50\text{ V}}{5\ \Omega} = 10 \text{ Amperes}$$

(4) Find current through R_2 (I_2) –

$$I_2 = \frac{V_2}{R_2} = \frac{50\text{ V}}{10\ \Omega} = 5 \text{ Amperes}$$

(5) Find current through (R_3) (I_3)

$$I_3 = \frac{V_3}{R_3} = \frac{50\text{ V}}{20\ \Omega} = 2.5 \text{ Amperes}$$

(6) Verify that the sum of the branch currents is equal to the total current (I_T) -

$$I_T = I_1 + I_2 + I_3$$

$$= 10\text{ A} + 5\text{ A} + 2.5\text{ A}$$

$$I_T = 17.5 \text{ Amperes}$$

Figure 2-3b. A parallel DC circuit problem-solving procedure.

Combination DC Circuits

Problems for combination circuits are solved by combining rules for series and parallel circuits.

Look at the sample circuit of Figure 2-4. The value that should first be calculated is the resistance of R_2 and R_3 in parallel. When this quantity is found, it can be added to the value of the series resistor (R_1) to find the total resistance of the circuit.

$$R_T = R_1 + (R_2 \parallel R_3)$$

$$= 30\ \Omega + \frac{10 \times 20}{10 + 20}$$

$$30\ \Omega + 6.67\ \Omega$$

$$= 36.67\ \Omega$$

(The symbol || means R_2 is in parallel with R_3). When the total resistance (R_T) is found, the total current (I_T) may be found:

$$I_T = \frac{V_T}{R_T} = \frac{40\ V}{36.67\ \Omega} = 1.09\ A$$

Notice that the total current flows through resistor R_1, since it is in series with the voltage source. The voltage drop across resistor R_1 is

$$V_{1'} = I_T \times R_1 = 1.09\ \text{A} \times 30\ \Omega = 32.7\ \text{V}$$

Figure 2-4. Combination-circuit example

The applied voltage is 40 V, and 32.7 V is dropped across resistor R_1'. The remaining voltage is dropped across the two parallel resistors R_2 and R_3; 40 V – 32.7 V = 7.3 V across R_2 and R_3. The currents through R_2 and R_3 are

$$I_2 = \frac{V_2}{R_2} = \frac{7.3\ \text{V}}{10\ \Omega} = 0.73\ \text{A}$$

$$I_3 = \frac{V_3}{R_3} = \frac{7.3\ \text{V}}{10\ \Omega} = 0.365\ \text{A}$$

Combination circuit problems may be solved by using the following step-by-step procedure:

1. Combine series and parallel parts to find the total resistance of the circuit.

2. Find the total current that flows through the circuit.

3. Find the voltage across each part of the circuit.

4. Find the current through each resistance of this circuit.

Steps 3 and 4 must often be done in combination with each other, rather than doing one and then the other. Some types of combination circuits are shown in Figure 2-5, while Figure 2-6 shows another example of a combination DC circuit problem.

REVIEW OF ALTERNATING CURRENT (AC) CIRCUITS

The following discussion provides a review of the three basic types of alternating current (AC) circuits. These basic circuits are: (1) resistive, (2) inductive, and (3) capacitive. The basic characteristics of each of these circuits should be reviewed to gain a fundamental understanding of electrical power systems.

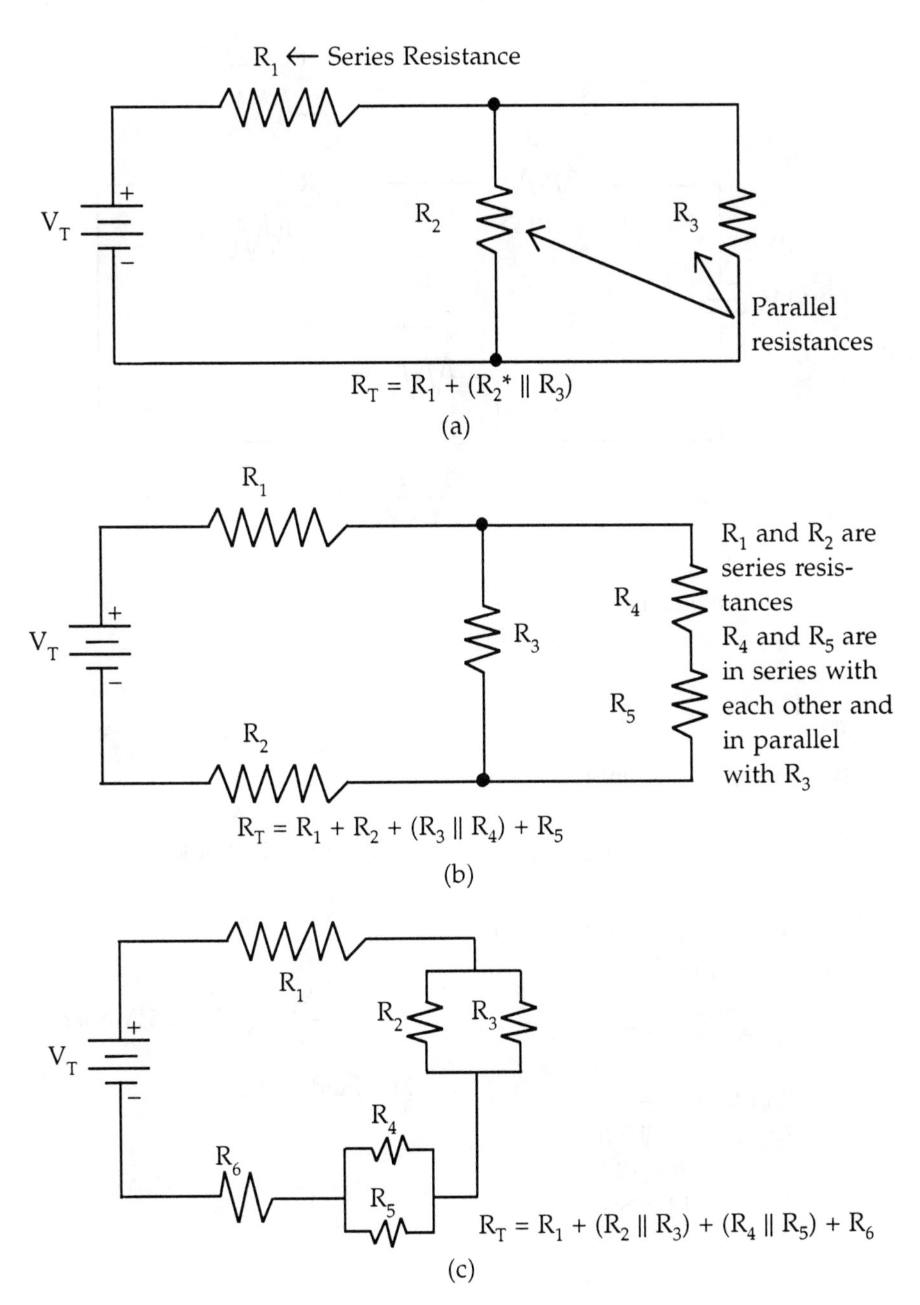

*This symbol means "in parallel with."

Figure 2-5. Types of combination circuits.

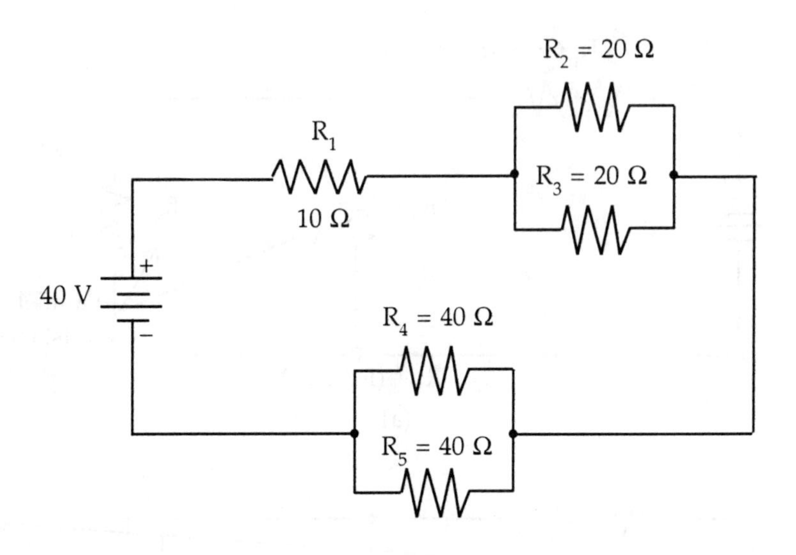

(1) Find total resistance (R_T) –

$$R_T = R_1 + (R_2 \parallel R_3) + (R_4 \parallel R_5)$$
$$= 10\ \Omega + 10\ \Omega + 20\ \Omega$$
$$R_T = 40\ \Omega$$

(2) Find total current (I_T) –

$$I_T = \frac{V_T}{R_T} = \frac{40\ \text{V}}{40\ \Omega} = 1\ \text{Ampere}$$

(3) Find voltage across R_1 (V_1) –

$$V_1 = I_1 \times R_1$$
$$= 1\ \text{A} \times 10\ \Omega$$
$$V_1 = 10\ \text{Volts}$$

(4) Find $V_2 = V_3$ –

$$V_2 = I_T \times (R_2 \parallel R_3)$$
$$= 1\ \text{A} \times 10\ \Omega$$
$$V_2 = 10\ \text{Volts}$$

(5) Find $V_4 = V_5$ –

$$V_4 = I_T \times (R_4 \parallel R_5)$$
$$= 1\ \text{A} \times 20\ \Omega$$
$$V_4 = 20\ \text{Volts}$$

(6) Find $I_2 = I_3$ –

$$I_2 = \frac{V_2}{R_2} = \frac{10\ \text{V}}{20\ \Omega} = 0.5\ \text{Ampere}$$

(7) Find $I_4 = I_5$ –

$$I_4 = \frac{V_4}{R_4} = \frac{20\ \text{V}}{40\ \Omega} = 0.5\ \text{Ampere}$$

Figure 2-6. A combination DC circuit problem-solving procedure.

Resistive AC Circuits

The simplest type of AC electrical circuit is a resistive circuit, such as the one shown in Figure 2-7a. The purely resistive circuit offers the same type opposition to alternating-current power sources as it does to pure direct-current power sources. In DC circuits,

$$\text{Voltage (V)} = \text{Current (I)} \times \text{Resistance (R)}$$

$$\text{Current (I)} = \frac{\text{Voltage (V)}}{\text{Resistance (R)}}$$

$$\text{Resistance (R)} = \frac{\text{Voltage (V)}}{\text{Current (I)}}$$

$$\text{Power (P)} = \text{Voltage (V)} \times \text{Current (I)}$$

These basic electrical relationships show that when voltage is increased, the current in the circuit increases proportionally. Also, as resistance is increased, the current in the circuit decreases. By looking at the waveforms of Figure 2-7b, we can see that the voltage and current in a purely resistive circuit, with AC applied, are in phase. An in-phase relationship exists when the minimum and maximum values of both voltage and current occur at the same time interval. Also, the power converted by the circuit is a product of voltage times current ($P = V \times I$). The power curve is also shown in Figure 2-7b. Thus, when an AC circuit contains only resistance, its behavior is very similar to a DC circuit. Purely resistive circuits are seldom encountered in the design of electrical power systems, although some devices are primarily resistive in nature.

Inductive AC Circuits

The property of inductance (L) is very commonly encountered in electrical power systems. This circuit property, shown in Figure 2-8 adds more complexity to the relationship between voltage and current in an AC circuit. All motors, generators, and transformers exhibit the property of inductance. This property is evident due to a counter-electromotive force (cemf) which is produced when a magnetic field is developed around a coil of wire. The magnetic flux produced around the coils affects circuit action. Thus, the inductive property (cemf) produced by a magnetic field offers an opposition to change in the current flow in a circuit.

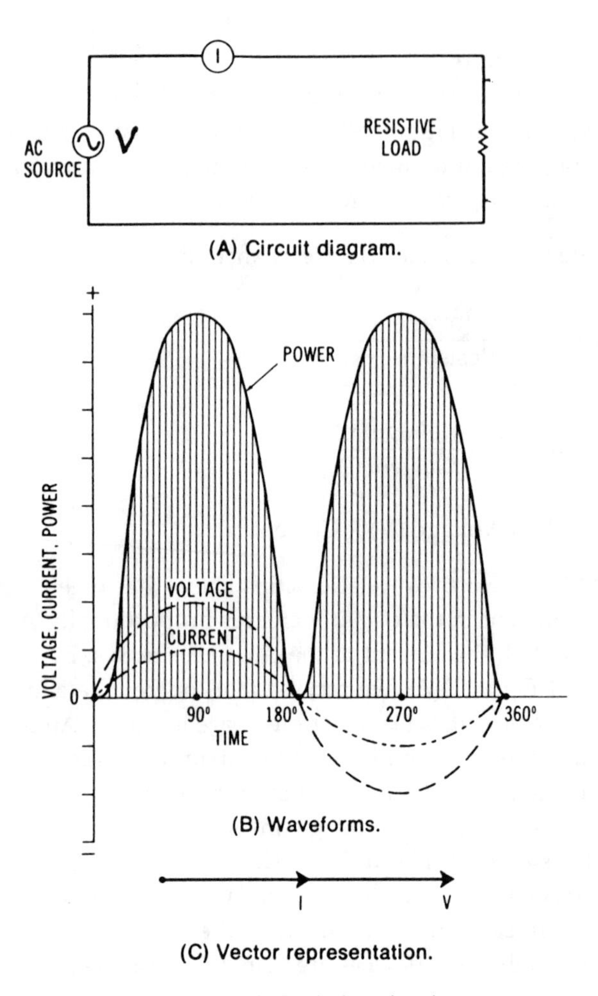

Figure 2-7. Resistive circuit.

The opposition to change of current is evident in the diagram of Figure 2-8b. In an inductive circuit, we can say that voltage leads current or that current lags voltage. If the circuit were purely inductive (contains no resistance), the voltage would lead the current by 90° (Figure 2-8b) and no power would be converted in the circuit. However, since all actual circuits have resistance, the inductive characteristic of a circuit might typically cause the condition shown in Figure 2-9 to exist. Here the voltage is leading the current by 30°. The angular separation between voltage and current is called the phase angle. The phase angle in-

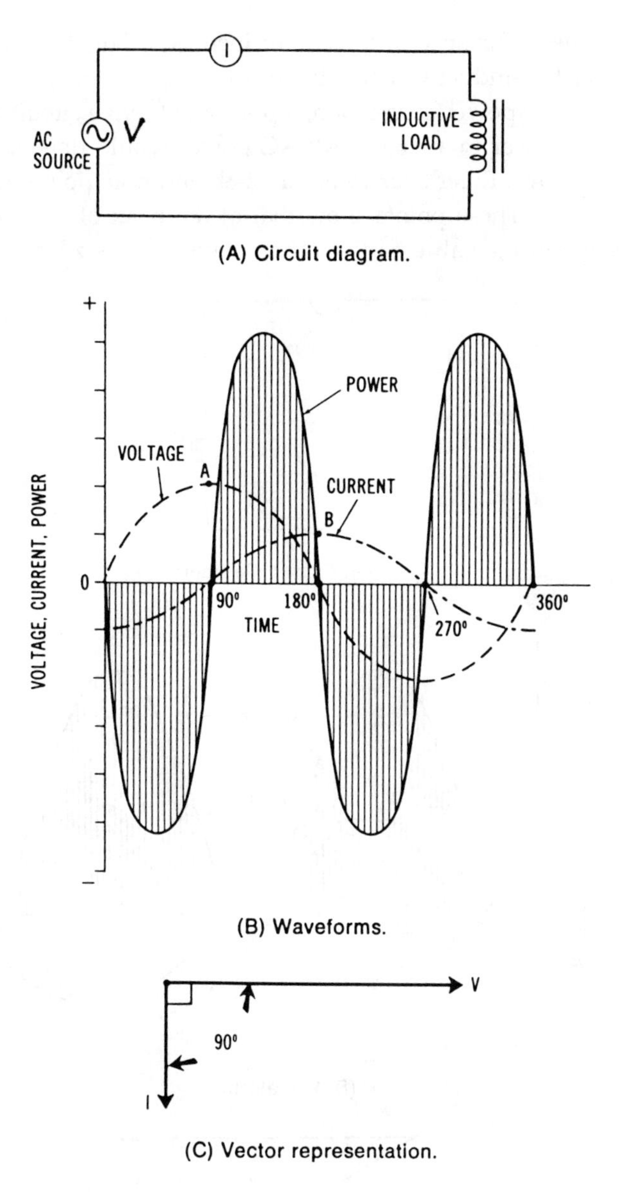

(A) Circuit diagram.

(B) Waveforms.

(C) Vector representation.

Figure 2-8. Inductive circuit.

creases as the inductance of the circuit increases. This type of circuit is called a resistive-inductive (RL) circuit.

In terms of power conversion, a purely inductive circuit would not convert any power in a circuit. All AC power would delivered back to the power source. Refer back to Figure 2-8b and note points A and B on the waveforms. These points show that at the peak of each waveform, the corresponding value of the other waveform is zero. The power

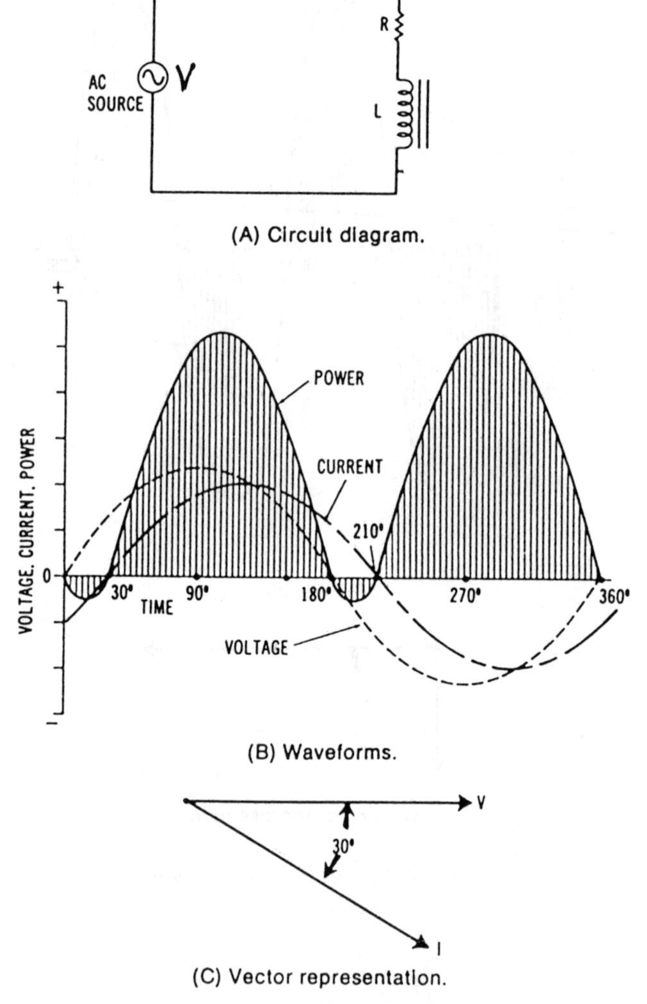

Figure 2-9. Resistive-inductive (RL) circuit.

curves shown are equal and opposite in value and will cancel each other out. Where both voltage and current are positive, the power is also positive since the product of two positive values is positive. When one value is positive and the other is negative, the product of the two values is negative; therefore, the power converted is negative. Negative power means that electrical energy is being returned from the load device to the power source without being converted to another form of energy. Therefore, the power converted in a purely inductive circuit (90° phase angle) would be equal to zero.

Compare the purely inductive waveforms to those of Figure 2-9b. In the practical resistive-inductive (RL) circuit, part of the power supplied from the source is converted in the circuit. Only during the intervals from 0° to 30° and from 180° to 210° does negative power result. The remainder of the cycle produces positive power; therefore, most of the electrical energy supplied by the source is converted to another form of energy.

Any inductive circuit exhibits the property of inductance (L) which is the opposition to a change in current flow in a circuit. This property is found in coils of wire (which are called inductors) and in rotating machinery and transformer windings. Inductance is also present in electrical power transmission and distribution lines to some extent. The unit of measurement for inductance is the henry (H). A circuit has a 1-henry inductance if a current changing at a rate of 1 ampere per second produces an induced counter-electromotive force (cemf) of 1 volt.

In an inductive circuit with AC applied, an opposition to current flow is created by the inductance. This type of opposition is known as inductive reactance (X_L). The inductive reactance of an AC circuit depends upon the inductance (L) of the circuit and the rate of change of current. The frequency of the applied alternating current establishes the rate of change of the current. Inductive reactance (X_L) may be expressed as:

$$X_L = 27\pi fL$$

where,

X_L is the inductive reactance in ohms,

2π is 6.28, the mathematical expression for one sine wave of alternating current (0°–360°),

f is the frequency of the AC source in hertz,

L is the inductance of the circuit in henrys.

Capacitive AC Circuits

Figure 2-10a shows a capacitive device connected to an AC source. We know that whenever two conductive materials (plates) are separated by an insulating (dielectric) material, the property of capacitance is exhibited. Capacitors have the capability of storing an electrical charge. They also have many applications in electrical power systems.

The operation of a capacitor in a circuit is dependent upon its ability to charge and discharge. When a capacitor charges, an excess of electrons (negative charge) is accumulated on one plate and a deficiency of electrons (positive charge) is created on the other plate. Capacitance (C) is determined by the size of the conductive material (plates) and by their separation (determined by the thickness of the dielectric or insulating material). The type of insulating material is also a factor in determining capacitance. Capacitance is directly proportional to the plate size and inversely proportional to the distance between the plates. The unit of capacitance is the farad (F). A capacitance of 1 farad results when a potential of 1 volt causes an electrical charge of 1 coulomb (a specific mass of electrons) to accumulate on a capacitor. Since the farad is a very large unit, microfarad (μF) values are ordinarily assigned to capacitors.

If a direct current is applied to a capacitor, the capacitor will charge to the value of that DC voltage. After the capacitor is fully charged, it will block the flow of direct current. However, if AC is applied to a capacitor, the changing value of current will cause the capacitor to alternately charge and discharge. In a purely capacitive circuit, the situation shown in Figure 2-10 would exist. The greatest amount of current would flow in a capacitive circuit when the voltage changes most rapidly. The most rapid change in voltage occurs at the 0° and 180° positions where the polarity changes. At these positions, maximum current is developed in the circuit. When the rate of change of the voltage value is slow, such as near the 90° and 270° positions, a small amount of current flows. In examining Figure 2-10b, we can observe that current leads voltage by 90° in a purely capacitive circuit or the voltage lags the current by 90°. Since a 90° phase angle exists, no power would be convened in this circuit, just as no power was developed in the purely inductive circuit. As shown in Figure 2-10b, the positive and negative power waveforms will cancel one another out.

Since all circuits contain some resistance, a more practical circuit is the resistive-capacitive (RC) circuit shown in Figure 2-11a. In an RC circuit, the current leads the voltage by some phase angle between 0° and

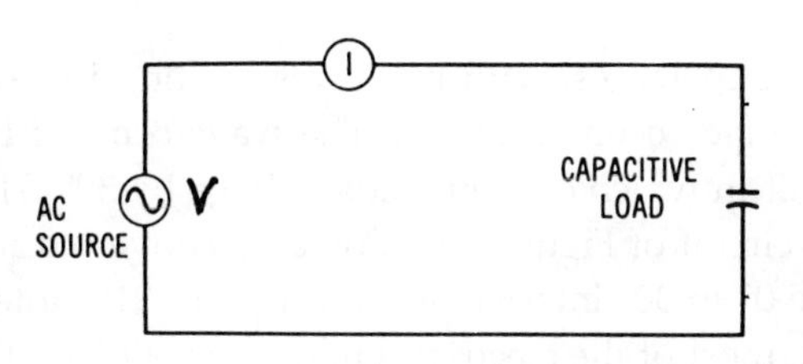

(A) Circuit diagram.

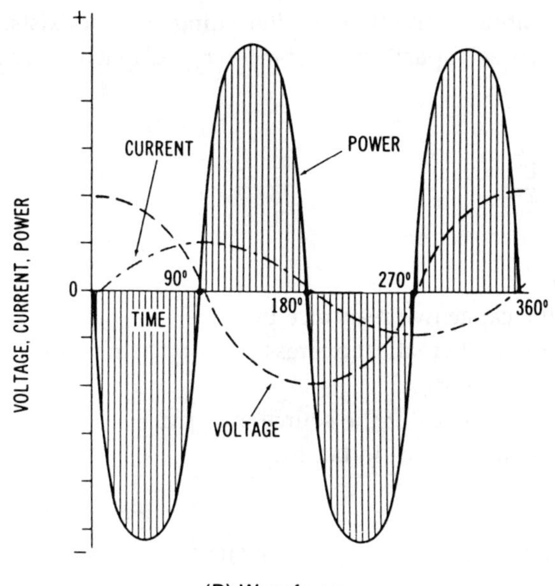

(B) Waveforms.

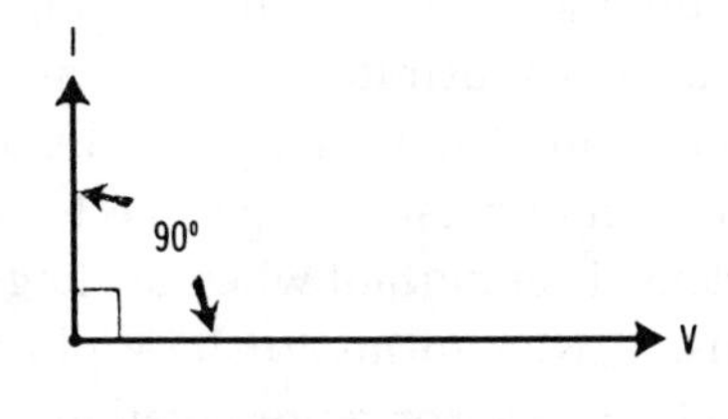

(C) Vector representation.

Figure 2-10. Capacitive circuit.

90°. As capacitance increases, with no corresponding increase in resistance, the phase angle becomes greater. The waveforms of Figure 2-11b show an RC circuit in which current leads voltage by 30°. This circuit is similar to the RL circuit of Figure 2-11. Power is converted in the circuit except during the 0° to 30° interval and the 180° to 210° interval. In the RC circuit shown, most of the electrical energy supplied by the source is convened to another form of energy in the load.

Due to the electrostatic field which is developed around a capacitor, an opposition to the flow of alternating current exists. This opposition is known as capacitive reactance (X_c). Capacitive reactance is expressed as:

$$X_c = \frac{1}{2\pi fC}$$

where,
X_c is the capacitive reactance in ohms.
2π is the mathematical expression of one sine wave (0° to 360°),
f is the frequency of the source in hertz,
C is the capacitance in farads.

VECTOR DIAGRAMS FOR AC CIRCUITS

In Figures 2-8 through 2-11, a vector diagram was shown for each circuit condition that was illustrated. Vectors are straight lines which have a specific direction and length. They may be used to represent voltage or current values. An understanding of vector diagrams (sometimes called phasor diagrams) is important when dealing with alternating current. Rather than using waveforms to show phase relationships, it is possible to use a vector or phasor representation.

Ordinarily, when beginning a vector diagram, a horizontal line is drawn with its left end as the reference point. Rotation in a counterclockwise direction from the reference point is then considered to be a positive direction. Note that in the preceding diagrams, the voltage vector was the reference. For the inductive circuits, the current vector was drawn in a clockwise direction, indicating a lagging condition. A leading

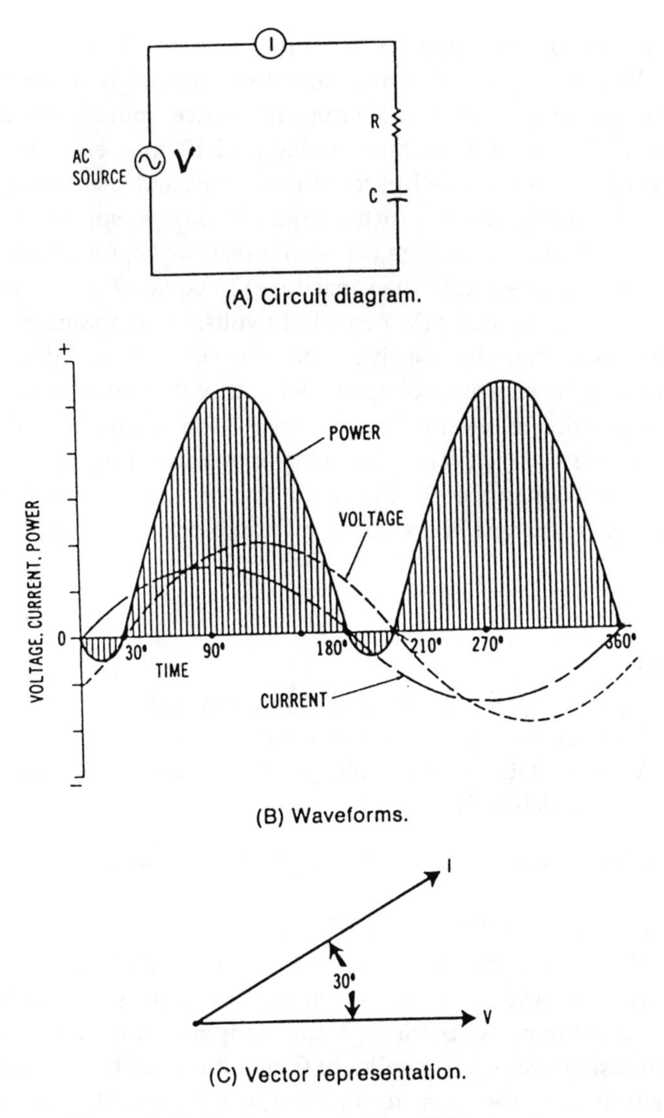

Figure 2-11. Resistive-capacitive (RC) circuit.

condition is shown for the capacitive circuits by the use of a current vector drawn in a counterclockwise direction from the voltage vector. You may now want to study Appendix C—Trigonometry for Electrical Power Systems.

Use of Vectors for Series AC Circuits

Vectors may be used to compare voltage drops across the components of a series circuit containing resistance, inductance and capacitance (an RLC circuit) as shown in Figure 2-12. In a series AC circuit, the current is the same in all parts of the circuit and the voltages must be added by using vectors. In the example shown, specific values have been assigned. The voltage across the resistor (V_R) is equal to 4 volts, while the voltage across the capacitor (V_C) equals 7 volts, and the voltage across the inductor (V_L) equals 10 volts. We diagram the capacitive voltage as leading the resistive voltage by 90° and the inductive voltage as lagging the resistive voltage by 90°. Since these two values are in direct opposition to one another, they may be subtracted to find the resultant reactive voltage (V_x). By drawing lines parallel to V_R and V_x, the resultant voltage applied to the circuit can be found. Since these vectors form a right triangle, the value of VT can be expressed as:

$$V_T = \sqrt{V_R^2 + V_x^2}$$

where,

V_T is the total voltage applied to the circuit,
V_R is the voltage across the resistance,
V_x is the total reactive voltage ($V_L - V_C$ or $V_C - V_L$, depending on which is the larger).

The solution to this problem is shown in Figure 2-12.

Use of Vectors for Parallel AC Circuits

Vector representation is also useful for parallel AC-circuit analysis. Voltage in a parallel AC circuit remains the same across all the components and the currents through the components of the circuit can be shown using vectors. A parallel RLC circuit is shown in Figure 2-13. The current through the capacitor (I_C) is shown leading the current through the resistor (I_R) by 90°. The current through the inductor (I_L) is shown lagging I_R by 90°. Since I_L and I_C are 180° out of phase, they are subtracted to find the total reactive current (I_x). By drawing lines parallel to I_R and I_x, the total current of the circuit (I_T) may be found. These vectors form a right triangle; therefore, total current can be expressed as:

$$I_T = \sqrt{I_R^2 + I_x^2}$$

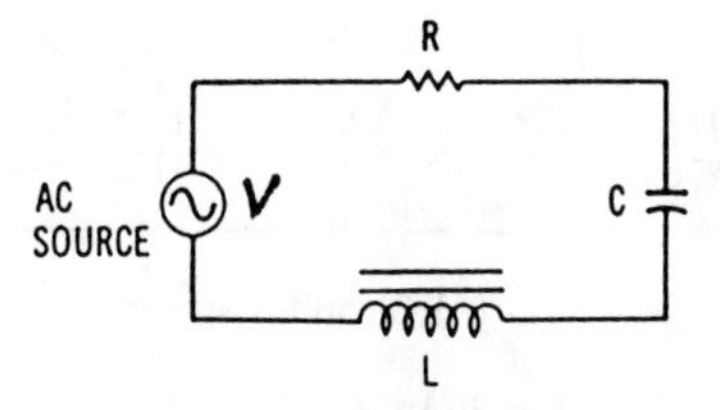

(A) Circuit diagram.

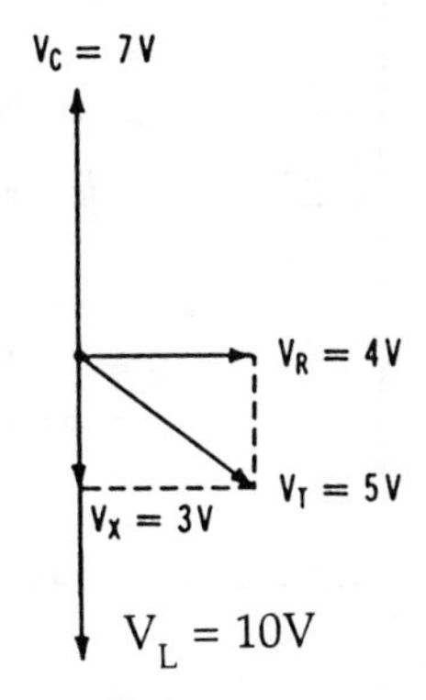

(B) Vector diagram.

$$V_L = 10\text{ V}$$
$$V_X = V_L - V_C = 10\text{ V} - 7\text{ V} = 3\text{ V}$$

$$V_T = \sqrt{V_R^2 + V_X^2} = \sqrt{4^2 + 3^2} = \sqrt{16 + 9} = \sqrt{25} = 5\text{V}$$

(C) Problem solution.

Figure 2-12. Voltage vector relationship in a series RLC circuit.

A similar method of vector diagramming can be used for voltages in RL and RC series circuits. This method may also be used for currents in RL and RC parallel circuits. RLC circuits were used in the examples to illustrate the method used to find a resultant reactive voltage or current in a circuit.

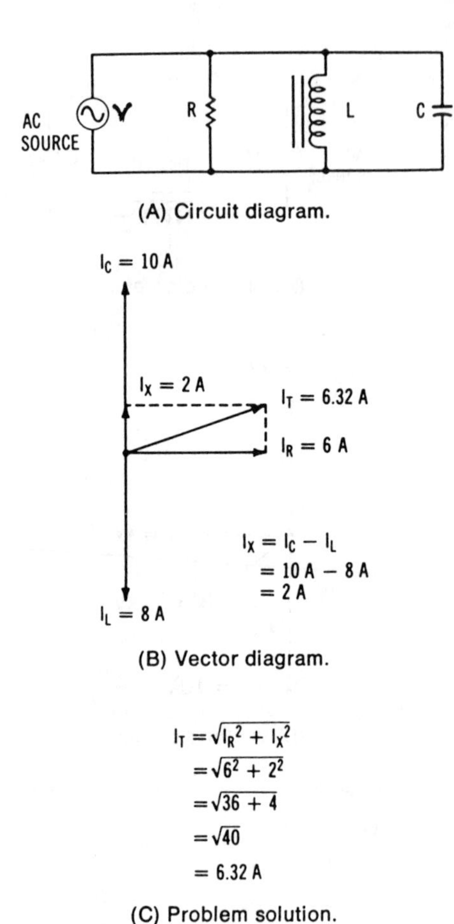

(A) Circuit diagram.

(B) Vector diagram.

(C) Problem solution.

Figure 2-13. Current vector relationship in a parallel RLC circuit.

IMPEDANCE IN AC CIRCUITS

Another application in the use of vectors is for determining the total opposition of an AC circuit to the flow of current. This total opposition is called impedance (Z).

Impedance In Series AC Circuits

Both resistances and reactances in AC circuits affect the opposition to current flow. Impedance (Z) of an AC circuit may be expressed as:

$$Z = \frac{V}{I}$$

or

$$Z = \sqrt{R^2 + (X_L - X_C)^2}$$

This formula may be clarified by using the vector diagram shown in Figure 1-9b. The total reactance (X_T) of an AC circuit may be found by subtracting the smallest reactance (X_L or X_C) from the largest reactance. The impedance of a series AC circuit is determined by using the preceding formula since a right triangle (called an impedance triangle) is formed by the three quantities which oppose the flow of alternating current. A sample problem for finding the total impedance of a series AC circuit is shown in Figure 2-14d.

Impedance in Parallel AC Circuits

When components are connected in parallel, the calculation of impedance becomes more complex. Figure 2-15a shows a simple RLC parallel circuit. Since the total impedance in the circuit is smaller than the resistance or reactance, an impedance triangle such as the one shown in Figure 2-14c cannot be developed. A simple method used to find impedance in parallel circuits is the *admittance* triangle shown in Figure 2-15b. The following quantities may be plotted on the triangle:

$$\text{admittance} = \frac{1}{Z} \quad \text{conductance} = \frac{1}{R}, \quad \text{inductive susceptance} = \frac{1}{X_L}.$$

Notice that these quantities are the reciprocals of each type of opposition to alternating current. Therefore, since total impedance (Z) is the smallest quantity in a parallel AC circuit, it becomes the largest value on the admittance triangle. The sample problem of Figure 2-15c shows the procedure used to find total impedance of a parallel RC circuit.

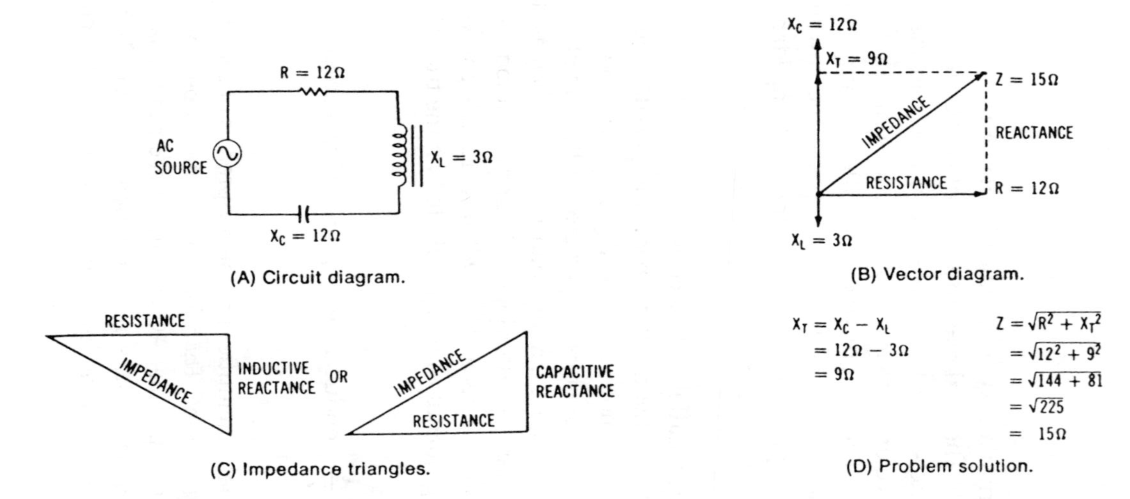

Figure 2-14. Impedance in series AC circuits: (a) circuit diagram; (b) vector diagram; (c) impedance triangles, (d) problem solution.

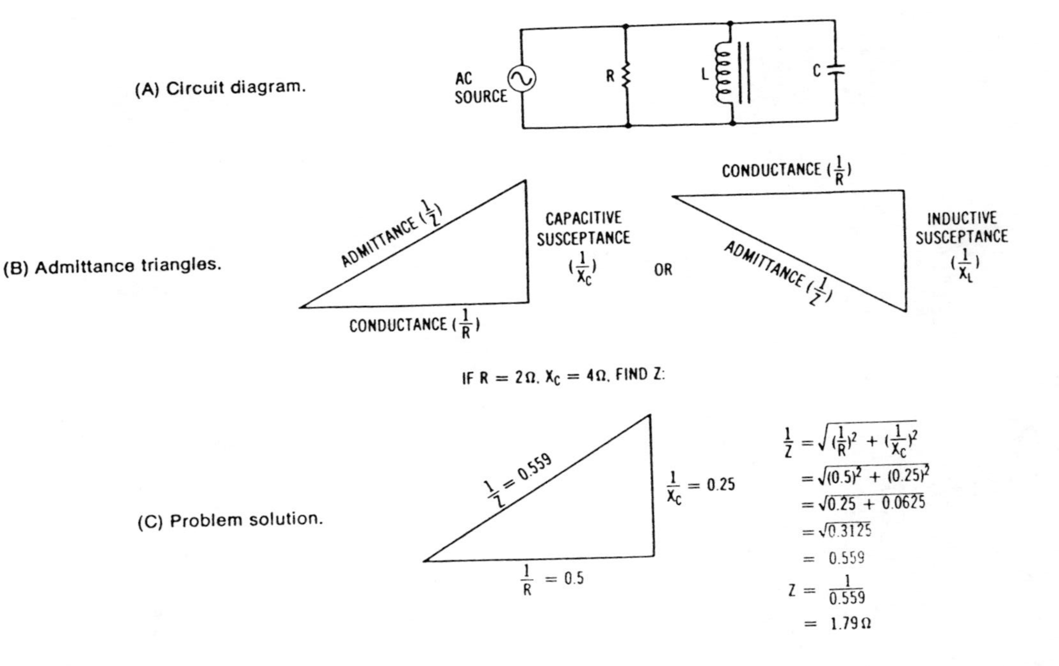

Figure 2-15. Impedance in parallel AC circuits; (a) circuit diagram; (b) vector diagram; (c) impedance triangles; (d) problem solution.

Chapter 3

Power Relationships in Electrical Circuits

An understanding of basic power relationships in electrical circuits is very important when studying electrical power systems. In Chapter 2, resistive, inductive, and capacitive circuits were discussed. Also, power converted in these circuits was discussed in terms of power waveforms which were determined by the phase angle (θ) between voltage and current. In a DC circuit, power is equal to the product of voltage and current ($P = V \times I$). This formula is true also for purely resistive AC circuits.

POWER IN DC ELECTRICAL CIRCUITS

In terms of voltage and current, power (P) in watts (W) is equal to voltage (in volts) multiplied by current (in amperes). The formula is $P = V \times I$. For example, a 120-V electrical outlet with 4 A of current flowing from it has a power value of

$$P = V \times I = 120\ V \times 4\ A = 480\ W$$

The unit of electrical power is the *watt*. In the example, 480 W of power is converted by the load portion of the circuit. Another way to find power is

$$P = \frac{V^2}{R}$$

This formula is used when voltage and resistance are known but current is not known. The formula $P = I^2 \times R$ is used when current and resistance are known.

Formulas and units of measurement are summarized in Figure 3-1. The quantity in the center of the circle may be found by any of the three formulas along the outer part of the circle in the same part of the circle. This circle is handy to use for making electrical calculations for voltage, current, resistance, or power.

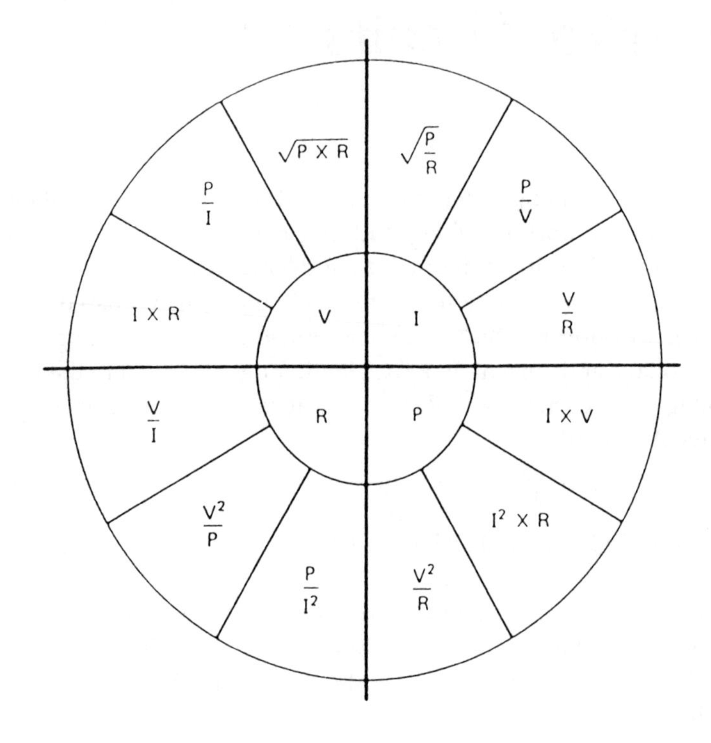

Figure 3-1. Formulas for finding voltage, current, resistance, or power.

It is easy to find the amount of power converted by each of the resistors in a series circuit, such as the one shown in Figure 3-2. In the circuit shown, the amount of power converted by each of the resistors and the total power is found as follows:

1. Power converted by resistor R_1:

$$P_1 = I^2 \times R_1 = 2^2 \times 20\ \Omega = 80\text{W}$$

2. Power converted by resistor R_2:

$$P_2 = I^2 \times R_2 = 2^2 \times 30\ \Omega = 120\ \text{W}$$

3. Power converted by resistor R_3:

$$P_3 = I^2 \times R^3 = 2^2 \times 50\ \Omega = 200\ \text{W}$$

4. Power converted by the circuit:

$$P_T = P_1 + P_2 + P_3 = 80\ \text{W} + 120\ \text{W} + 200\ \text{W} = 400\ \text{W, or}$$

$$P_T = V_T \times I = 200\ \text{V} \times 2\ \text{A} = 400\ \text{W}$$

When working with electrical circuits it is possible to check your results by using other formulas.

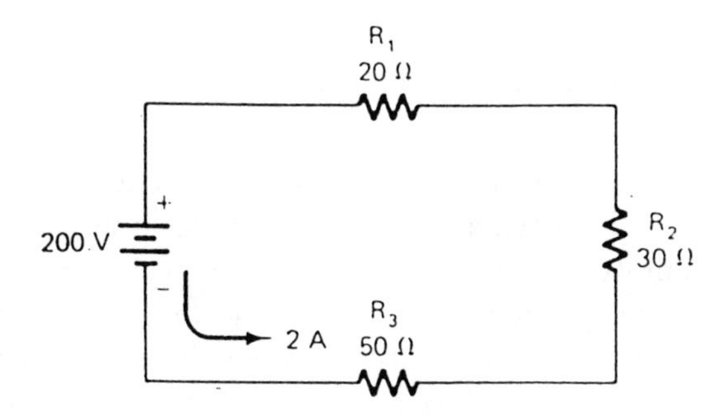

Figure 3-2. Finding power in a series circuit.

Power in parallel circuits is found in the same way as for series circuits. In the example of Figure 3-3 the power converted by each of the resistors and the total power of the parallel circuit is found as follows:

1. Power converted by resistor R_1:

$$P_1 = \frac{V^2}{R_1} = \frac{30^2}{5} = \frac{900}{5} = 180\ \text{W}$$

2. Power converted by resistor R_2:

$$P_2 = \frac{V^2}{R_2} = \frac{30^2}{10} = \frac{900}{10} = 90\ \text{W}$$

3. Power converted by resistor R_3:

$$P_3 = \frac{V^2}{R_3} = \frac{30^2}{20} = \frac{900}{20} = 45 \text{ W}$$

4. Total power converted by the circuit:

$$P_T = P_1 = P_2 + P_3 = 180 \text{ W} + 90 \text{ W} + 45 \text{ W} = 315 \text{ W}$$

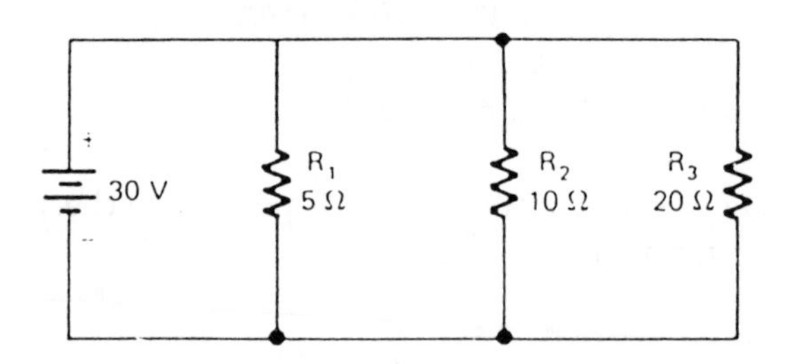

Figure 3-3. Finding power values in a parallel circuit.

The *watt* is the basic unit of electrical power. To determine an actual quantity of electrical energy, a factor that indicates how long a power value continued must be used. Such a unit of electrical energy is called a watt-second. It is the product of watts (W) and time (in seconds). The watt-second is a very small quantity of energy. It is more common to measure electrical energy in *kilowatt-hours (kWh)*. It is the kWh quantity of electrical energy which is used to determine the amount of electric utility bills. A kilowatt-hour is 1000 W in 1 h of time or 3,600,000 W per second.

As an example, if an electrical heater operates on 120 V, and has a resistance of 20 Q, what is the cost to use the heater for 200 h at a cost of 5 cents per kWh?

1. $P = \frac{V^2}{R} = \frac{120^2}{20\ \Omega} = \frac{14{,}400}{20\ \Omega} = 720 \text{ W} = 0.72 \text{ kW}.$

2. There are 1000 W in a kilowatt (1000 W = 1 kW).

3. Multiply the kW that the heater has used by the hours of use:
   kW × 200 h = kilowatt-hours (kWh)
   0.72 × 200 h = 144 kWh

4. Multiply the kWh by the cost:
   kWh × cost = 144 kWh × 0.05 = $7.20

Some simple electrical circuit examples have been discussed in this chapter. They become easy to understand after practice with each type of circuit. It is very important to understand the characteristics of series, parallel, and combination circuits.

Maximum Power Transfer in Circuits

An important consideration in electrical power systems is called *maximum power transfer*. Maximum power is transferred from a voltage source to a load when the load resistance (R_L) is equal to the internal resistance of the source (R_s). The source resistance limits the amount of power that can be applied to a load. Electrical circuits may be considered as shown in Figure 3-4.

For example, as a flashlight battery gets older, its internal resistance increases. This increase in the internal resistance causes the battery to supply less power to the lamp load. Thus the light output of the flashlight is reduced.

Figure 3-5 shows an example that illustrates maximum power transfer. The source is a 100-V battery with an internal resistance of 5 Ω. The values of I_L, V_{out}, and power output (P_{out}) are calculated as follows:

$$I_L = \frac{V_T}{R_S + R_L} \qquad V_{out} = I_L \times R_L \qquad P_{out} = I_L \times V_{out}$$

Notice the graph shown in Figure 3-5. This graph shows that maximum power is transferred from the source to the load when $R_L = R_S$. This is an important circuit design consideration for power sources.

POWER IN AC ELECTRICAL CIRCUITS

When a reactance (either inductive or capacitive) is present in an AC circuit, power is no longer a product of voltage and current. Since reactive circuits cause changes in the method used to compute power, the following described techniques express the basic power relationships in AC circuits. The product of voltage and current is expressed in *volt-amperes (VA)* or *kilovolt-amperes (kVA)*, and is known as *apparent power*. When using meters to measure power in an AC circuit, apparent power is the voltage reading multiplied by the current reading. The actual power which is converted to another form of energy by the circuit is measured with a wattmeter. This actual power is referred to as *true power*. Ordinarily, it is desirable to know the ratio of true power converted in a circuit to apparent power. This ratio is called the *power factor* and is expressed as:

$$pf = \frac{P}{VA}$$

or

$$\%\,pf = \frac{P}{VA} \times 100$$

where

pf is the power factor of the circuit,
P is the true power in watts,
VA is the apparent power in volt-amperes.

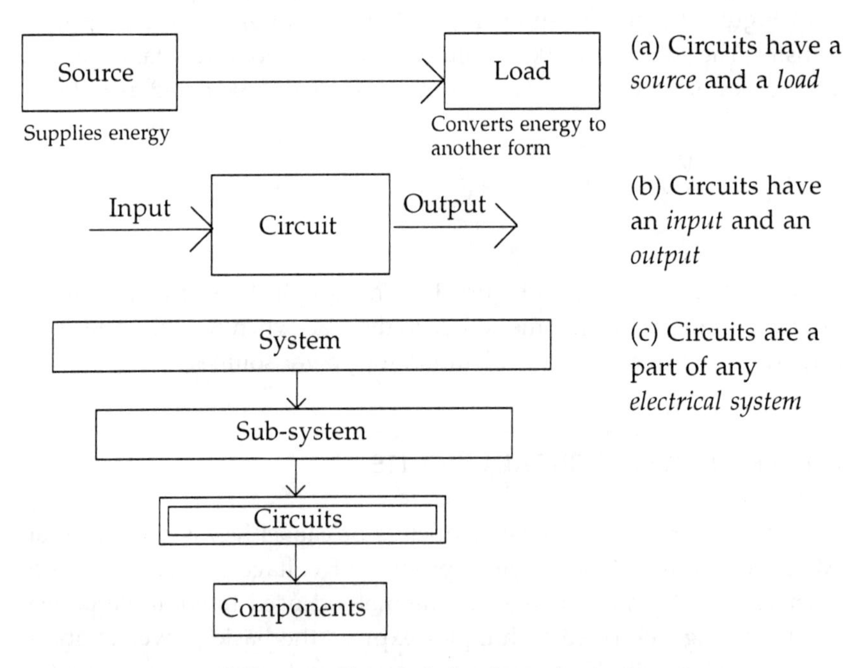

Figure 3-4. Electrical circuits and systems.

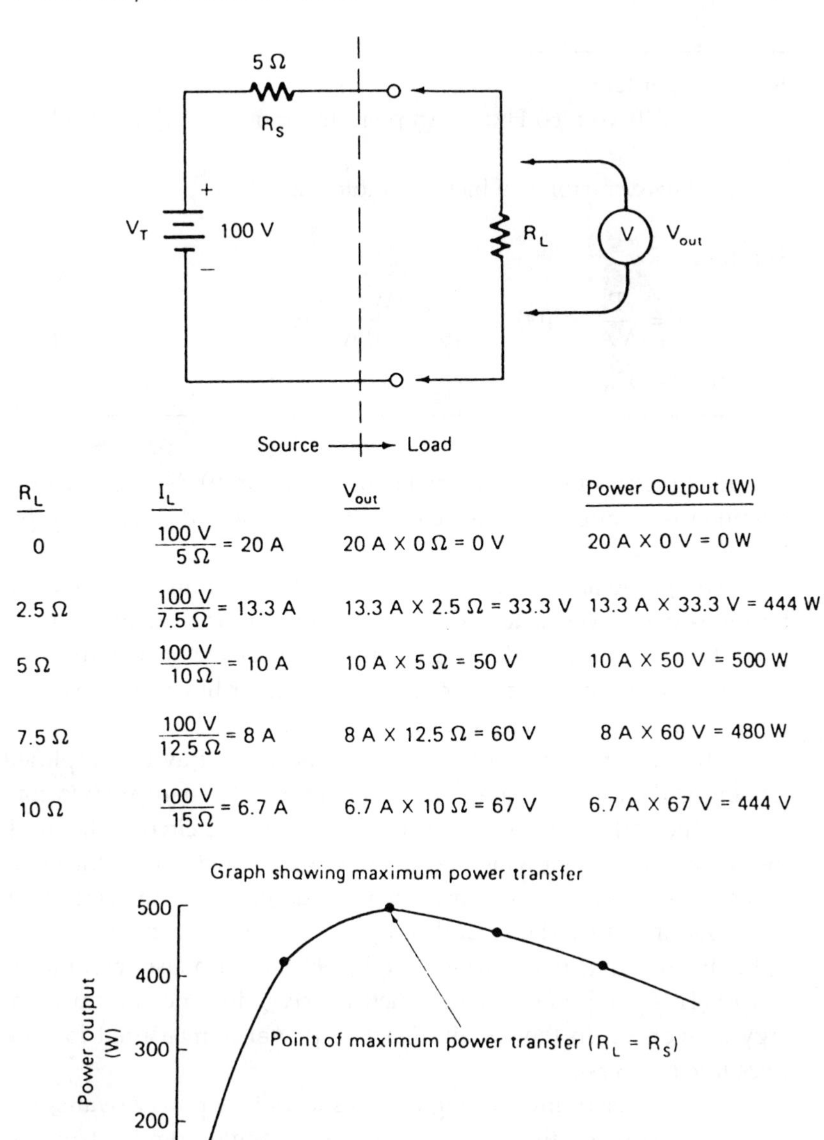

R_L	I_L	V_{out}	Power Output (W)
0	$\frac{100\text{ V}}{5\ \Omega}$ = 20 A	20 A X 0 Ω = 0 V	20 A X 0 V = 0 W
2.5 Ω	$\frac{100\text{ V}}{7.5\ \Omega}$ = 13.3 A	13.3 A X 2.5 Ω = 33.3 V	13.3 A X 33.3 V = 444 W
5 Ω	$\frac{100\text{ V}}{10\ \Omega}$ = 10 A	10 A X 5 Ω = 50 V	10 A X 50 V = 500 W
7.5 Ω	$\frac{100\text{ V}}{12.5\ \Omega}$ = 8 A	8 A X 12.5 Ω = 60 V	8 A X 60 V = 480 W
10 Ω	$\frac{100\text{ V}}{15\ \Omega}$ = 6.7 A	6.7 A X 10 Ω = 67 V	6.7 A X 67 V = 444 V

Figure 3-5. Problem that shows maximum power transfer.

Sample Problem:

Given: A 240 Volt, 60 Hz, 30 Amp electric motor rated at 6000 Watts

Find: Power factor at which the motor operates

Solution:

$$\% \text{ pf} = \frac{P}{VA} \times 100 = \frac{6000 \text{ W}}{240 \text{ V} \times 30 \text{ A}} \times 100$$

$$\% \text{ pf} = 83.3\%$$

The maximum value of power factor is 1.0, or 100% which would be obtained is a purely resistive circuit. This is referred to as unity power factor.

The phase angle between voltage and current in an AC circuit determines the power factor. If a purely inductive or capacitive circuit existed, the 90° phase angle would cause a power factor of zero to result. In practical circuits, the power factor varies according to the relative values of resistance and reactance.

The power relationships we have discussed may be simplified by looking at the power triangle shown in Figure 3-6. There are two components which affect the power relationship in an AC circuit. The in-phase (resistive) component which results in power conversion in the circuit is called *active power*. Active power is the true power of the circuit and is measured in watts. The second component is that which results from an inductive or capacitive reactance and is 90° out of phase with the active power. This component, called *reactive power*, does not produce an energy conversion in the circuit. Reactive power is measured in *volt-amperes reactive (vars)*.

The power triangle of Figure 3-6 shows true power (watts) on the horizontal axis, reactive power (var) at a 90° angle from the true power, and volt-amperes (VA) as the longest side (hypotenuse) of the right triangle. Note the similarity between this right triangle, the voltage triangle for series AC circuits of Figure 2-14b, the current triangle for parallel AC circuits of Figure 2-15b, the impedance triangles of Figure 2-14c, and the admittance triangles of Figure 2-15b. Each of these right triangles has a horizontal axis that corresponds to the resistive component of the circuit, while the vertical axis corresponds to the reactive compo-

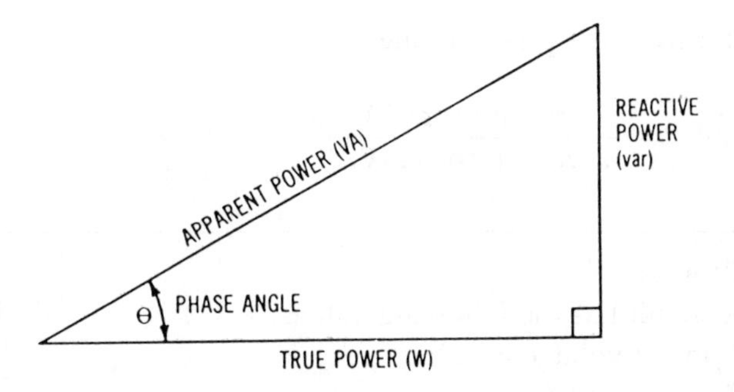

Figure 3-6. Power triangle.

nent. The hypotenuse represents the resultant which is based on the relative values of resistance and reactance in the circuit. We can now see how important vector representation and an understanding of the right triangle are in analyzing AC circuits.

We can further examine the power relationships of the power triangle by expressing each value mathematically, based on the value of apparent power (VA) and the phase angle (θ). Remember that the phase angle is the amount of phase shift, in degrees, between voltage and current in the circuit. Trigonometric ratios, which are discussed in Appendix C, show that the sine of an angle of a right triangle is expressed as:

$$\text{sine } \theta = \frac{\text{opposite side}}{\text{hypotenuse}}$$

Since this is true, the phase angle can be expressed as:

$$\text{sine } \theta = \frac{\text{reactive power (var)}}{\text{apparent power (VA)}}.$$

Therefore,

$$\text{var} = \text{VA} \times \text{sine } \theta.$$

We can determine either the phase angle or the var value by using trigonometric functions.

We also know that the cosine of an angle of a right triangle is expressed as:

$$\text{cosine } \theta = \frac{\text{adjacent side}}{\text{hypotenuse}}.$$

Thus, in terms of the power triangle:

$$\text{cosine } \theta = \frac{\text{true power (W)}}{\text{apparent power (VA)}}.$$

Sample Problem:

Given: A circuit has the following values:
Applied voltage = 240;
Current = 12 Amperes;
Power factor = 0.83

Find: True power of the circuit

Solution:

$$W = VA \times \text{Cosine } \theta$$
$$= 240 \times 12 \times 0.83$$
$$W = 2390 \text{ Watts}$$

Note that the expression

$$\frac{\text{true power}}{\text{apparent power}}$$

is the power factor of a circuit; therefore, the power factor is equal to the cosine of the phase angle (pf = cosine θ).

Right triangle relationships can also be expressed to determine the value of any of the sides of the power triangle when the other two values are known. These expressions are as follows:

$$VA = \sqrt{W^2 + var^2}$$

$$W = \sqrt{VA^2 - var^2}$$

$$var = \sqrt{VA^2 - W^2}$$

Sample Problem:

Given: Total reactive power = 54 var, applied voltage = 120 volts, current - 0.5 amps

Find: True Power of the circuit

Solution:

$$W = \sqrt{VA^2 - V^2}$$

$$= \sqrt{(120 \times 0.5)^2 - 54^2}$$

$$= \sqrt{3600 - 2916}$$

$$W = 26.15 \text{ Watts}$$

Appendix C should be review thoroughly in order to gain a better understanding of the use of right triangles and trigonometric ratios for solving AC circuit problems.

POWER RELATIONSHIPS IN THREE-PHASE AC CIRCUITS

To illustrate the basic concepts of three-phase power systems, we will use the example of a three-phase AC generator. A simplified pictorial diagram of a three-phase generator is shown in Figure 3-7. A three-phase voltage diagram is shown in Figure 3-8.

In Figure 3-7, poles A′, B′, and C′ represent the beginnings of each of the phase windings, while poles A, B, and C represent the ends of each of the windings. These windings may be connected in either of two ways. These methods of connection, called the *wye* configuration and the *delta* configuration, are the basic types of three-phase power systems. These three-phase connections are shown schematically in Figure 3-9. Keep in mind that these methods apply not only to three-phase AC generators, but also to three-phase transformer windings and three-phase motor windings.

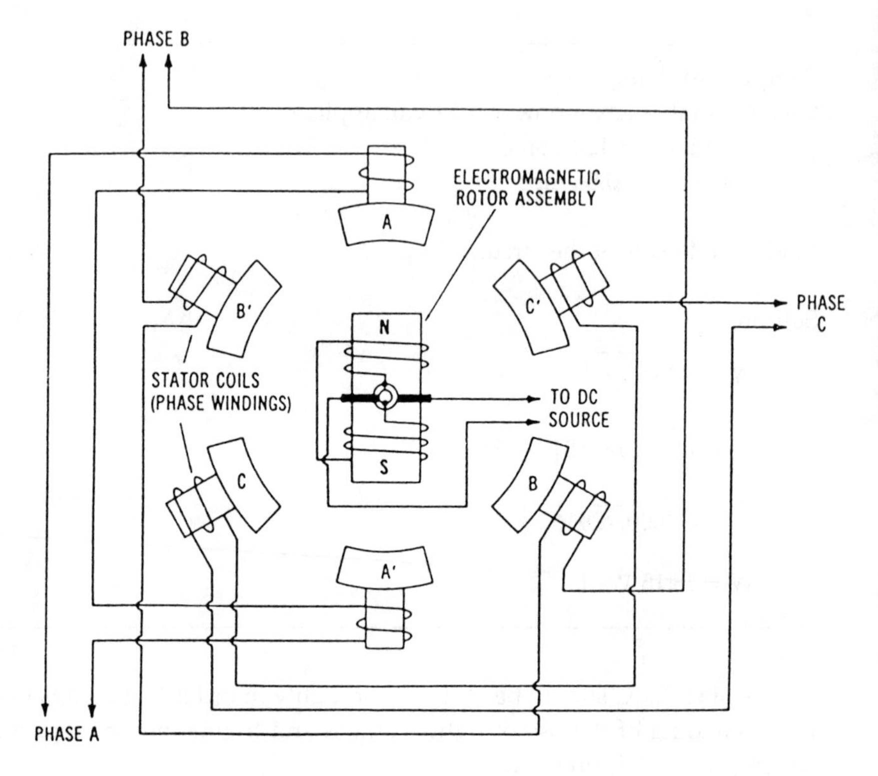

Figure 3-7. Simplified drawing showing the basic construction of a three-phase AC alternator.

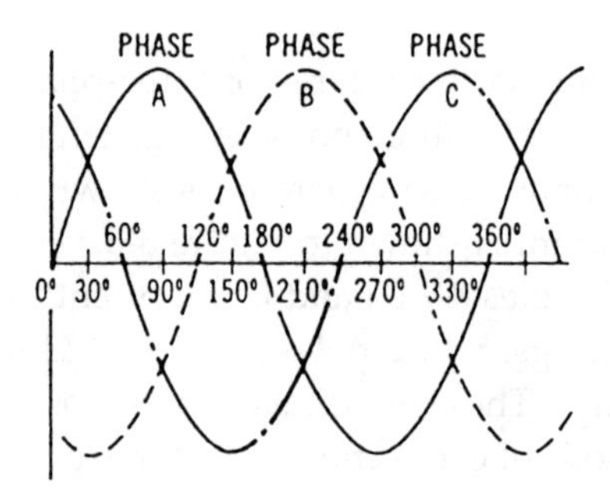

Figure 3-8. Three-phase output waveform developed by an alternator.

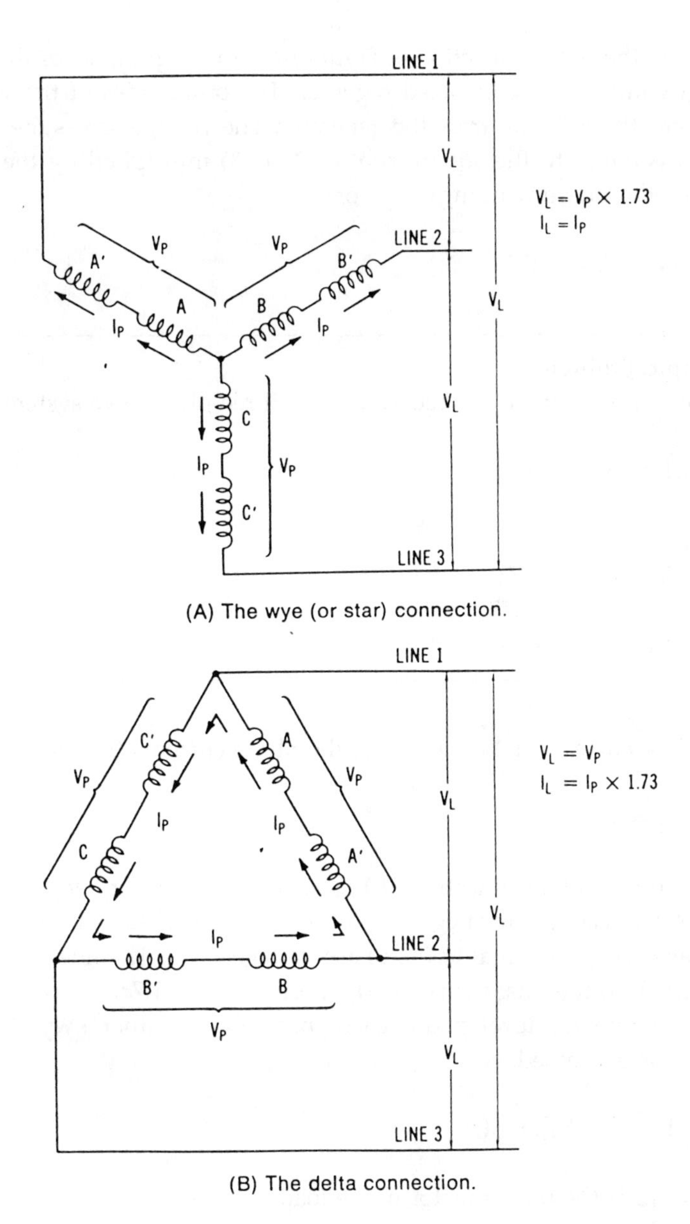

(A) The wye (or star) connection.

(B) The delta connection.

Figure 3-9. The two methods of connecting three-phase stator coils together.

In the wye connection of Figure 3-9a, the beginnings or the ends of each winding are connected together. The other sides of the windings become the AC lines from the generator. The voltage across the AC lines (V_L,) is equal to the square root of 3 (1.73) multiplied by the voltage across the phase windings (V_P), or

$$V_L = V_P \times 1.73.$$

Sample Problem:
Given: Phase voltage = 120 Volts for a three-phase wye system

Find: Line voltage

Solution:

$$V_L = V_p \times 1.73$$
$$= 120 \times 1.73$$
$$V_L = 208 \text{ volts}$$

The line currents (I_L) are equal to the phase currents (I_P), or

$$I_L = I_P.$$

In the delta connection of Figure 3-9b, the end of one phase winding is connected to the beginning of the adjacent phase winding The line voltages (V_L) are equal to the phase voltages (V_P). The line currents (I_L) are equal to the phase currents (I_P) multiplied by 1.73.

The power developed in each phase (P_P) for either a wye or a delta circuit is expressed as:

$$P_P = V_P \times I_P \times pf,$$

where pf is the power factor of the load.

Sample Problem:
Given: A three-phase delta system has a phase voltage of 240 volts, phase current of 20 amperes, and a power factor of 0.75

Find: Power per Phase

Solution:

$$P_P = V_p \times I_p \times pf$$
$$P_p = 240 \times 20 \times 0.75$$
$$P_p = 3600 \text{ Watts}$$

The total power (P_T) developed by all three phases of a three-phase system is expressed as:

$$P_T = 3 \times P_P$$
$$= 3 \times V_P \times I_P \times pf$$
$$= 1.73 \times V_L \times I_L \times pf.$$

Sample Problem:
Given: A three-phase wye system has a phase voltage of 277 volts, phase current of 10 amperes, and a power factor of 0.85

Find: Total three-phase Power

Solution:

$$P_T = 3 \times 277V \times 10A \times 0.85$$
$$P_T = 7063.5 \text{ Watts}$$

We can summarize three-phase power relationships as follows:

Volt-amperes per phase (VA_P) $= V_P I_P$

Total volt-amperes (VA_T) $= 3V_P I_P$

$= 1.73 V_L I_L$

Sample Problem:
Given: A three-phase delta system has a line voltage of 208 volts and a line current of 4.86 amps

Find: Total three-phase volt-amperes

Solution:

$VA = 1.73 \times 200V \times 4.86A$

$VA = 1748.8$ Volt-amperes

$$\text{Power factor (pf)} = \frac{\text{true power (W)}}{1.73\ V_L I_L}$$

$$= \frac{W}{3V_P I_P}$$

Power per phase (P_P) $= V_P I_P \times pf$

Total power (P_T) $= 3V_P I_P \times pf$

$= 1.73 V_L I_L \times pf$

where,

V_L is the line voltage in volts,

I_L is the line current in amperes

V_P is the phase voltage in volts,

I_P is the phase current in amperes.

Sample Problem:

Given: A three-phase wye system has the following values—

Phase voltage = 120 V;

Phase current = 18.5 A;

Power factor = 0.95

Find: Total Three-phase Power of the circuit

Solution:

$P_T = 3 \times V_P \times I_P \times pf$

$= 3 \times 120 \times 18.5 \times 0.95$

$P_T = 6{,}327\ W = 6.327\ kW$

Calculations involving three-phase power are somewhat more complex than for single-phase power calculations. We must keep in mind the difference between phase values and line values to avoid making mistakes.

LOAD CHARACTERISTICS

In order to plan for electrical power system load requirements it is necessary to understand the electrical characteristics of all the loads connected to the power system. The types of power supplies and distribution systems which a building uses are determined by the load characteristics. All loads may be considered as either resistive, inductive capacitive, or a combination of these. Examples of these loads are shown in Figure 3-10. We should be aware of the effects which various types of loads will have on the power system. The nature of alternating current causes certain electrical circuit properties to exist.

One primary factor which affects the electrical power system is the presence of inductive loads. These are mainly electric motors. To counteract the inductive effects utility companies use power-factor corrective capacitors as part of the power system design. Capacitor units are located at substations to improve the power factor of the system. The inductive effect, therefore increases the cost of a power system and reduces the actual amount of power which is converted to another form of energy.

Load (Demand) Factor

One electrical load relationship that is important to understand is the *load or demand factor*. Load factor expresses the ratio between the average power requirement and the peak power requirement or:

$$\text{load (demand) factor} = \frac{\text{average demand (kW)}}{\text{peak demand (kW)}}.$$

Sample Problem:

Given: A factory has a peak demand of 12 MW and an average power demand of 9.86 MW.

Find: The load (demand) factor for the factory.

Solution:

$$D\ F = \frac{\text{Avg. Demand}}{\text{Peak Demand}} = \frac{9.86\ \text{MW}}{12\ \text{MW}}$$

$$D\ F = 0.82$$

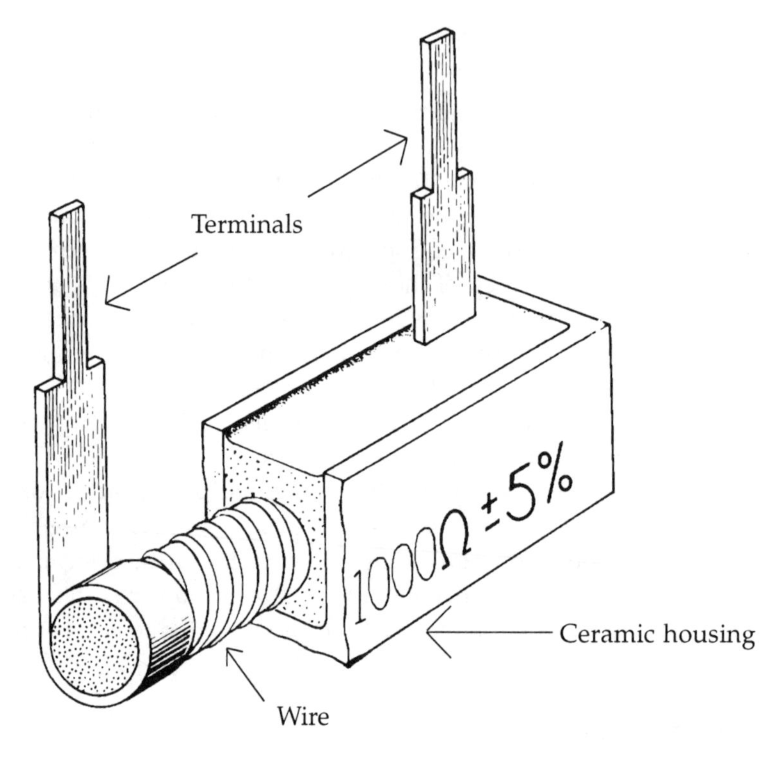

Figure 3-10a. Resistive-load, high-wattage wire-wound resistor.

The average demand for an industry or commercial building is the average electrical power used over a specific time period. The peak demand is the maximum amount of power (kW) used during that time period. The load profile shown in Figure 3-11 shows a typical industrial demand versus time curve for a working day. Demand peaks that far exceed the average demand cause a decrease in the load factor ratio. Low load factors result in an additional billing charge by the utility company.

Utility companies must design power distribution systems to meet the peak demand time and insure that their generating capacity will be able to meet this peak power demand. Therefore it is inefficient electrical design for an industry to operate at a low load factor, since this represents a significant difference between peak power demand and average power demand. Every industry should attempt to raise its load factor to the maximum level it can. By minimizing the peak demands of industrial plants power demand control systems and procedures can help increase the efficiency of our nation's electrical power systems.

Figure 3-10b. Inductive load—filter circuits (inductors). (Courtesy TRW/UTC Transformers)

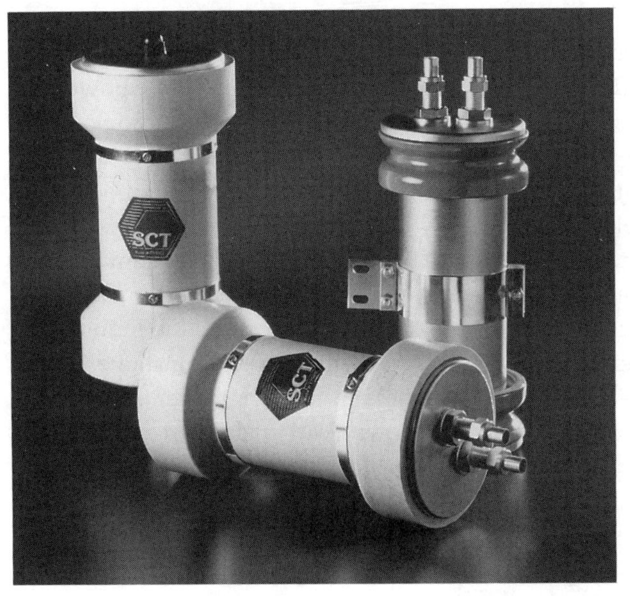

Figure 3-10c. Capacitive load—high power capacitors. (Courtesy Sprague-Goodman Electronics)

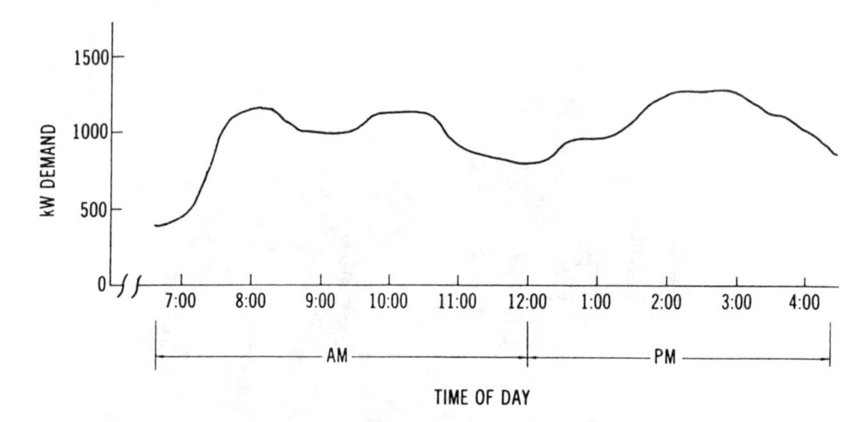

Figure 3-11. Load profile for an industrial plant.

Be careful not to confuse the load factor of a power system with the power factor. The power factor is the ratio of power converted (true power) to the power delivered to a system (apparent power).

POWER FACTOR CORRECTION

Most industries use a large number of electric motors; therefore, industrial plants represent highly inductive loads. This means that industrial power systems operate at a power factor of less than unity (1.0). However, it is undesirable for an industry to operate at a low power factor, since the electrical power system will have to supply more power to the industry than is actually used.

A given value of volt-amperes (voltage × current) is supplied to an industry by the electrical power system. If the power factor (pf) of the industry is low, the current must be higher since the power converted by the total industrial load equals VA × pf. The value of power factor decreases as the reactive power (unused power) drawn by the industry increases. This is shown in Figure 3-12. We will assume a constant value of true power to see the effect of increases in reactive power drawn by a load. The smallest reactive power shown (var_1) results in the volt-ampere value of VA_1. As reactive power is increased, as shown by the var_2 and var_3 values, more volt-amperes (VA_2 and VA_3) must be drawn from the source. This is true since the voltage component of the supplied volt-amperes remains constant. This example represents the same effect as a

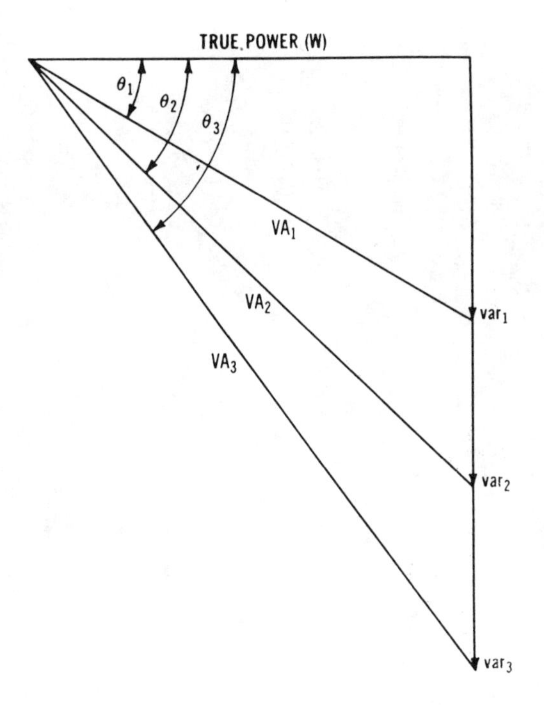

Figure 3-12. Effect of increases in reactive power (var) on apparent power (VA).

decrease in the power factor, since pf = W/VA, and as VA increases, the pf will decrease if W remains constant.

Utility companies usually charge industries for operating at power factors below a specified level. It is desirable for industries to "correct" their power factor to avoid such, charges and to make more economical use of electrical energy. Two methods may be used to cause the power factor to increase: (1) power factor-corrective capacitors, and (2) three-phase synchronous motors. Since the effect of capacitive reactance is opposite to that of inductive reactance, their reactive effects will counteract one another. Either power factor-corrective capacitors or three-phase synchronous motors may be used to add the effect of capacitance to an AC power line. Some power factor-corrective capacitors are shown in Figure 3-13.

An example of power factor correction is shown in Figure 3-14. We will assume from the example that both true power and inductive reactive power remain constant at values of 10 kW and 10 kvar. In Figure 3-14a, the formulas show that the power factor equals 70%. However, if 5-kvar capacitive reactive power in introduced into the electrical power

Figure 3-13a. Three-phase power factor correction capacitors manufactured by Sangamo Corporation.

Figure 3-13b. Pole-mounted power factor correction capacitors. (Courtesy McGraw-Edison Co.)

system, the net reactive power becomes 5 kvar (10-kvar inductive minus 5-kvar capacitive), as shown in Figure 3-14b. With the addition of 5-kvar capacitive to the system, the power factor is increased to 89%. Now, in Figure 3-14c, if 10-kvar capacitive is added to the power system, the total reactive power (kvar) becomes zero. The true power is now equal to the apparent power; therefore, the power factor is 1.0, or 100%, which is characteristic of a purely resistive circuit. The effect of the increased capacitive reactive power in the system is to increase or "correct" the power factor and thus reduce the current drawn from the power distribution lines that supply the loads. In many cases it is beneficial for industries to invest in either power factor-corrective capacitors or three-phase synchronous motors to correct their power factor. Calculations may be simplified by using the chart of Table 3-1.

Utility companies also attempt to correct the power factor of the power distribution system. A certain quantity of inductance is present in most of the power distribution system including the generator windings the transformer windings and the power lines. To counteract the inductive effects, utilities use power factor-corrective capacitors such as the units shown in Figure 3-13.

CAPACITORS FOR POWER FACTOR CORRECTION

Static capacitors are used for power factor correction in the system. They are constructed similar to the smaller capacitors used in electrical equipment that have metal-foil plates separated by paper insulation. Ordinarily, static capacitors are housed in metal tanks, so that the plates can be immersed in an insulating oil to improve high-voltage operation. The usual operating voltages of static capacitors is from 230 volts to 13.8 kilovolts. These units are connected in parallel with power lines usually at the industrial plants, to increase the system power factor. Their primary disadvantage is that their capacitance cannot be adjusted to compensate for changing power factors.

Power factor correction can also be accomplished by using *synchronous capacitors* connected across the power lines. Three-phase synchronous motors (see Chapter 10) are also called synchronous capacitors. The advantage of synchronous capacitors over static capacitors is that their capacitive effect can be adjusted as the system power factor increases or decreases. The capacitive effect of a synchronous capacitor is

Table 3-1. Kilowatt (kW) multipliers for determining capacitor kilovars (kVAR).

	Desired Power Ractor in Percentage																				
	80	81	82	83	84	85	86	87	88	89	90	91	92	93	94	95	96	97	98	99	1.0
50	0.982	1.008	1.034	1.060	1.086	1.112	1.139	1.165	1.192	1.220	1.248	1.276	1.306	1.337	1.369	1.403	1.440	1.481	1.529	1.589	1.732
51	0.937	0.962	0.989	1.015	1.041	1.067	1.094	1.120	1.147	1.175	1.203	1.231	1.261	1.292	1.324	1.358	1.395	1.436	1.484	1.544	1.686
52	0.893	0.919	0.945	0.971	0.997	1.023	1.050	1.076	1.103	1.131	1.159	1.187	1.217	1.248	1.280	1.314	1.351	1.392	1.440	1.500	1.643
53	0.850	0.876	0.902	0.928	0.954	0.980	1.007	1.033	1.060	1.088	1.116	1.144	1.174	1.205	1.237	1.271	1.308	1.349	1.397	1.457	1.600
54	0.809	0.835	0.861	0.887	0.913	0.939	0.966	0.992	1.019	1.047	1.075	1.103	1.133	1.164	1.196	1.230	1.267	1.308	1.356	1.416	1.559
55	0.769	0.795	0.821	0.847	0.873	0.899	0.926	0.952	0.979	1.007	1.035	1.063	1.093	1.124	1.156	1.190	1.227	1.268	1.316	1.376	1.519
56	0.730	0.756	0.782	0.808	0.834	0.860	0.887	0.913	0.940	0.968	0.996	1.024	1.054	1.085	1.117	1.151	1.188	1.229	1.277	1.337	1.480
57	0.692	0.718	0.744	0.770	n.796	0.822	0.849	0.875	0.902	0.930	0.958	0.986	1.016	1.047	1.079	1.113	1.150	1.191	1.239	1.299	1.442
58	0.655	0.681	0.707	0.733	0.759	0.785	0.812	0.838	0.865	0.893	0.921	0.949	0.979	1.010	1.042	1.076	1.113	1.154	1.202	1.262	1.405
59	0.619	0.645	0.671	0.697	0.723	0.749	0.776	0.802	0.829	0.857	0.885	0.913	0.943	0.974	1.006	1.040	1.077	1.118	1.166	1.226	1.369
60	0.583	0.609	0.635	0.661	0.687	0.713	0.740	0.766	0.793	0.821	0.849	0.877	0.907	0.938	0.970	1.004	1.041	1.082	1.130	1.190	1.333
61	0.549	0.575	0.601	0.627	0.653	0.679	0.706	0.732	0.759	0.787	0.815	0.843	0.873	0.904	0.936	0.970	1.007	1.048	1.096	1.156	1.299
62	0.516	0.542	0.568	0.594	0.620	0.646	0.673	0.699	0.725	0.754	0.782	0.810	0.840	0.871	0.903	0.937	0.974	1.015	1.063	1.123	1.266
63	0.483	0.509	0.535	0.561	0.587	0.613	0.640	0.666	0.693	0.721	0.749	0.777	0.807	0.838	0.870	0.904	0.941	0.982	1.030	1.090	1.233
64	0.451	0.474	0.503	0.529	0.555	0.581	0.608	0.634	0.661	0.689	0.717	0.745	0.775	0.806	0.838	0.872	0.909	0.950	0.998	1.068	1.201
65	0.419	0.445	0.471	0.497	0.523	0.549	0.576	0.602	0.629	0.657	0.685	0.713	0.743	0.774	0.805	0.840	0.877	0.918	0.966	1.026	1.169
66	0.388	0.414	0.440	0.466	0.492	0.518	0.545	0.571	0.598	0.626	0.654	0.682	0.712	0.743	0.775	0.809	0.846	0.887	0.935	0.995	1.138
67	0.358	0.384	0.410	0.436	0.462	0.488	0.515	0.541	0.568	0.596	0.624	0.652	0.682	0.713	0.745	0.779	0.816	0.857	0.905	0.965	1.108
68	0.328	0.354	0.380	0.406	0.432	0.458	0.485	0.511	0.538	0.56'6	0.594	0.622	0.652	0.683	0.715	0.749	0.786	0.827	0.875	0.935	1.078
69	0.299	0.325	0.351	0.377	0.403	0.429	0.456	0.482	0.509	0.537	0.565	0.593	0.623	0.654	0.686	0.720	0.757	0.798	0.846	0.906	1.049
70	0.270	0.296	0.322	0.348	0.374	0.400	0.427	0.453	0.480	0.508	0.536	0.564	0.594	0.625	0.657	0.691	0.728	0.769	0.817	0.877	1.020
71	0.242	0.268	0.294	0.320	0.346	0.372	0.399	0.425	0.452	0.480	0.508	0.536	0.566	0.597	0.629	0.663	0.700	0.741	0.789	0.849	0.992
72	0.214	0.240	0.266	0.292	0.318	0.344	0.371	0.397	0.424	0.452	0.480	0.508	0.538	0.569	0.601	0.635	0.672	0.713	0.761	0.821	0.964

73	0.186	0.212	0.238	0.264	0.290	0.316	0.343	0.369	0.396	0.424	0.452	0.480	0.510	0.541	0.573	0.607	0.644	0.685	0.733	0.793	0.936
74	0.159	0.185	0.211	0.237	0.263	0.289	0.316	0.342	0.369	0.397	0.425	0.453	0.483	0.514	0.546	0.580	0.617	0.658	0.706	0.766	0.909
75	0.132	0.158	0.184	0.210	0.236	0.262	0.289	0.315	0.342	0.370	0.398	0.426	0.456	0.487	0.519	0.553	0.590	0.631	0.679	0.739	0.882
76	0.105	0.131	0.157	0.183	0.209	0.235	0.262	0.288	0.315	0.343	0.371	0.399	0.429	0.460	0.492	0.526	0.563	0.604	0.652	0.712	0.855
77	0.079	0.105	0.131	0.157	0.183	0.209	0.236	0.262	0.289	0.317	0.345	0.373	0.403	0.434	0.466	0.500	0.537	0.578	0.626	0.686	0.829
78	0.052	0.078	0.104	0.130	0.156	0.182	0.209	0.235	0.262	0.290	0.318	0.346	0.376	0.407	0.439	0.473	0.510	0.554	0.599	0.569	0.802
79	0.026	0.052	0.078	0.104	0.130	0.156	0.183	0.209	0.236	0.264	0.292	0.320	0.350	0.381	0.413	0.447	0.484	0.525	0.573	0.633	0.776
80	0.000	0.026	0.052	0.078	0.104	0.130	0.157	0.183	0.210	0.238	0.266	0.294	0.324	0.355	0.387	0.421	0.458	0.499	0.547	0.609	0.750
81		0.000	0.026	0.052	0.078	0.104	0.131	0.157	0.184	0.212	0.240	0.268	0.298	0.329	0.361	0.395	0.432	0.473	0.521	0.581	0.724
82			0.000	0.026	0.052	0.078	0.105	0.131	0.158	0.186	0.214	0.242	0.272	0.303	0.335	0.369	0.406	0.447	0.495	0.555	0.698
83				0.000	0.026	0.052	0.079	0.105	0.132	0.160	0.188	0.216	0.246	0.277	0.309	0.343	0.380	0.421	0.469	0.529	0.672
84					0.000	0.026	0.053	0.079	0.106	0.134	0.162	0.190	0.220	0.251	0.283	0.317	0.354	0.395	0.443	0.503	0.646
85						0.000	0.027	0.053	0.080	0.108	0.136	0.164	0.194	0.225	0.257	0.291	0.328	0.369	0.417	0.477	0.620
86							0.000	0.026	0.053	0.081	0.109	0.137	0.167	0.198	0.230	0.264	0.301	0.342	0.390	0.450	0.593
87								0.000	0.027	0.055	0.083	0.111	0.141	0.172	0.204	0.238	0.275	0.316	0.364	0.424	0.567
88									0.000	0.028	0.056	0.084	0.114	0.145	0.177	0.211	0.248	0.289	0.337	0.397	0.540
89										0.000	0.028	0.056	0.086	0.117	0.149	0.183	0.220	0.261	0.309	0.369	0.512
90											0.000	0.028	0.058	0.089	0.121	0.155	0.192	0.233	0.281	0.341	0.484
91												0.000	0.030	0.061	0.093	0.127	0.164	0.205	0.253	0.313	0.456
92													0.000	0.031	0.063	0.097	0.134	0.175	0.223	0.283	0.426
93														0.000	0.032	0.066	0.103	0.144	0.192	0.252	0.395
94															0.000	0.034	0.071	0.112	0.160	0.220	0.363
95																0.000	0.037	0.079	0.126	0.186	0.329
96																	0.000	0.041	0.089	0.149	0.292
97																		0.000	0.048	0.108	0.254
98																			0.000	0.060	0.203
99																				0.000	0.143

Example: Total kW input of load from wattmeter reading 100 kW at a power factor of 60%. The leading reactive kVAR necessary to raise the power factor to 90% is found by multiplying the 100 kW by the factor found in the table, which is .849. Then 100 kW $\times$ 0.849 = 84.9 kVAR. Use 85 kVAR.

$$\text{kVA} = \sqrt{\text{kW}^2 + \text{kvar}^2}$$
$$= \sqrt{10^2 + 10^2}$$
$$= \sqrt{100 + 100}$$
$$= \sqrt{200}$$
$$= 14.14$$
$$\text{pf} = \frac{\text{TRUE POWER}}{\text{APPARENT POWER}}$$
$$= \frac{10\text{ kW}}{14.14\text{ kVA}}$$
$$= 0.7\ (70\%)$$

(A) Reactive power = 10-kvar inductive.

$$\text{kVA} = \sqrt{\text{kW}^2 + \text{kvar}^2}$$
$$= \sqrt{10^2 + 5^2}$$
$$= \sqrt{100 + 25}$$
$$= \sqrt{125}$$
$$= 11.18$$
$$\text{pf} = \frac{\text{TRUE POWER}}{\text{APPARENT POWER}}$$
$$= \frac{10\text{ kW}}{11.18\text{ kVA}}$$
$$= 0.89\ (89\%)$$

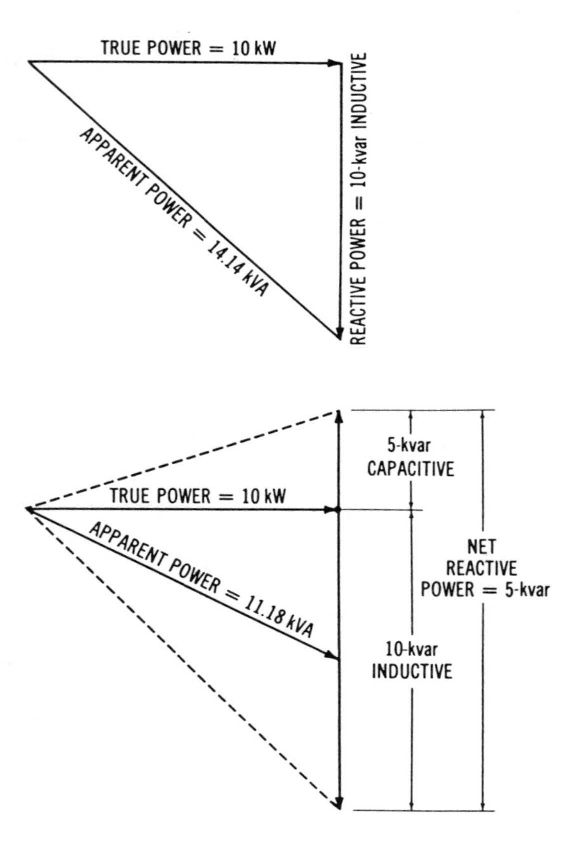

(B) Reactive power = 10-kvar inductive, 5-kvar capacitive.

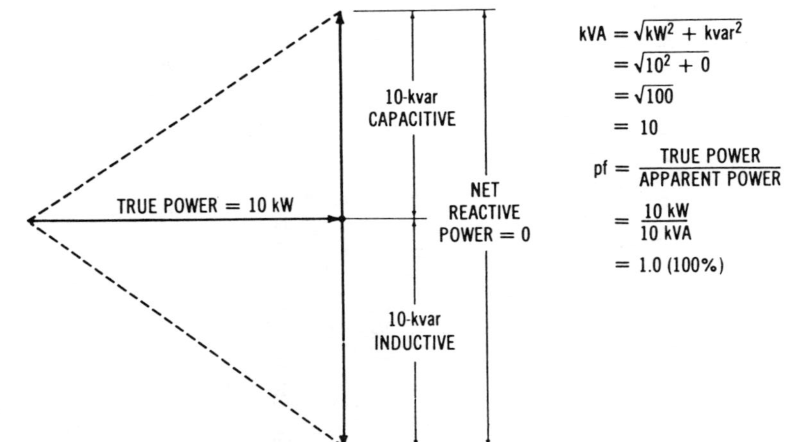

(C) Reactive power = 10-kvar inductive, 10-kvar capacitive.

Figure 3-14. Illustration of the effect of capacitive reactance on an inductive load.

easily changed by varying the DC excitation voltage applied to the rotor of the machine. Industries considering the installation of either static or synchronous capacitors should first compare the initial equipment cost and the operating cost against the savings brought about by an increased system power factor.

BALANCED THREE-PHASE LOADS

A balanced three-phase load means that the resistance or impedances connected across each phase of The system are equal. Three-phase motors are one type of balanced load. The following relationships exist in a balanced three-phase delta system. The line voltages (V_L) are equal to the phase voltages (V_P). The line currents (I_L) are equal to the phase currents (I_P) multiplied by 1.73. Thus:

$$V_L = V_P$$
$$I_L = I_P \times 1.73$$

For a balanced three-phase wye system, the method used to find the voltages and currents is similar. The voltage across the AC lines (V_L) is equal to the square root of 3 (1.73) multiplied by the voltage across the phase windings (V_P), or

$$V_L = V_P \times 1.73.$$

The line currents (I_L) are equal to the phase currents (I_P), or

$$I_L = I_P$$

The power developed in each phase (P_P) for either a wye or a delta circuit is expressed as:

$$P_P = V_P \times I_P \times pf,$$

where,

pf is the power factor (phase angle between voltage and current) of the load.

The total power developed by all three phases of a three-phase generator (P_T) is expressed as:

$$\begin{aligned} P_T &= 3 \times P_P \\ &= 3 \times V_P \times I_P \times pf \\ &= 1.73 \times V_L \times I_L \times pf. \end{aligned}$$

As an example, we can calculate the phase current (I_P) line current (I_L) and total power (P_T) for a 240-volt three-phase delta-connected system with 1000-watt load resistances connected across each power line. The phase current is:

$$I_P = \frac{P_P}{V_P}$$

$$= \frac{1000 \text{ watts}}{240 \text{ volts}}$$

$$= 4.167 \text{ amperes}$$

The line current is calculated as:

$$\begin{aligned} I_L &= I_P \times 1.73 \\ &= 4.167 \text{ x } 1.73 \\ &= 7.21 \text{ amperes.} \end{aligned}$$

Thus the total power is determined as:

$$\begin{aligned} P_T &= 3 \times P_P \\ &= 3 \times 1000 \text{ watts} \\ &= 3000 \text{ watts.} \end{aligned}$$

Another way to calculate the total power is:

$$\begin{aligned} P_T &= 1.73 \times V_L \times I_L \times pf \\ &= 1.73 \times 34() \text{ volts} \times 7.21 \text{ amperes} \times 1 \\ &= 2993.59 \text{ watts} \end{aligned}$$

(Note that your answer depends upon the number of decimal places to which you carry your calculations. Round off I_L at 7.2 amperes and your answer for the last calculation is 2989.44 watts instead of 2993.59 watts.)

The following is an example of the power calculation for a three-phase wye system. For the problem, "find the line current (I_L) and power per phase (P_P) of a balanced 20,000-watt 277/480-volt three-phase wye system operating: at a 0.75 power factor," we use the formula

$$P_T = V_L \times I_L \times 1.73 \times pf.$$

Transposing and substituting, we have:

$$I_L = \frac{P_T}{V_L \times 1.73 \times pf}$$

$$= \frac{20{,}000 \text{ watts}}{480 \text{ volts} \times 1.73 \times 0.75}$$

$$= 32.11 \text{ amperes}$$

Then, for the power per phase (P_P), divide the 20,000 watts of the system by 3 (the number of phases) for a value of 6,666.7 watts (6.66 kW).

UNBALANCED THREE-PHASE LOADS

Often, three-phase systems are used to supply power to both three-phase and single-phase loads. If three identical single-phase loads were connected across each set of power lines, the three-phase system would still be balanced. However, this situation is usually difficult to accomplish, particularly when the loads are lights. Unbalanced loads exist when the individual power lines supply loads that are not of equal resistances or impedances.

The total power converted by the loads of an unbalanced system must be calculated by looking at each phase individually. Total power of a three-phase unbalanced system is:

$$P_T = P_{P-A} + P_{P-B} + P_{P-C}$$

where the power-per-phase (P_P) values are added. Power per phase is found in the same way as when dealing with balanced loads:

$P_P = V_P \times I_P \times$ power factor.

The current flow in each phase may be found if we know the power per phase and the phase voltage of the system. The phase currents are found in the following manner:

1. The following 120-volt single-phase loads are connected to a 120/208-volt wye system. Phase A has 2000 watts at a 0.75 power factor, Phase B has 1000 watts at a 0.85 power factor, and Phase C has 3000 watts at a 1.0 power factor. Determine the total power of the three-phase system and the current flow through each line.

2. To find the phase currents, use the formula $P_P = V_P \times I_P \times$ power factor, and transpose. Thus, we have $I_P = P_P/V_P \times$ pf. Substitution of the values for each leg of the system gives us:

   a.

$$*I_{P\text{-}A} = \frac{P_{P\text{-}A}}{V_P \times pf}$$

$$= \frac{2000 \text{ watts}}{120 \text{ volts} \times 0.75}$$

$$= 22.22 \text{ amperes}$$

   b.

$$I_{P\text{-}B} = \frac{P_{P\text{-}B}}{V_P \times pf}$$

$$= \frac{1000 \text{ watts}}{120 \text{ volts} \times 0.85}$$

$$= 9.8 \text{ amperes}$$

---

*This notation means phase current (I_P) of "A" power line.

c.

$$I_{P-C} = \frac{P_{P-C}}{V_P \times pf}$$

$$= \frac{3000 \text{ watts}}{120 \text{ volts} \times 1.0}$$

$$= 25.0 \text{ amperes}$$

3. To determine the total power, use the total power formula for a three-phase unbalanced system that was just given.

$$P_T = P_{P-A} + P_{P-B} + P_{P-C}$$
$$= 2000 + 1\ 000 + 3000$$
$$= 6000 \text{ watts (6 kW)}$$

Both single-phase and three-phase AC power systems are used extensively. Electrical power systems involve:

1. Electrical power *sources*_such as generators.

2. *Distribution* of electrical power.

3. *Control* of electrical power by various methods.

4. Electrical power *conversion* systems or loads such as electrical lights and motors.

5. *Measurement* of electrical power-related quantities with specialized equipment.

Chapter 4

Electrical Power Production Systems

Most electrical power distributed in the United States is produced at power plants that are either fossil-fuel steam plants, nuclear-fission steam plants, or hydroelectric plants. A steam turbine generator unit is shown in Figure 4-1. Fossil-fuel and nuclear-fission plants utilize steam turbines to deliver the mechanical energy needed to rotate the large three-phase alternators which produce massive quantities of electrical power. Hydroelectric plants ordinarily use vertically mounted hydraulic turbines. One is shown in Figure 4-2. These units convert the force of flowing water into mechanical energy to rotate three-phase alternators.

Power plants may be located near the energy sources, near cities, or near the large industries where great amounts of electrical power are consumed. The generating capacity of power plants in the United States is greater than the combined capacity of the next four leading countries of the world. Thus, we can see how dependent we are upon the efficient production of electrical power.

There are many residential, commercial and industrial customers of electricity in the United States today. To meet this vast demand for electrical power, power companies combine to produce tremendous quantities of electrical power. This vast quantity of electrical power is supplied by power-generating plants. Individual generating units which supply over 1000 megawatts of electrical power are now in operation at some power plants.

Electrical power can be produced in many ways, such as from chemical reactions, heat, light, or mechanical energy. The great majority of our electrical power is produced by power plants located throughout our country which convert the energy produced by burning coal, oil, or natural gas, the falling of water, or from nuclear reactions into electrical energy. Electrical generators at these power plants are driven by steam

Figure 4-1. Steam turbine generator unit used to produce electrical power; (top) shows the three-phase generator with DC excitation unit in front; (bottom) shows the steam turbine unit which is connected to the generator.

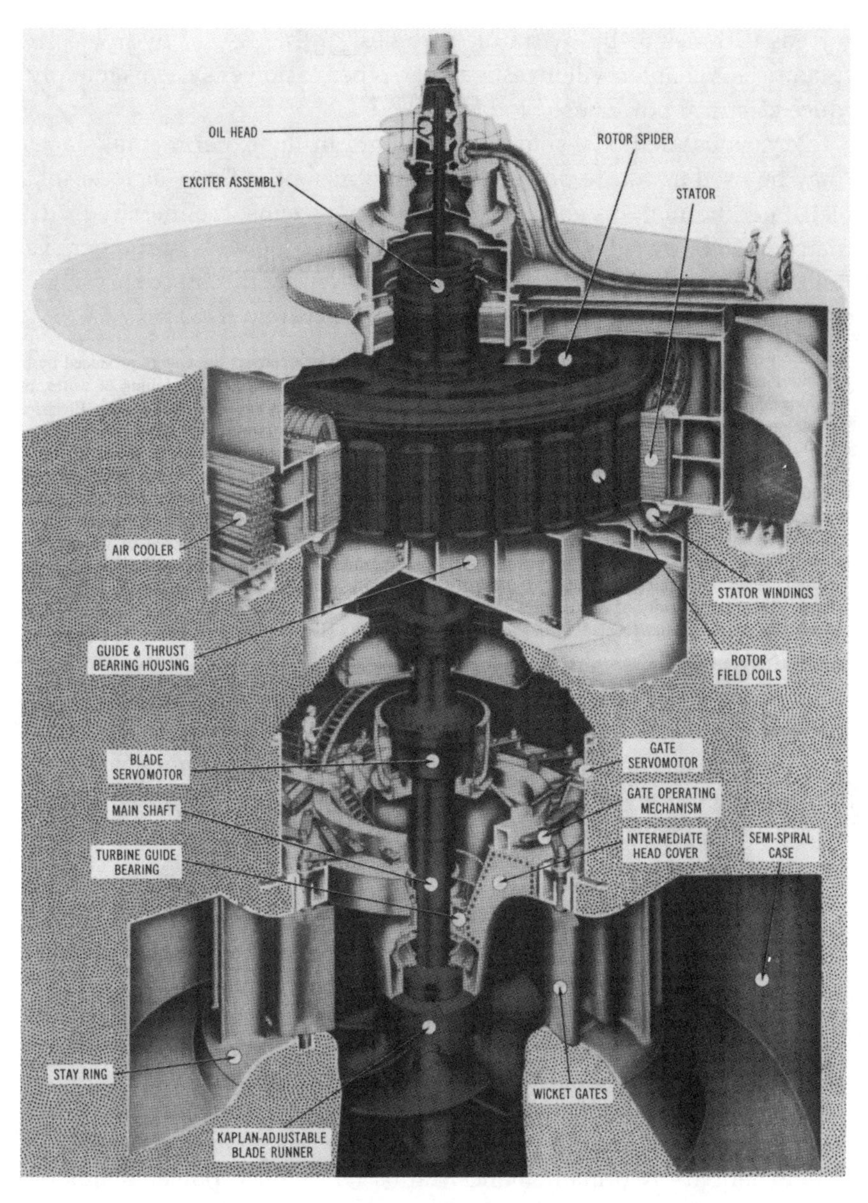

Figure 4-2. Hydraulic generator unit.

or gas turbines or by hydraulic turbines, in the case of hydroelectric plants. This chapter will investigate the types of power systems that produce electrical power used today.

Various methods, some of which are in the experimental stages, may be used as future power production methods. These include solar cells, geothermal systems, wind-powered systems, magnetohydrodynamic (MHD) systems, nuclear-fusion systems, fuel cells and other systems. The three major types of electrical power are discussed: (a) single-phase AC; (2) three-phase AC; and (3) direct current (DC).

SUPPLY AND DEMAND

The supply and demand situation for electrical energy is much different from other products which are produced by an organization and, then later, sold to consumers. Electrical energy must be supplied at the same time that it is demanded by consumers. There is no simple storage system which may be used to supply additional electrical energy at peak demand times. This situation is quite unique and necessitates the production of sufficient quantities of electrical energy to meet the demand of the consumers at any time. Accurate forecasting of load requirements at various given times must be maintained by utilities companies in order that they may recommend the necessary power plant output for a particular time of the year, week, or day.

PLANT LOAD AND CAPACITY FACTORS

There is a significant variation in the load requirement that must be met at different times. Thus, the power plant generating capacity is subject to a continual change. For the above reasons, much of the generating capacity of a power plant may be idle during low demand times. This means that not all the generators at the plant will be in operation.

There are two mathematical ratios with which power plants are concerned. These ratios are called load factor (demand factor) and capacity factor. They are expressed as:

$$\text{Load (demand) factor} = \frac{\text{Average load for a time period}}{\text{Peak load for a time period}}$$

$$\text{Capacity factor} = \frac{\text{Average load for a time period}}{\text{Output capacity of a power plant}}$$

It would be ideal, in terms of energy conservation, to keep these ratios as close to unity as possible.

FOSSIL FUEL SYSTEMS

Millions of years ago, large deposits of organic materials were formed under the surface of the earth. These deposits, which furnish our coal, oil, and natural gas, are known as fossil fuels. Of these, the most abundant fossil fuel is coal and coal-fired electrical power systems produce about one-half of the electrical power used in the United States. Natural-gas-fired systems are used for about one-fourth of our electrical power, while oil-fired systems produce around 1096 of the power at the present time. These relative contributions of each system to the total electrical power produced in the United States are subject to change due to the addition of new power generation facilities and fuel availability. At the present time, over 80% of our electrical energy is produced by fossil-fuel systems. It is important to note that these percentages vary from year to year.

A basic fossil-fuel power system is shown in Figure 4-3. In this type of system, a fossil fuel (coal, oil, or gas) is burned to produce heat energy. The heat from the combustion process is concentrated within a boiler where circulating water is converted to steam. The high-pressure steam is used to rotate a turbine. The turbine shaft is connected directly to the electrical/generator and provides the necessary mechanical energy to rotate the generator. The generator then converts the mechanical energy into electrical energy. Figures 4-4 and 4-5 show the layout and cross-section view of a typical coal-fired fossil fuel system.

HYDROELECTRIC SYSTEMS

The use of water power goes back to ancient times. It has been developed to a very high degree, but is now taking a secondary role due to the emphasis on other power sources that are being developed in our country today. Electrical power production systems using water power

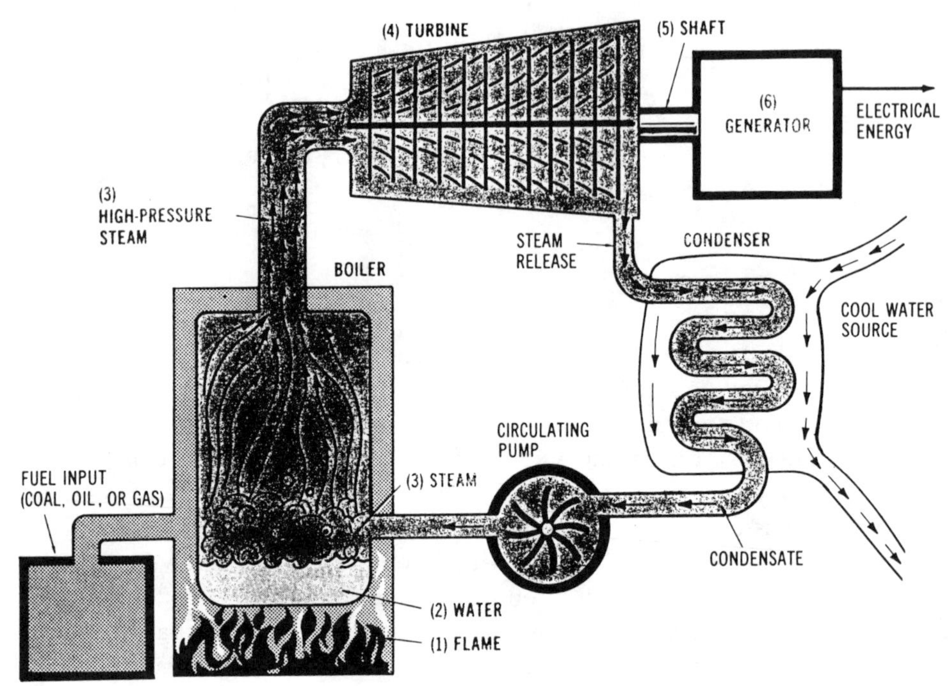

Heat from burning fuel (1) changes water in boiler (2) into steam (3), which spins turbine (4) connected by shaft (5) to generator (6), producing electrical energy.

Figure 4-3. A basic fossil fuel power system.

were developed for use in the early 20th Century.

The energy of flowing water may be used to generate electrical power. This method of power production is used in hydroelectric power systems as shown by the simple system illustrated in the diagram of Figure 4-6. Water, which is confined in a large reservoir, is channeled through a control gate which adjusts the flow rate. The flowing water passes through the blades and control vanes of a hydraulic turbine which produces rotation. This mechanical energy is used to rotate a generator that is connected directly to the turbine shaft. Rotation of the alternator causes electrical power to be produced. However, hydroelectric systems are limited by the availability of large water supplies. Many hydroelectric systems are part of multipurpose facilities. For instance, a hydroelectric power system may be part of a project planned for flood control, recreation, or irrigation. Some hydroelectric power systems are shown in Figures 4-7 through 4-9.

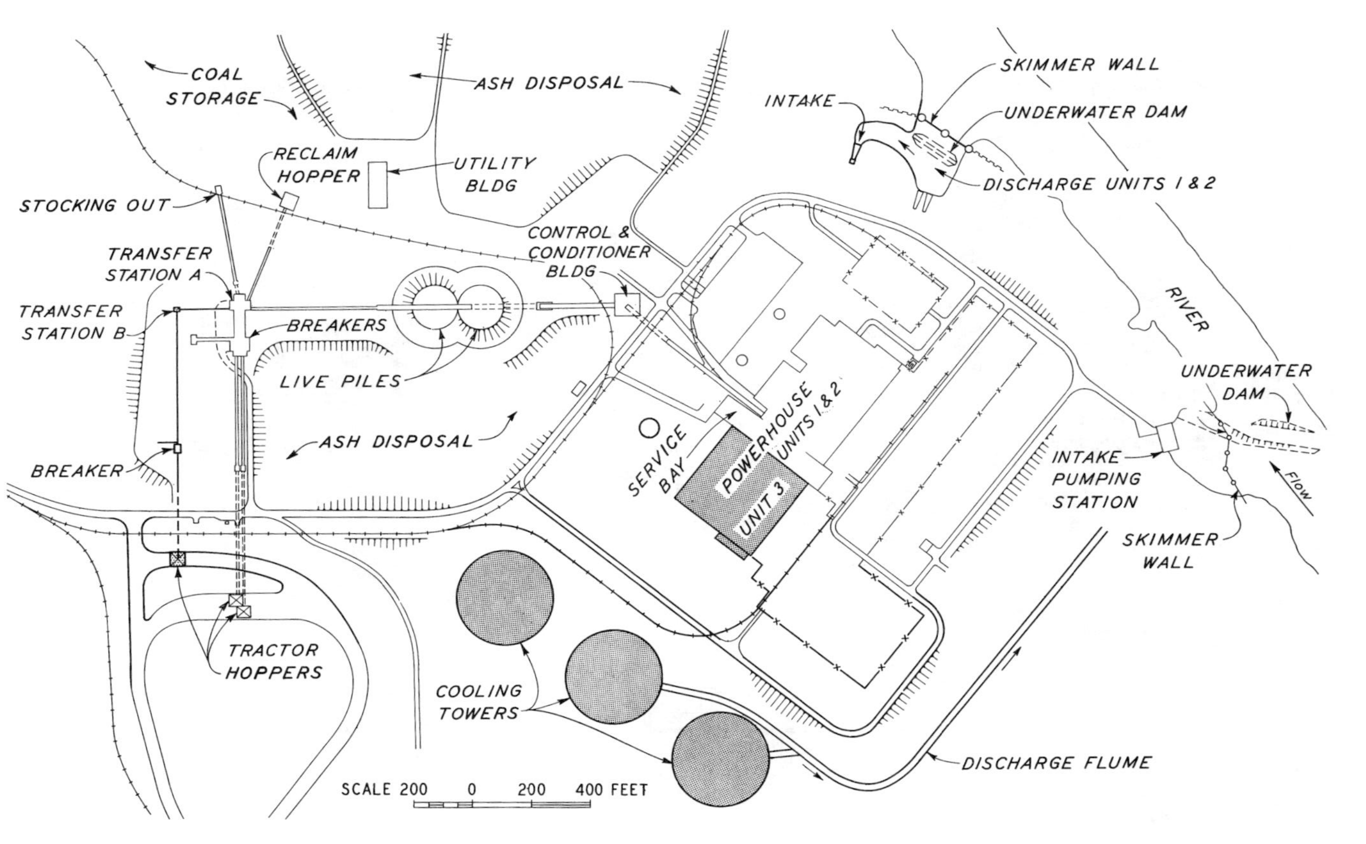

Figure 4-4. Layout of a typical coal-fired electrical power plant. (Courtesy Tennessee Valley Authority)

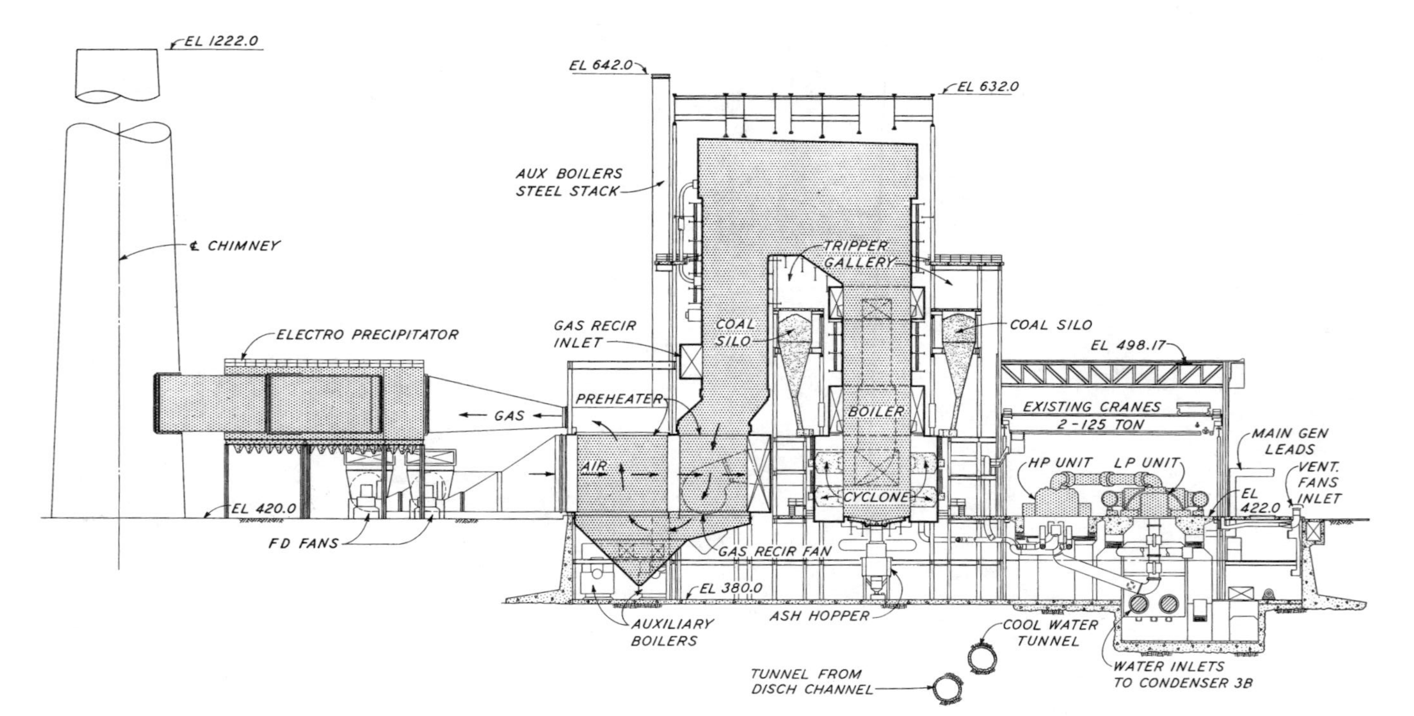

Figure 4-5. Cross-section view of a typical coal-fired power plant (Courtesy Tennessee Valley Authority)

PUMPED-STORAGE HYDROELECTRIC SYSTEMS

Several megawatts of electrical power are developed in the United States by pumped-storage hydroelectric systems. This type of system, shown in Figure 4-10, operates by pumping water to a higher elevation and storing it in a reservoir until it is released to drop to a lower elevation to drive the hydraulic turbines of a hydroelectric power-generating plant.

The variable nature of the electrical load demand makes pumped-storage systems desirable to operate. During low-load periods, the hydraulic turbines may be used as pumps to pump water to a storage reservoir of a higher elevation from a water source of a lower elevation. The water in the upper reservoir can be stored for long periods of time, if necessary. When the electrical load demand on the power system increases; the water in the upper reservoir can be allowed to flow (by gravity feed) through the hydraulic turbines which will then rotate the three-phase generators in the power plant. Thus, electrical power can be generated without any appreciable consumption of fuel. The pump-turbine and motor-generator units are constructed so that they will operate in two ways: (1) as a pump and motor, and (2) as a turbine and generator. In either case, the two machines are connected by a common shaft and operate together.

NUCLEAR-FISSION SYSTEMS

Nuclear power plants in operation today utilize reactors which function due to the nuclear-fission process. Nuclear fission is a complex reaction which results in the division of the nucleus of an atom into two nuclei. This splitting of the atom is brought about by the bombardment of the nucleus with neutrons, gamma rays, or other charged particles and is referred to as induced fission. When an atom is split, it releases a great amount of heat. A nuclear-fission power plant steam turbine system is shown in Figure 4-11.

In recent years, several nuclear-fission power plants have been put into operation. A nuclear-fission power system relies upon heat produced during a nuclear reaction process. Nuclear reactors "burn" nuclear material whose atoms are split causing the release of heat. This reaction is referred to as nuclear fission. The heat from the fission process is used to change circulating water into steam. The high-pressure steam rotates a turbine which is connected to an electrical generator. This is shown in the diagram of Figure 4-12.

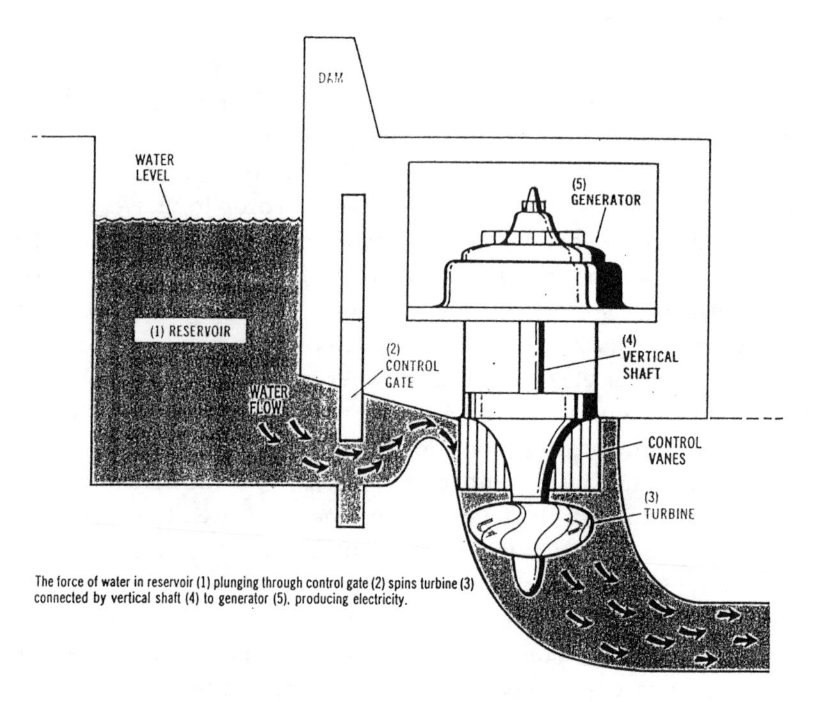

Figure 4-6. Drawing of a basic hydroelectric power system

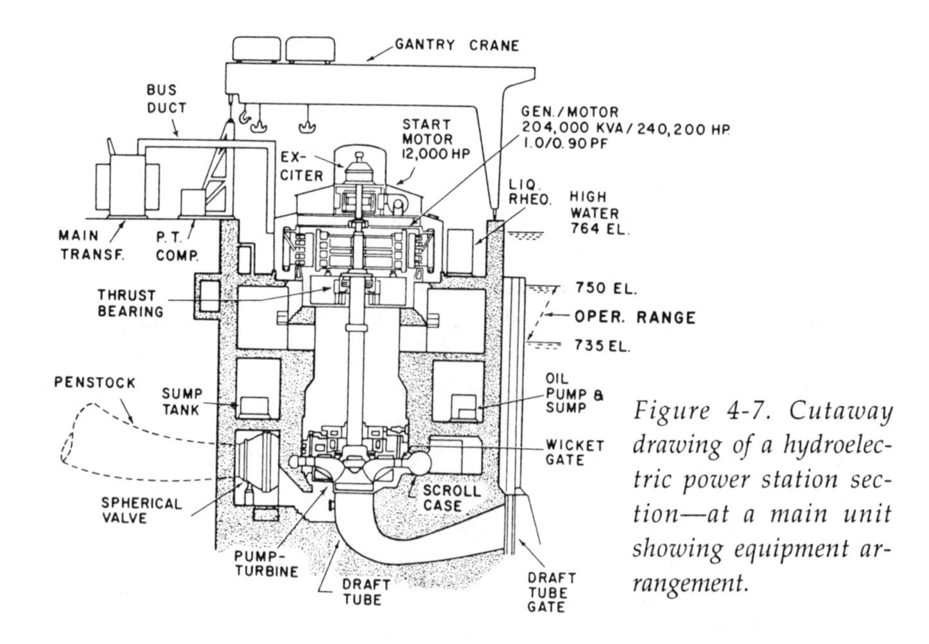

Figure 4-7. Cutaway drawing of a hydroelectric power station section—at a main unit showing equipment arrangement.

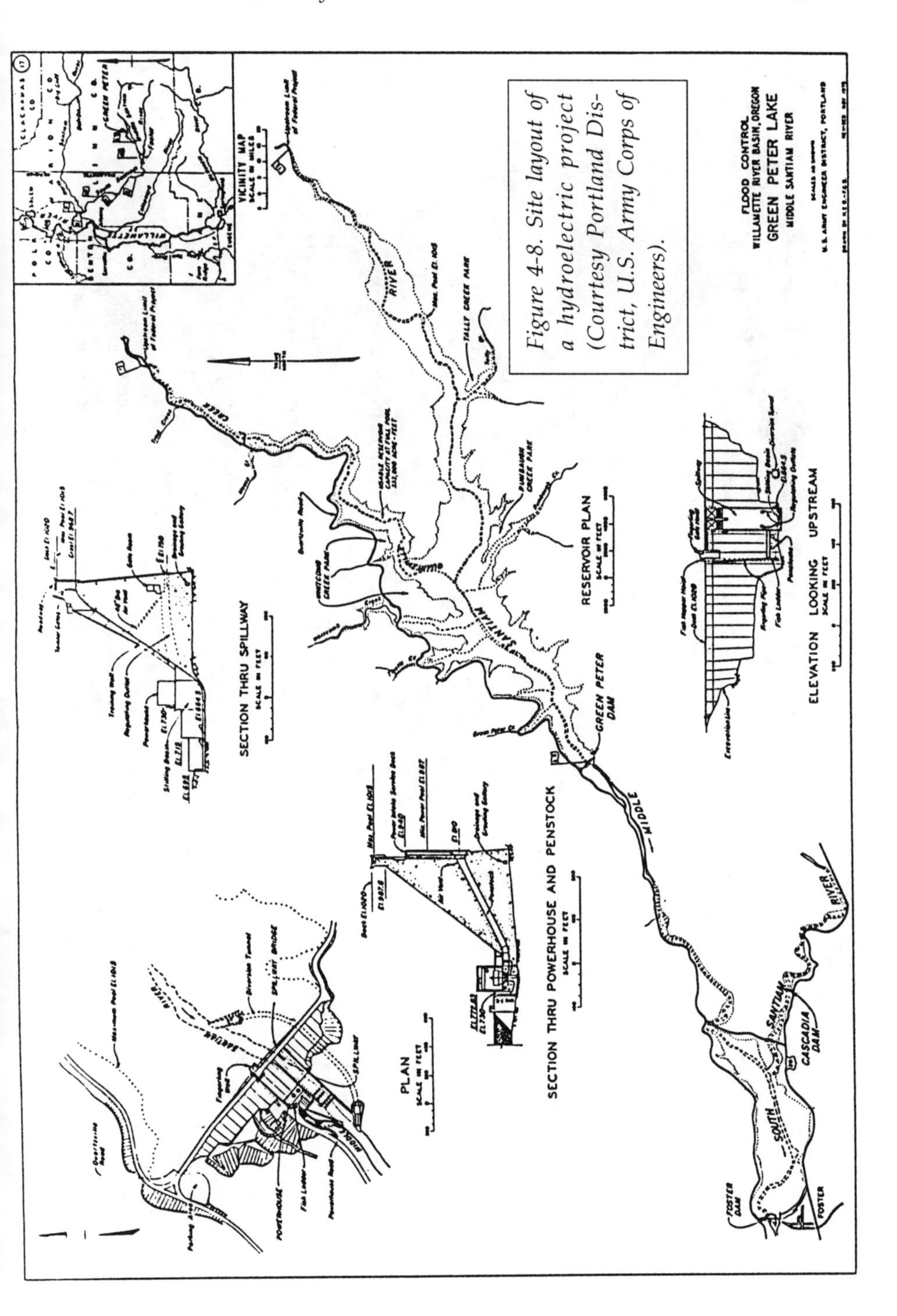

Figure 4-8. Site layout of a hydroelectric project (Courtesy Portland District, U.S. Army Corps of Engineers).

Figure 4-9. John Day Lock & Dam project, on the Columbia River between Oregon and Washington, has a power-generating capacity equaled by few hydroelectric dams in the world. The project, with its 16-generator powerhouse, has a 2.1 million kilowatt capacity. (Courtesy, Portland District, U.S. Army Corps of Engineers.)

The nuclear-fission system is very similar to fossil-fuel systems in that heat is used to produce high-pressure steam which rotates a turbine. The source of heat in the nuclear-fission system is a nuclear reaction while, in the fossil-fuel system, heat is developed by a burning fuel. At the present time, less than 10% of the electrical power produced in the United States comes from nuclear-fission sources. However, this percentage is also subject to rapid change.

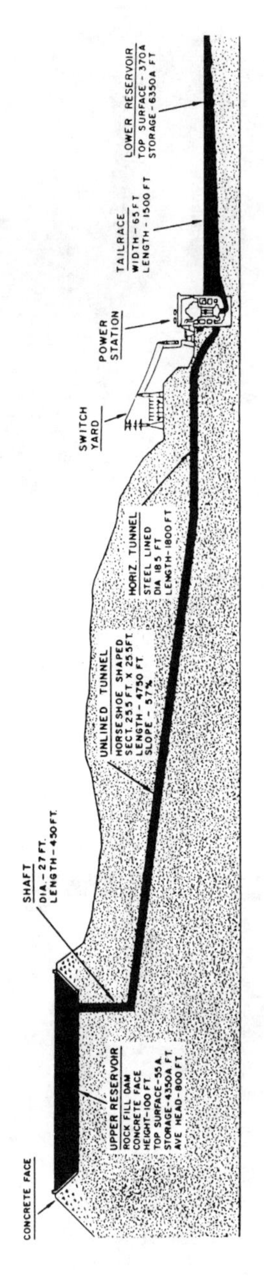

Figure 4-10. Pumped-storage hydroelectric power plant.

Figure 4-11. Nuclear Fission Electrical Power Plant (Courtesy of General Electric Company, Schenectady, NY).

Figure 4-12. Drawing illustrating the principles of a nuclear-fission power system.

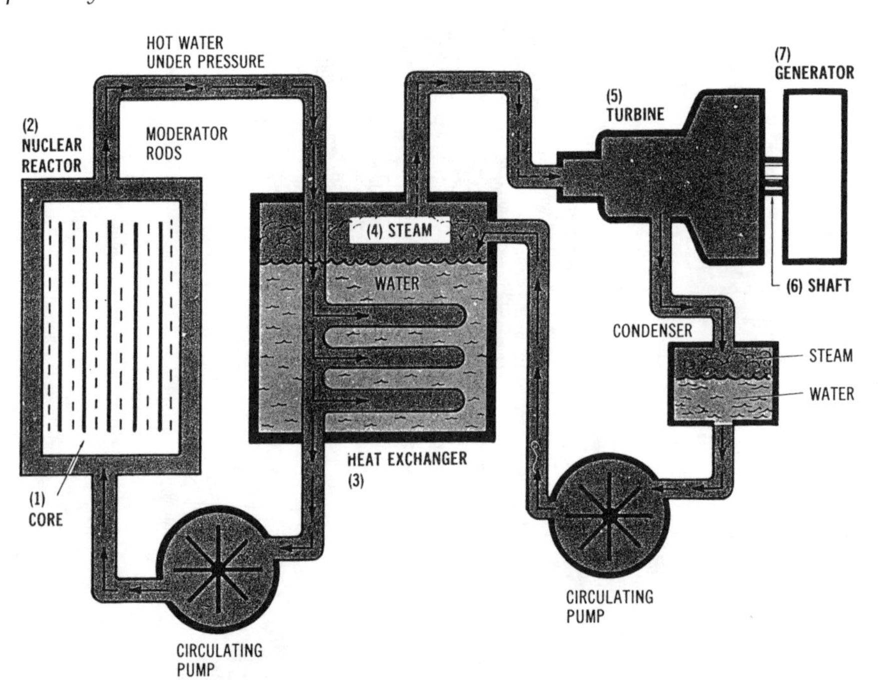

LOCATION OF ELECTRICAL POWER PLANTS

A critical issue which now faces those involved in the production of electrical power is the location of power plants. Federal regulations associated with the National Environmental Policy Act (NEPA) have made the location of power plants more difficult. At present, there is a vast number of individual power plants throughout the country. However, the addition of new generating plants involves such current issues as air pollution, water pollution, materials handling (particularly with nuclear plants), fuel availability and federal, state, and municipal regulations.

These issues have brought about some recent thought about the construction of "energy centers." Such systems would be larger and more standardized than the power plants of today. This concept would reduce the number of plants that are needed to produce a specific quantity of electrical power. Other advantages of this concept include better use of land resources, easier environmental control management, and a more economic construction and management of facilities. These advantages may make centralized power production the best alternative, socially, economically, and technically, for meeting future electrical power requirements.

ELECTRICAL LOAD REQUIREMENTS

The electrical power which must be produced by our power systems varies greatly according to several factors such as the time of the year, the time of the week, and the time of the day. Electrical power supply and demand is much more difficult to predict than most quantities that are bought and sold. Electrical power must be readily available and in sufficient quantity whenever it is required. The overall supply and demand problem is something most of us take for granted until our electrical power is interrupted. Electrical power systems in the United States must be interconnected on a regional basis so that power stations can support one another in meeting the variable load demands.

The use of electrical power is forecast to increase every ten years at a rate that will cause a doubling of the kilowatt hours required. Some forecasts, however, show the rate of electrical power demand to have a "leveling-off" period in the near future. This effect may be due to a saturation of the possible uses of electrical power for home appliances, industrial processes, and commercial use. These factors, combined with greater conservation efforts, and social and economic factors, support the idea that the electrical power demand will increase at a lesser rate in future years. The forecasting of the present demand by the electrical utilities companies must be based on an analysis by regions. The demand varies according to the type of consumer that is supplied by the power stations (which comprise the system). A different type of load is encountered when residential, industrial, and commercial systems are supplied by the utilities companies.

Industrial use of electrical power accounts for over 40% of the total kilowatt-hour (kWh) consumption and the industrial use of electrical power is projected to increase at a rate similar to its present rate in the near future. The shortage of natural gas should not significantly affect electrical power consumption by industry. Most of the conversions of gas systems will be to the usage of oil systems in their place.

The major increases in residential power demand have been due to an increased use by customers. A smaller increase was accounted for by an increase in the number of customers. Such variables as the type of heating used, the use or nonuse of air conditioning and the use of major appliances (freezers, dryers, ranges), affect the residential electrical power demand. At present, residential use of electrical power accounts for over 30% of the total consumption. The rate of increase will probably taper off in the near future.

Commercial use of electrical power accounts for less than 25% of the total kWh usage. Commercial power consumption includes usage by office buildings, apartment complexes, school facilities, shopping establishments, and motel or hotel buildings. The prediction of the electrical power demand by these facilities is somewhat similar to the residential demand. Commercial use of electrical power is also expected to increase at a declining rate in the future. These percentages are subject to change over time.

ELECTRICAL LOAD-DEMAND CONTROL

As the costs of producing power continue to rise, power companies must search for ways to limit the maximum rate of energy consumption. To cut down on power usage, industries have begun to initiate programs which will cut down on the load during peak operating periods. The use of certain machines may be limited while other large power-consuming machines are operating. In larger industrial plants and at power-production plants, it would be impossible to manually control the complex regional switching systems, so computers are being used to control loads.

To prepare the computers for power-consumption control, the peak demand patterns of local industries and the surrounding region supplied by a specific power station must be determined. The load of an industrial plant may then be balanced according to area demands with the power station output. The computer may be programmed to act as a switch, allowing only those processes to operate which are within the load calculated for the plant for a specific time period. If the load drawn by an industry exceeds the limit, the computer may deactivate part of the system. When demand is decreased in one area, the computer can cause the power system to increase power output to another part of the system. Thus, the industrial load is constantly monitored by the power company to insure a sufficient supply of power at all times.

ALTERNATIVE POWER SYSTEMS

Several methods of producing electrical power are either in limited use or in the experimental stage at the present time. Some of these methods show promise as a possible electrical power production method for the future.

Solar Energy Systems

For many years, many have regarded the sun as a possible source of electrical power. However, few efforts to use this cheap source of energy have been made. We are now faced with the problem of finding alternative sources of energy and solar energy is one possible alternative source. The sun delivers a constant stream of radiant energy. The amount of solar energy coming toward the earth through sunlight in one day equals the energy produced by burning many millions of tons of coal. It is estimated that enough solar energy is delivered to the United States in less than one hour to meet the power needs of the country for one year. This is why solar energy is a potential source for our ever-growing electrical power needs. However, there are still several problems to be solved prior to using solar energy. One major problem is in developing methods for controlling and utilizing the energy of the sun. There are two methods for collecting and concentrating solar energy presently in use. Both of these methods involve a mirrorlike reflective surface.

The first method uses parabolic mirrors to capture the energy of the sun. These mirrors concentrate the energy from the sun by focusing the light onto an opaque receiving surface. If water could be made to circulate through tubes, the heat focused onto the tubes could turn the water into steam. The steam, then, could drive a turbine to produce mechanical energy. The mirrors could be rotated to keep them in proper position for the best light reflection.

The second method uses a flat-plate solar collector. Layers of glass are laid over a blackened metal plate, with an air space between each layer. The layers of glass act as a heat trap. They let the rays of the sun in, but keep most of the heat from escaping. The heated air could be used to warm a home.

The first widespread use of solar energy will probably be to heat homes and other buildings. Experiments in doing this are already underway in many areas. To heat a home, a flat-plate collector may be mounted on the part of the roof that slopes in a southward direction. It should be tilted at an angle to receive the greatest amount of sunlight possible. The sun would be used to heat a liquid (or air) which would be circulated through the collector. The heated liquid (air) would be stored in an insulated tank and, then, pumped into the house through pipes and radiators (air ducts). The system could be as shown in the top portion of Figure 4-13. By adding a steam turbine, a generator, etc., to the solar collector just discussed, the heat could be used to drive the steam turbine to generate electrical energy. This solar power system is illustrated in Figure 4-13.

Another major problem in solar heating is in the storage of the heat produced by the sun. In areas that have several cloudy days each year, an auxiliary heating system is required. However, solar energy is being used today on a limited basis. For instance, flashlights and radios can be powered by solar cells. The main advantage of this usage is that a solar battery can be used for an infinite period of time. Considering all of these aspects of solar energy and its potential, many, feel that solar energy will be the next major form of energy to be utilized extensively in the United States. The development of large-scale power production systems which use solar energy, however, remains questionable.

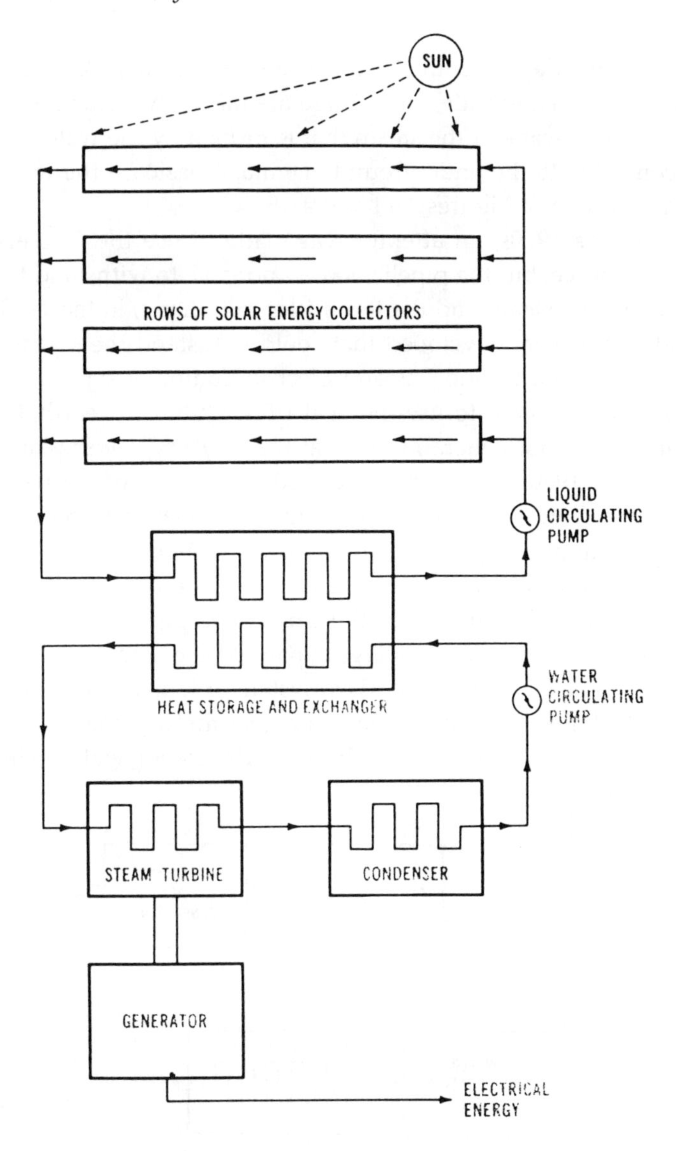

Figure 4-13. Block diagram of a solar power system.

Geothermal Power Systems

About 20 miles below the crust of the earth is a molten mass of liquid and gaseous matter called magma, which is still cooling from the time that the earth was formed. When this magma comes close to the crust of the earth, possibly through a rupture, a volcano could be formed and erupt. Magma could also cause steam vents, like the ones at the "Geysers" area in California. These are naturally occurring vents which permit the escape of the steam that is formed by the water which comes in contact with the underground magma. A basic geothermal power system is shown in Figures 4-14 and 4-15.

In the 1920s, an attempt was made to use the Geysers area as a power source, but the pipelines were not able to withstand the corrosive action of the steam and the impurities in it. Later, in the 1950s, stainless steel alloys were developed that could withstand the steam and its impurities, so the Pacific Gas and Electric Company started development of a power system to use the heat from within the earth as an energy source. The first generating unit at the Geysers power plant began operation in 1960. At present, over 500 megawatts of electrical power is available from the generating units in the Geysers area. The Geysers geothermal power plant is shown in Figure 4-16.

In the geothermal system, steam enters a path (through a pipeline or a vent) to the surface of the earth. The pipelines which carry the steam are constructed with large expansion loops that cause small pieces of rock to be left in the loop. This system of loops avoids damage to the steam turbine blades. After the steam goes through the turbine, it goes to a condenser, where it is combined with cooler water. This water is

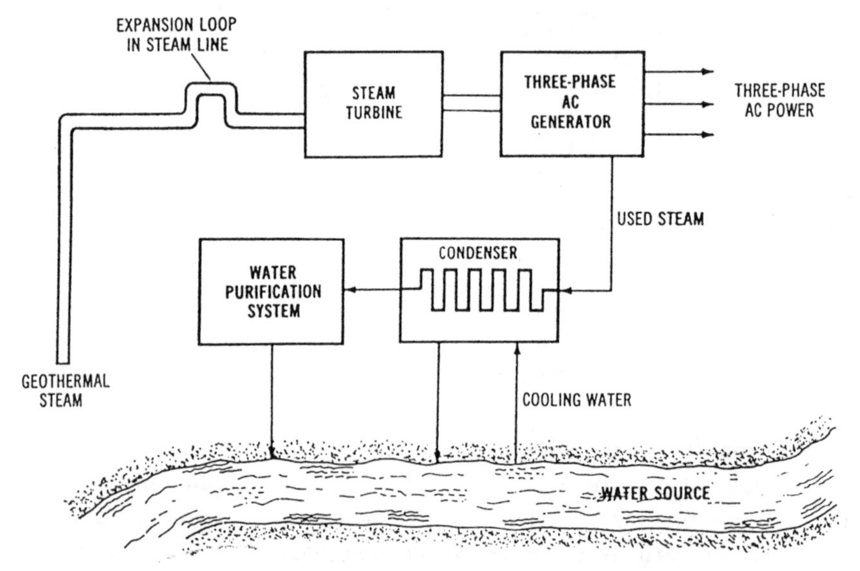

Figure 4-14. Drawing of a basic geothermal power system.

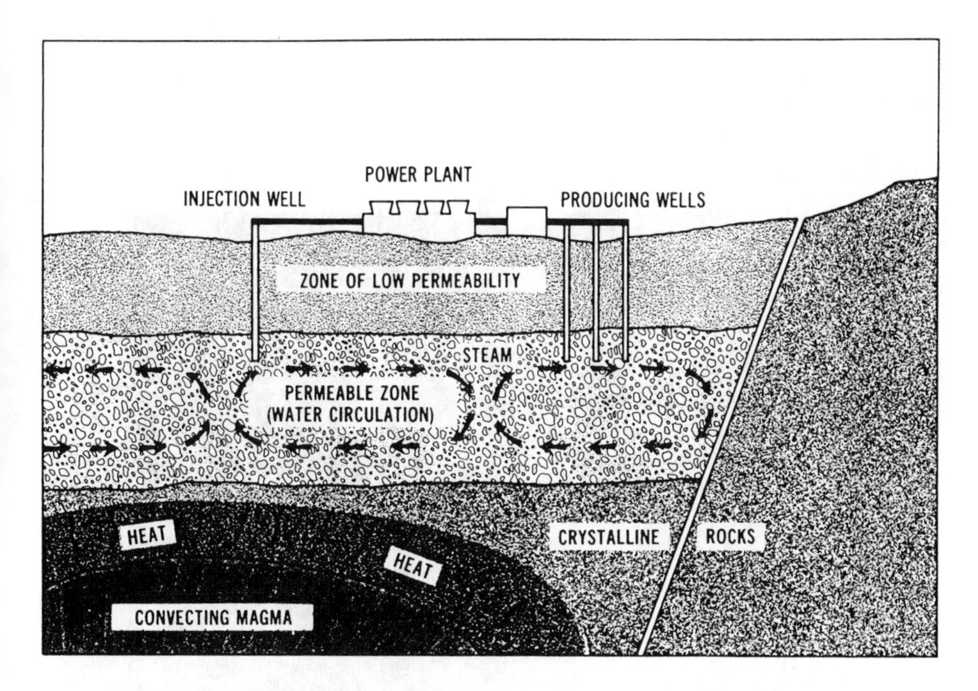

Figure 4-15. Drawing of a basic geothermal power system that illustrates the underground structure.

pumped to cooling towers where the water temperature is reduced. This part of the geothermal system is similar to conventional steam systems.

Wind Systems

Nonpolluting, inexpensive power systems have been desired for many years. Now, technology has advanced to the point where there may be some inexpensive methods of producing electrical power that we have not used significantly before.

In the early 1900s, in many farm areas, particularly in the Midwest, electrical power had not been distributed to homes. It was conceived that the wind could be used to provide a mechanical energy not only for water pumps but also for generators with which to provide electrical power. The major problem was that the wind did not blow all the time and, then, when it did, it often blew so hard that it destroyed the windmill. Most of these units used fixed-pitch propellers as fans. They also used low-voltage direct-current electrical systems and there was no way of storing power during periods of nonuse or during times of low wind speeds. These early units used a rotating-armature system (see Single-phase AC

Figure 4-16. The Geysers geothermal power plant (Courtesy Pacific Gas and Electric Co.)

Generators, later in this chapter) and, therefore, along with other maintenance problems, had to have the brushes replaced very often.

The wind generating plants used today have a system of storage batteries, rectifiers, and other components that provides a constant power output even when the wind is not blowing. They also have a 2-blade or 3-blade propeller system that can be "feathered" during periods of high winds so that the mill will not destroy itself. A simplified wind-power system is shown in Figure 4-17. Most of the systems that are in use today are individual units capable of producing 120-volt alternating current with constant power outputs in the low-kilowatt range.

Magnetohydrodynamic (MHD) Systems

MHD stands for magnetohydrodynamic, a process of generating electricity by moving a conductor of small particles suspended in a su-

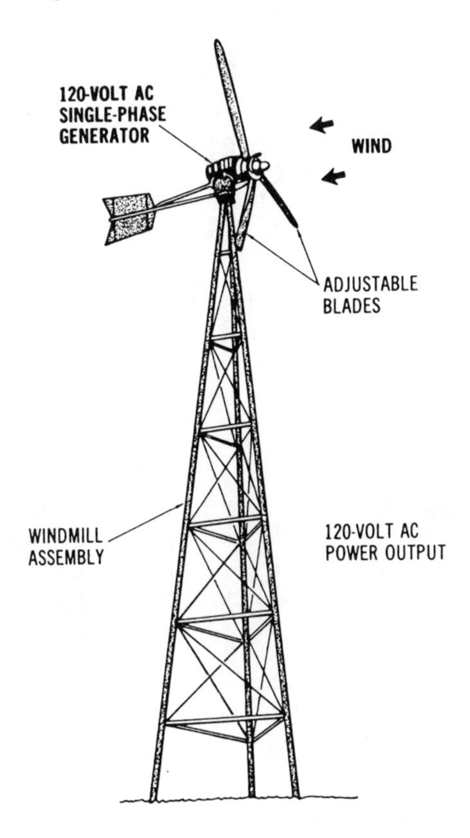

Figure 4-17. A simplified wind power system.

perheated gas through a magnetic field. The process is illustrated in Figure 4-18. The metallic conductors are made of metals such as potassium or cesium and can be recovered and used again. The gas is heated to a temperature much hotter than the temperature to which steam is heated in conventional power plants. This superheated gas is in what is called a plasma state at these high temperatures. This means that the electrons of many of the gas atoms have been stripped away, thus making the gas a good electrical conductor. The combination of metal and gas is forced through an electrode-lined channel which is under the influence of a superconducting magnet that has a tremendous field strength. The magnet must be of the superconducting type since a regular electromagnet of that strength would require too much power. A superconducting magnet, therefore, is one of the key parts to this type of generation system.

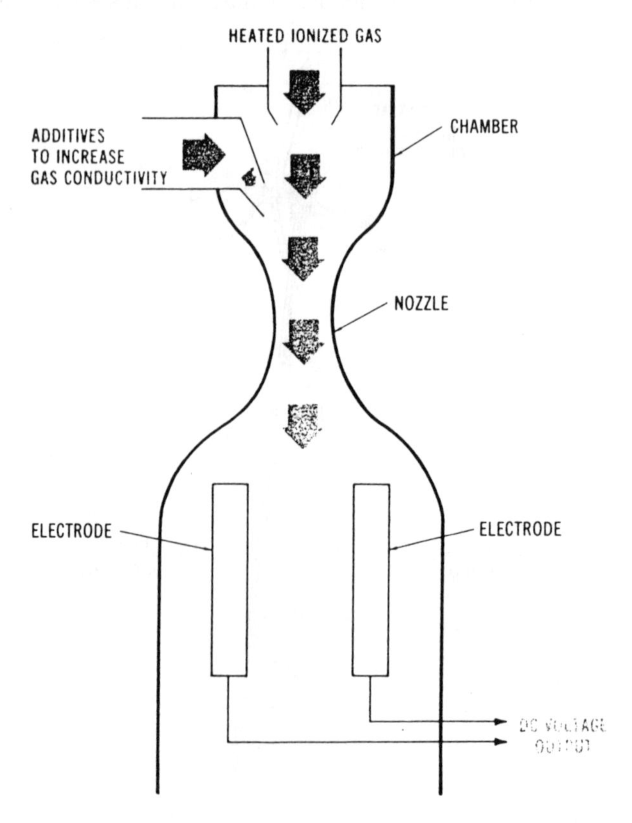

Figure 4-18. Simplified drawing of a magnetohydrodynamic (MHD) generator system.

Nuclear-fusion Power Systems

Another alternative power production method which has been considered is nuclear fusion. Deuterium, the type of fuel used for this process, is very abundant. The supply of deuterium is considered to be unlimited since it exists as heavy hydrogen in sea water. The use of such an abundant fuel, however, could solve some of our problems that are related to the depletion of fossil fuels. Another outstanding advantage of this system is that its radioactive waste products would be minimal.

The fusion process results when two atomic nuclei collide under controlled conditions to cause a rearrangement of their inner structure and, thus, release a large amount of energy during the reaction. These nuclear reactions or fusing of atoms must take place under tremendously high temperatures. The energy released through nuclear fusion would be thousands of times greater per unit than the energy from typical chemical reactions and considerably greater than that of a nuclear-fission reaction.

The fusion reaction involves the fusing together of two light elements to form a heavier element, with heat energy being released during the reaction. This reaction could occur when a deuterium ion and a tritium ion are fused together. A deuterium ion is a hydrogen atom with one additional neutron and a tritium ion is a hydrogen atom with two additional neutrons. A temperature in the range of 100,000,000°C is needed for this reaction to produce a great enough velocity for the two ions to fuse together. Sufficient velocity is needed to overcome the forces associated with the ions. The deuterium-tritium-fusion reaction produces a helium atom and a neutron. The neutron, with a high enough energy level, could cause another deuterium-tritium reaction of nearby ions, providing the time of the original reaction is long enough. A much higher amount of energy would be produced by a nuclear-fusion reaction than by a fission reaction. There are several different techniques being investigated for producing nuclear fusion. One method is called magnetic confinement. The proposed design of a magnetic-confinement power system is shown in Figure 4-19.

Fuel-cell Systems

Another alternative energy system which has been researched is the fuel cell. Fuel cells convert the chemical energy of fuels into electrical energy. An advantage of this method is that its efficiency is greater than steam turbine-driven production systems, since the conversion of energy

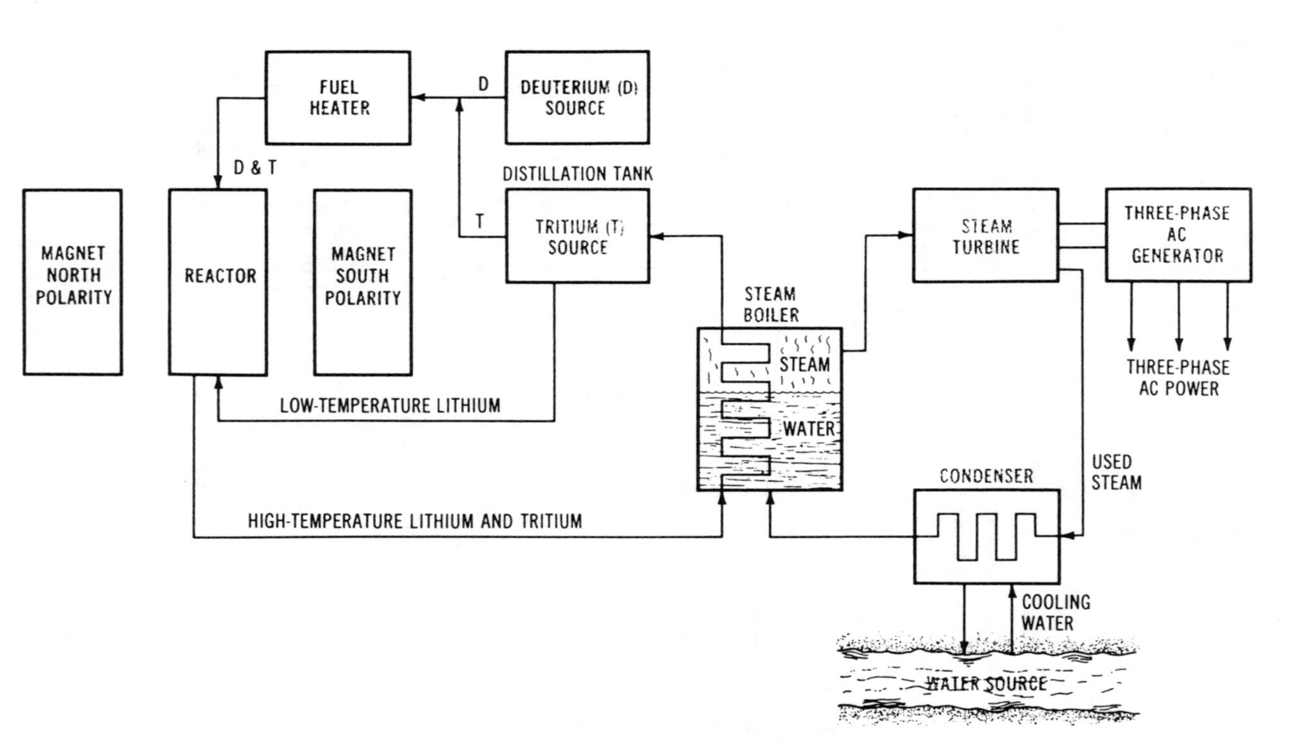

Figure 4-19. Proposed design of a magnetic-confinement nuclear-fusion power system.

is directly from chemical to electrical. Ordinarily, fuel cells use oxygen and hydrogen gas as fuels, as shown in Figure 4-20. Instead of consuming its electrodes, as ordinary batteries do, a fuel cell is supplied with chemical reactants from an external source. Ordinarily, hydrogen gas is delivered to the anode of the cell and oxygen gas or air is delivered to the cathode.

Fuel cells were developed long ago. It was known early that electrochemical reactions could be used to convert chemical energy directly to electrical energy. The first commercial fuel cells were used as auxiliary power sources for United States space vehicles. These, were hydrogen-oxygen fuel cells which produce low-voltage direct current. Several cells may be connected in series-parallel configurations to produce greater voltage and current levels.

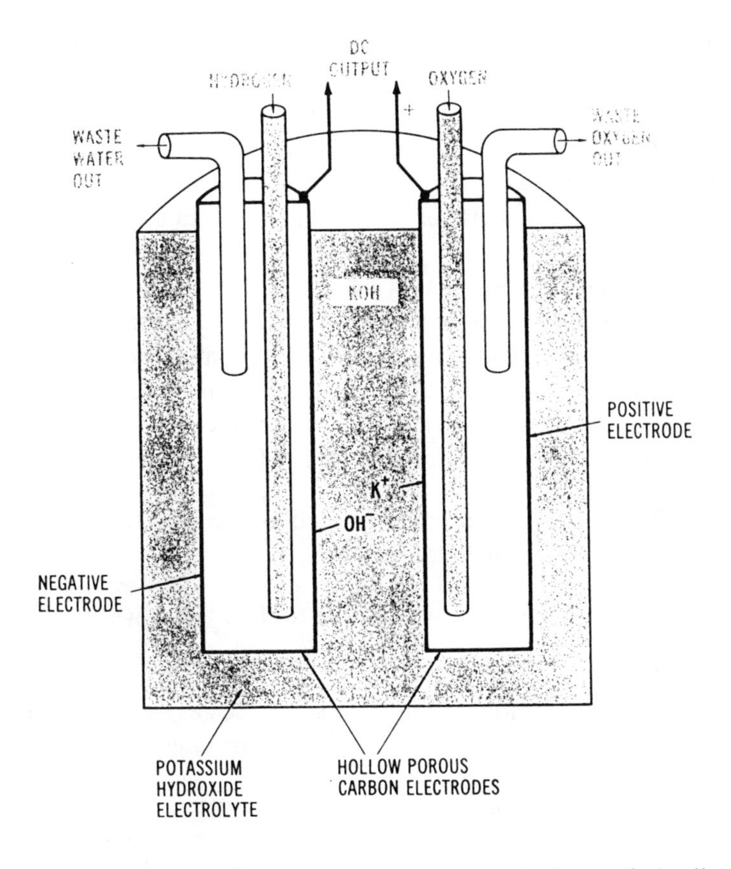

Figure 4-20. Simplified drawing of an oxygen-hydrogen fuel cell.

Tidal Power Systems

The rise and fall of waters along coastal areas that is caused by gravitational forces is the basis for the tidal electrical power production method. There are presently some tidal power systems in operation. Tidal systems would be desirable since they do not pollute the atmosphere, do not consume any natural resource, and do not drastically change the surrounding environment as some conventional hydroelectric systems do.

The depth of tidal water varies greatly at different times of the year. These depths are determined by changes in the sun and moon in relation to the earth. Tides are readily predictable since the same patterns are established year after year. A tidal power system would have to be constructed where water in sufficient quantity could be stored with a minimum amount of dam construction. A tidal system could be made to operate during the rise of tides and the fall of tides. Also, the pumped-storage method could be used in conjunction with tidal systems to assure power output during peak load times. A potential tidal system site along the United States-Canada border has been studied; however, the economic feasibility of tidal systems at this time is not too promising. A tidal system that is in operation is at Normandy, France.

Coal-gasification Fuel Systems

The process of coal gasification has aroused interest in recent years. This process involves the conversion of coal or coke to a gaseous state through a reaction with air, oxygen, carbon dioxide, or steam. Many people feel that this process will be able to produce a natural gas substitute. Some of the methods of producing gas from coal include:

1. The BI-GAS process in which a reaction of coal, steam, and hydrogen gas produces methane gas.

2. The COGAS process in which a liquid fuel and a gaseous fuel are produced.

3. The CSG process (or consolidated synthetic gas process) which develops a very slow reaction to produce methane.

4. The Hydrane process in which methane is produced by a direct reaction of hydrogen and coal, with no intermediate gas production.

5. The Synthane process in which methane is produced by a method involving several different steps.

Coal gasification may be one solution to the problem of our depleting natural gas supply. Since some electrical power systems use natural gas as the fuel, we should be concerned about coal gasification as an alternative method to aid in the production of electrical power.

Oil-shale Fuel-production Systems

Another method which should be mentioned as an alternative method that could aid in the production of electrical energy is fuel from oil shale. There is a potential fuel source located at the oil-shale deposits which exist primarily in Colorado, Wyoming, and Utah.

This oil shale was formed by a process similar to that which created crude petroleum. However, there was never enough underground heat or pressure to convert the organic sediment to the same consistency as oil. Instead, a waxy hydrocarbon called *kerogen* was produced. This compound is mixed with fine rock and is referred to as oil shale.

Extracting crude petroleum from this substance is a very complex process which starts with either underground or strip mining. The mined rock is crushed and then heated to produce a raw oil which, in turn, must be upgraded to a usable level. All of these operations could take place at the mining site. In addition, large amounts of shale waste must be removed. The resulting waste substance is about the consistency and color of fireplace soot. The waste occupies more volume than the original oil-shale formations. The potential impact of oil-shale development is questionable. Vast quantities of land might be disturbed. There are definitely many factors to consider in addition to our energy needs.

Biomass Systems

Another alternative system being considered as a potential method of producing electrical power is biomass. Biomass sources of energy for possible use as fuel sources for electrical power plants are wood, animal wastes, garbage, food processing wastes, grass, and kelp from the ocean. Many countries in the world use biomass sources as primary energy sources. In fact, the United States used some of the biomass sources of energy almost exclusively for many years. The potential amount of energy which could be produced in the United States by biomass sources is substantial at this time also. There are still several questions about using biomass sources of energy for producing electrical power; however, as fossil fuels become less abundant, biomass sources may receive more active consideration.

This discussion is included to stimulate thought about potential "alternative" systems for producing electrical power. Each of the systems discussed have potential problems; however, serious experimentation must be conducted to assure that electrical power can be produced economically. Our technology is dependent upon low-cost electrical power.

SINGLE-PHASE AC POWER SYSTEMS

Electrical power can be produced by single-phase generators, commonly called alternators. The principle of operation of a single-phase alternator is shown in Figure 4-21. In order for a generator to convert mechanical energy into electrical energy, three conditions must exist:

1. There must be a magnetic field developed.

2. There must be a group of conductors adjacent to the magnetic field.

3. There must be relative motion between the magnetic field and the conductors.

Generator Construction

Generators used to produce electrical power require some form of mechanical energy. This mechanical energy is used to move electrical conductors through the magnetic field of the generator. Figure 4-21a shows the basic parts of a mechanical generator. A generator has a stationary part and a rotating part. The stationary part is called the *stator* and the rotating part is called the *rotor*. The generator has magnetic *field poles* of north and south polarities. Also, the generator must have a method of producing a rotary motion or a *prime mover* connected to the generator shaft. There must also be a method of electrically connecting the rotating conductors to an external circuit. This is done by a *slip ring/ brush assembly*. The stationary brushes are made of carbon and graphite.

The slip rings used on AC generators are made of copper. They are permanently mounted on the shaft of the generator. The two slip rings connect to the ends of a *conductor loop*. When a load is connected, a closed external circuit is made. With all of these generator parts functioning together, electromagnetic induction can take place and electrical power can be produced.

Generating AC Voltage

Figure 4-21 shows a magnetic field developed by a set of permanent magnets. Conductors that can be rotated are placed within the magnetic field and they are connected to a load device by means of a slip ring/brush assembly. Figure 4-21 simulates a single-phase alternator.

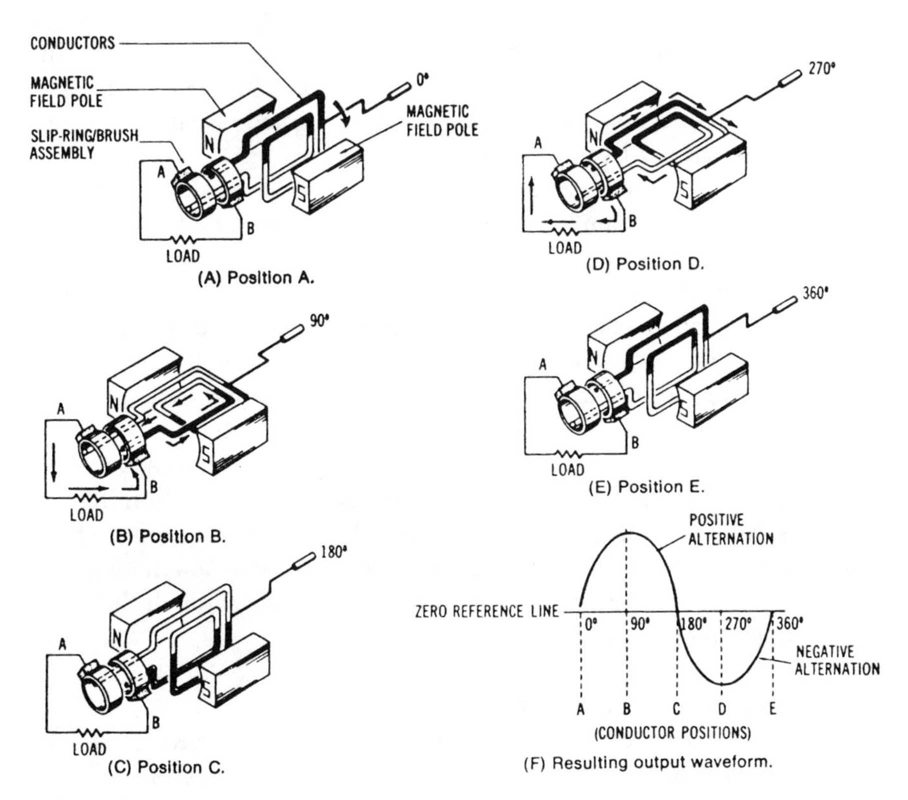

Figure 4-21. Basic principle of a single-phase alternator: (A) Position A, (B) Position B, (C) Position C, (D) Position D, (E) Position E, (F) Resultilng output waveform.

In position A (Figure 4-21a), the conductors are positioned so that the minimum amount of magnetic lines of force are "cut" by the conductors as they rotate. No current is induced into the conductors at position A and the resulting current flow through the load will be zero. If the conductors are rotated 90° in a clockwise direction to position B (Figure 4-21b) they will pass from the minimum lines of force to the most concentrated area of the magnetic field. At position B, the induced current will be maximum, as shown by the waveform diagram of Figure 4-21f. Note that the induced current rises gradually from the zero reference line to a maximum value at position B. As the conductors are rotated another 90° to position C (Figure 4-21c), the induced current becomes zero again. No current Rows through the load at this position. Note, in the diagram of Figure 4-21f, how the induced current drops gradually from maximum to zero. This part of the induced AC current (from 0° to 180°) is called the positive alternation. Each value of the induced current, as the conductors rotate from the 0° position to the 180° position, is in a positive direction. This action could be observed visually if a meter was connected in place of the load.

When the conductors are rotated another 90° to position D (Figure 4-21d), they once again pass through the most concentrated portion of the magnetic field. Maximum current is induced into the conductors at this position. However, the direction of the induced current is in the opposite direction as compared with position B. At the 270° position, the induced current is maximum in a negative direction. As the conductors are rotated to position E (same as at position A), the induced current is minimum once again. Note in the diagram of Figure 4-21e how the induced current decreases from its maximum negative value back to zero again (at the 360° position). The part of the induced current from 180° to 360° is called the negative alternation. The complete output, which shows the induced current through the load, is called an alternating-current waveform. As the conductors continue to rotate through the magnetic field, the cycle is repeated.

Alternating-Current Sine Wave

The induced current produced by the method discussed above is in the form of a sinusoidal waveform or sine wave. This waveform is referred to as a sine wave due to its mathematical origin, based on the trigonometric sine function. The current induced into the conductors, shown in Figure 4-21, varies as the sine of the angle of rotation between the conductors and the magnetic field. This induced current produces a voltage. The instantaneous voltage induced into a single conductor can be expressed as:

$$V_i = V_{max} \times \sin \theta$$

where,

V_i is the instantaneous induced voltage,
V_{max} is the maximum induced voltage,
θ is the angle of conductor rotation from the zero reference.

For example, at the 30° position (Figure 4-22), if the maximum voltage is 100 volts, then

$$V_i = 100 \times \sin \theta.$$

Refer to Appendix C if necessary and use a calculator to find that the sine of 30° = 0.5, then

$$\begin{aligned} V_i &= 100 \times 0.5 \\ &= 50 \text{ volts.} \end{aligned}$$

SINGLE-PHASE AC GENERATORS

Although much single-phase electrical power is used, particularly in the home, very little electrical power is produced by single-phase alternators. The single-phase electrical power used in the home is usually developed by three-phase alternators and then converted to single-phase, by the power distribution system. There are two basic methods which can be used to produce single-phase alternating current. One method is called the rotating-armature method and the other is the rotating-field method. These methods are illustrated in Figure 4-23.

Rotating Armature Method

In the rotating-armature method, shown in Figure 4-23a, an alternating-current voltage is induced into the conductors of the rotating part of the machine. The electromagnetic field is developed by a set of stationary pole pieces. Relative motion between the conductors and the mag-

TRIGONOMETRIC SINE VALUES

Degrees	Sine	Degrees	Sine
0/360	0.000	180	0.000
30	0.500	210	0.500
60	0.866	240	0.866
90	1.000	270	1.000
120	0.866	300	0.866
150	0.500	330	0.500

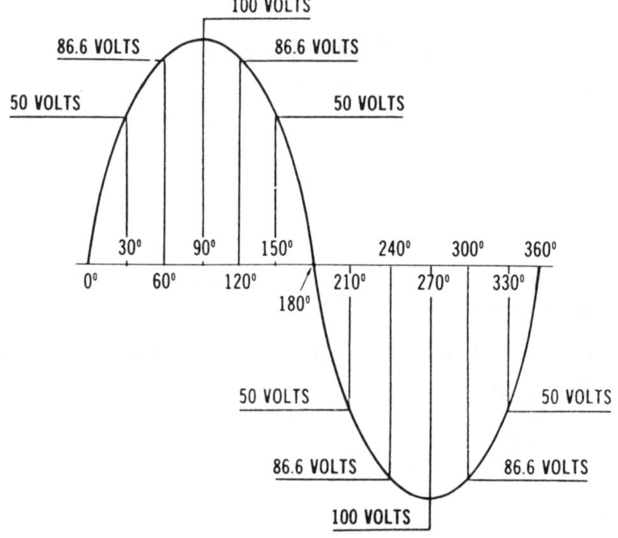

Figure 4-22. Mathematic origin of an AC sine wave.

netic field is provided by a prime mover or mechanical-energy source connected to the shaft (which is a part of the rotor assembly). Prime movers may be steam turbines, gas turbines, hydraulic turbines, gasoline engines, diesel engines, or possibly gas engines or electric motors. Remember that all generators convert mechanical energy into electrical energy, as shown in Figure 4-24. Only small power ratings can be used with the rotating-armature type of alternator. The major disadvantage of this method is that the AC voltage is extracted from a slip ring/brush assembly (see Figure 4-23a). A high voltage could produce tremendous sparking or arc-over between the brushes and the slip rings. The maintenance involved in replacing brushes and repairing the slip-ring commutator assembly would be very time consuming and expensive. Therefore, this method is used only for alternators with low power ratings.

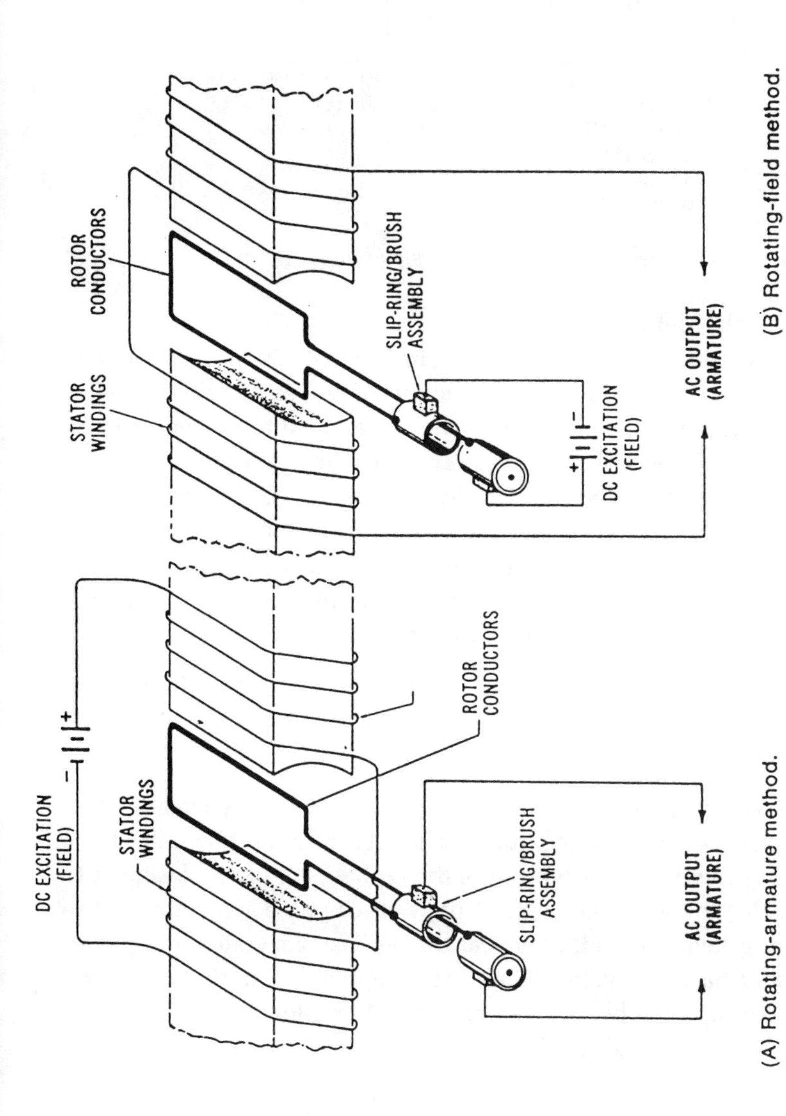

(A) Rotating-armature method.

(B) Rotating-field method.

Figure 4-23. The two basic methods of generating single-phase alternating current.

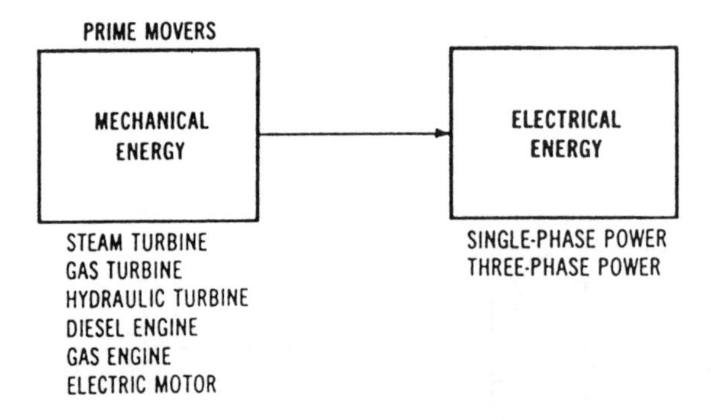

Figure 4-24. Energy conversion by generators.

Rotating-field Method

The rotating-field method, shown in Figure 4-23b, is used for alternators capable of producing larger amounts of power. The DC-excitation voltage which develops the magnetic field is applied to the rotating portion of the machine. The AC voltage is induced into the stationary conductors of the machine. Since the DC-excitation voltage is a much lower value than the AC voltage that is produced, maintenance problems associated with the slip ring/brush assembly are minimized. In addition, the conductors of the stationary portion of the machine may be larger so as to handle more current since they do not rotate.

THREE-PHASE AC GENERATORS

The vast majority of electrical power produced in the United States is three-phase power. A large, commercial generator is shown in Figure 4-25. Most generators which produce three-phase voltage look similar to this one. Due to their large power ratings, three-phase generators utilize the rotating field method. A typical three-phase generator in a power plant might have 250 volts DC excitation applied to the rotating field through the slip ring/brush assembly while 13.8 kilovolts AC is induced into the stationary conductors.

Commercial power systems use many three-phase alternators connected in parallel to supply their regional load requirements. Normally, industrial loads represent the largest portion of the load on our power systems. The residential (home) load is somewhat less. Due to the vast load that has to be met by the power systems, three-phase generators have high power ratings. Nameplate data for a typical commercial three-phase alternator is shown in Figure 4-26. Note the large values for each of the nameplate ratings.

Figure 4-25. A steam turbine generator unit (Courtesy Westinghouse Electric Corp.).

TURBINE-GENERATOR

STEAM TURBINE						
No. 128917						
Rating 66000 kW	3600 RPM	21 Stages				
Steam: Pressure 1250 PSIG	Temp 95 F	Exhaust Pressure 1.5" HG. ABS.				
GENERATOR						
No. 8287069	*Hydrogen Cooled*			*Rating*	*Capability*	*Capability*
Type ATB	2 Poles	60 Cycles	Gas Pressure	30 psig	15 psig	0.5 psig
3 PH. Y Connected for		13800 Volts	kVA	88235	81176	70588
Excitation		250 Volts	Kilowatts	75000	69000	60000
Temp rise guaranteed not to exceed			Armature Amp	3691	3396	2953
45 C on armature by detector			Field Amp	721	683	626
74 C on field by resistance			Power Factor	0.85	0.85	0.85

Figure 4-26. Nameplate data for a commercial three-phase alternator.

Generation of Three-Phase Voltage

The basic construction of a three-phase AC generator is shown in Figure 4-27 with its resulting output waveform given in Figure 4-28. Note that three-phase generators must have at least six stationary poles or two poles for each phase. The three-phase generator shown in the drawing is a rotating-field type generator. The magnetic field is developed electromagnetically by a direct-current voltage. The DC voltage is applied from an external power source through a slip ring/brush assembly to the windings of the rotor. The magnetic polarities of the rotor as shown are north at the top and south at the bottom of the illustration. The magnetic lines of force would be developed around the outside of the electromagnetic rotor assembly.

Through electromagnetic induction, a current can be induced into each of the stationary (stator) coils of the generator. Since the beginning of phase A is physically located 120° from the beginning of phase B, the induced currents will be 120° apart. Likewise, the beginning of phase B and phase C are located 120° apart. Thus, the voltages developed due to electromagnetic induction are 120° apart, as shown in Figure 4-28. Voltages are developed in each stator winding as the pectromagnetic field rotates within the enclosure which houses the stator coils.

Three-Phase Connection Methods

In Figure 4-27, poles A′, B′, and C′ represent the beginnings of each of the phase windings of the alternator. Poles A, B, and C represent the ends of each of the phase windings. There are two methods which may be used to connect these windings together. These methods are carted wye and delta connections.

Three-phase Wye-connected Generators

The windings of a three-phase generator can be connected in a wye configuration by connecting the beginnings or the ends of the windings together. The other ends of the windings become the three-phase power lines from the generator. A three-phase wye-connected generator is illustrated in Figure 4-29. Notice that the beginnings of the windings (poles A′, B′, and C′) are connected together. The other ends of the windings (poles A, B, and C) are the three-phase power lines that are connected to the load to which the generator will supply power.

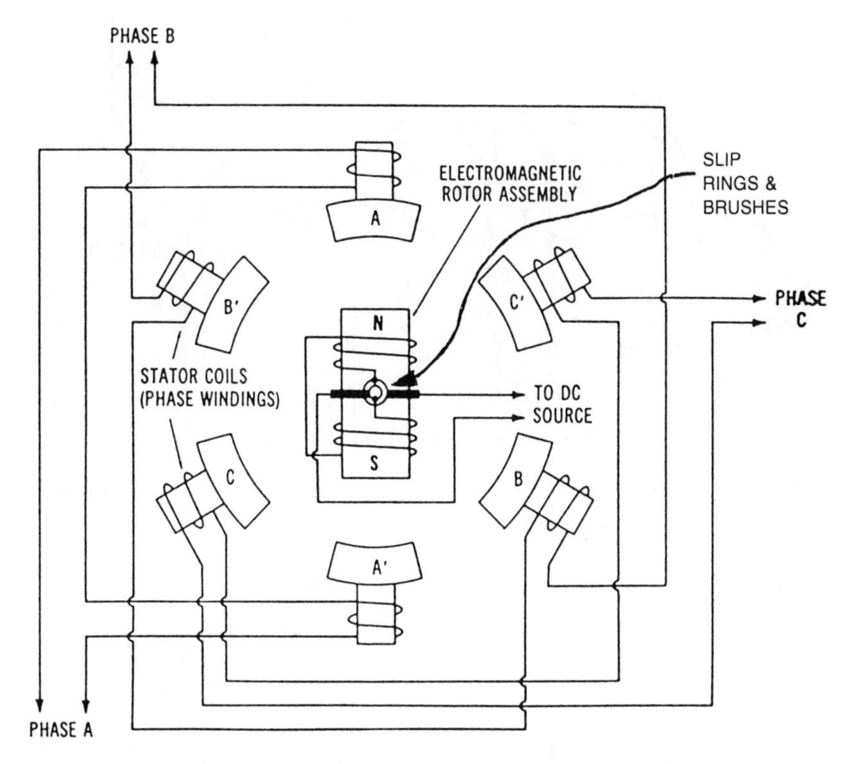

Figure 4-27. Simplified drawing showing the basic construction of a three-phase AC generator.

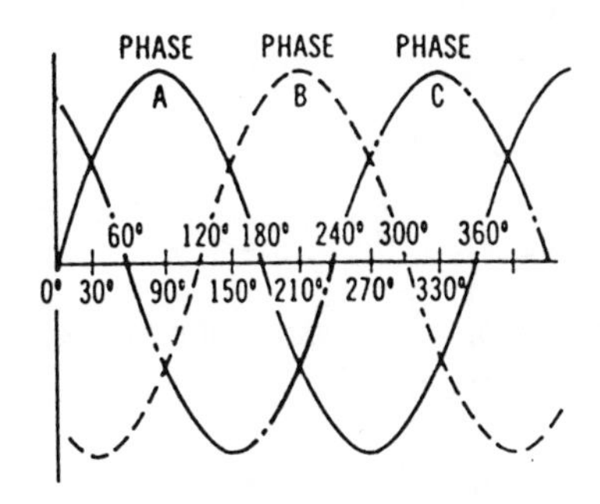

Figure 4-28. Output waveform of a three-phase AC generator.

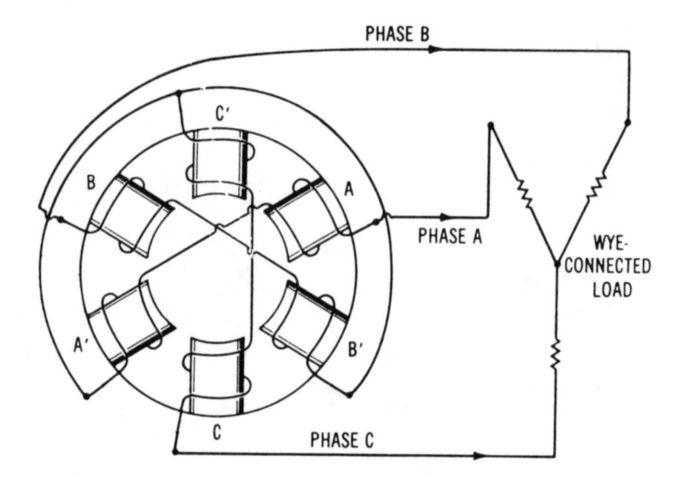

Figure 4-29. Simplified drawing of the stator of a three-phase generator that is connected in a wye configuration.

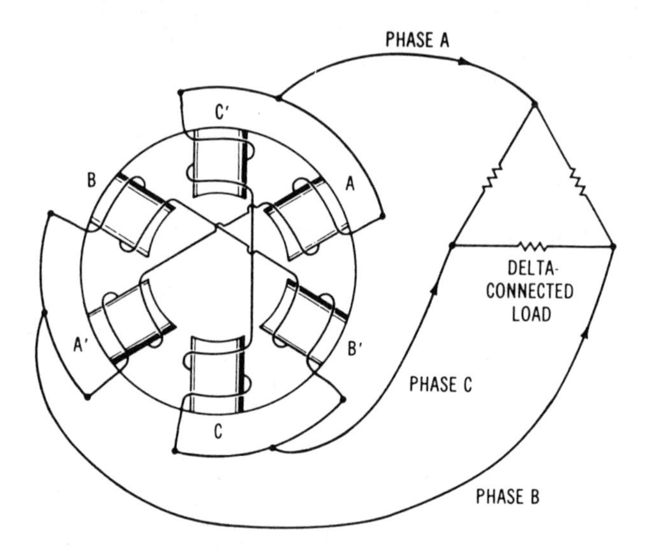

Figure 4-30. Simplified drawing of the stator of a three-phase generator that is connected in a delta configuration.

Three-phase Delta-connected Generators

The windings of a three-phase generator may also be connected in a delta arrangement, as shown in Figure 4-28. In the delta configuration, the beginning of one phase winding is connected to the end of the adjacent phase winding. Thus, the beginnings and ends of all adjacent phase windings are connected together. The voltage, current, and power characteristics of three-phase wye and delta connections were discussed earlier in Chapter 1.

Advantages of Three-phase Power

Three-phase power is used primarily for industrial and commercial applications. Many types of industrial equipment use three-phase alternating-current power because the power produced by a three-phase voltage source, as compared to single-phase power, is less pulsating. You can see this effect by observing that a peak voltage occurs every 120° in the three-phase waveform of Figure 4-28. A single-phase voltage has a peak voltage only once every 360° (Figure 4-22). This comparison is somewhat similar to comparing the power developed by an eight-cylinder engine to the power developed by a four-cylinder engine. The eight-cylinder engine provides smoother, less-pulsating power. The effect of smoother power development on electric motors (with three-phase voltage applied) is that it produces a more uniform torque in the motor. This factor is very important for the large motors that are used in industry.

Three separate single-phase voltages can be derived from a three-phase transmission line and it is more economical to distribute three-phase power from plants to consumers that are located a considerable distance away. Fewer conductors are required to distribute the three-phase voltage. Also, the equipment which uses three-phase power is physically smaller in size than similar single-phase equipment.

GENERATOR FREQUENCY

The frequency of the sinusoidal waveforms (sine waves) produced by an alternating-current generator is usually 60 hertz. One cycle of alternating current is generated when a conductor makes one complete revolution past a set of north and south field poles. A speed of 60 revolutions per second (3600 revolutions per minute) must be maintained to produce 60 hertz. The frequency of an alternating-current generator (alternator) may be expressed as:

$$f = \frac{\text{number of poles per phase} \times \text{speed of rotation (rpm)}}{120}$$

where,

f is the frequency in hertz.

Note that if the number of poles is increased, the speed of rotation may be reduced while still maintaining a 60-Hz frequency.

VOLTAGE REGULATION

As an increased electrical load is added to an alternator, it tends to slow down. The decreased speed causes the generated voltage to decrease. The amount of voltage change depends on the generator design and the type of load connected to its terminals. The amount of change in generated voltage from a no-load condition to a rated full-load operating condition is referred to as voltage regulation. Voltage regulation may be expressed as:

$$\%\,VR = \frac{V_{NL} - V_{FL}}{V_{FL}} \times 100$$

where,

VR is the voltage regulation in percent,
V_{NL} is the no-load terminal voltage,
V_{FL} is the rated full-load terminal voltage.

EFFICIENCY

Efficiency is the ratio of the power output of a generator in watts to the power input in horsepower. The efficiency of a generator may be expressed as:

$$\text{Efficiency}\,(\%) = \frac{P_{out}}{P_{in}} \times 100$$

where,

P_{in} is the power input in horsepower,
P_{out} is the power output in watts.

To convert horsepower to watts, remember that 1 horsepower = 746 watts. The efficiency of a generator usually ranges from 70% to 85%.

DIRECT-CURRENT (DC) POWER SYSTEMS

Alternating current is used in greater quantities than direct current; however, many important operations are dependent upon direct-current power. Industries use direct-current power for many specialized processes. Electroplating and DC variable-speed motor drives are only two examples which show the need for direct-current power to sustain industrial operations. We use direct-current power to cause our automobiles to start and many types of portable equipment in the home use direct-current power. Most of the electrical power produced in the United States is three-phase alternating current and this three-phase AC power may be easily converted to direct current for industrial or commercial use. Direct current is also available in the form of primary and secondary chemical cells. These cells are used extensively. In addition, direct-current generators are also used to supply power for specialized applications.

Batteries

Storage batteries are used in industry and commercial buildings to provide emergency power in the event of power failure. Such standby systems are necessary to sustain lighting and some critical operations when power is not available. Industrial trucks and loaders use storage batteries for their everyday operation. Many types of instruments and portable equipment rely upon batteries for power. Several of these instruments get their power from rechargeable secondary cells rather than primary cells, which cannot be recharged. Railway cars use batteries for lighting when they are not in motion. Of course; automotive systems of all kinds use secondary batteries to supply direct current power for starting, lighting, and other electrical systems.

Direct-current Generating Systems

Mechanical generators are used in many situations to produce direct current. These generators convert mechanical energy into direct-current electrical energy. The parts of a simple direct-current generator are shown in Figure 4-31. The principle of operation of a direct-current generator is similar to that of the alternating-current generator. A rotating armature coil passes through a magnetic field that is developed between the north and south polarities of permanent magnets or electromagnets. As the coil rotates, electromagnetic induction causes a current to be induced into the coil. The current produced is an *alternating current*. However, it is possible to convert the alternating current which is induced into the armature into a form of direct current. The conversion of AC to DC is accomplished through the use of a split-ring commutator. The conductors of the armature of a direct-current generator are connected to *split-ring commutator* segments. The split-ring commutator shown in Figure 4-31 has two segments which are insulated from one another and from the shaft of the machine on which it rotates. An end of each armature conductor is connected to each commutator segment. The purpose of the split-ring commutator is to reverse the armature-coil connection to the external load circuit at the same time that the current induced in the armature coil reverses. This causes direct current of the correct polarity to be applied to the load at all times.

Voltage Output of DC Generator

The voltage developed by the single-coil generator shown in Figure 4-32 would appear as illustrated in Figure 5-1a. This pulsating direct current is not suitable for most applications. However, by using many turns of wire around the armature and several split-ring commutator

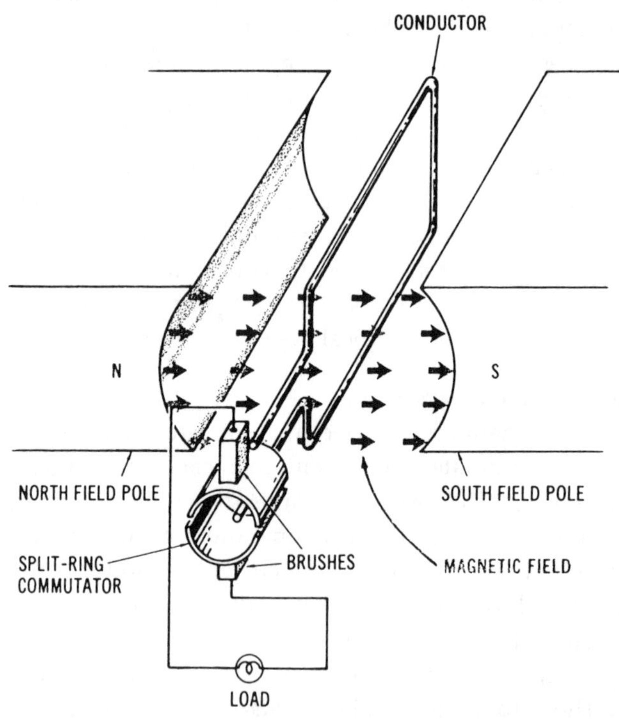

Figure 4-31. Simplified drawing of the basic parts of a direct-current generator.

segments, the voltage developed can be a smooth direct current such as that produced by a battery. This type of output is shown in Figure 4-32b. The voltage developed by a direct-current generator depends upon the strength of the magnetic field, the number of coils in the armature, and the speed of rotation of the; armature. By increasing any of these factors, the voltage output can be increased.

Types of Direct-Current Generators

Direct-current generators are classified according to the way in which a magnetic field is developed in the stator of the machine. One method is to use a permanent-magnet field. It is also possible to use electromagnets to develop a magnetic field by applying a separate source of direct current to the electromagnetic coils. However, the most common method of developing a magnetic field is for part of the generator output to be used to supply direct-current power to the field of the machine. Thus, there are three basic classifications of direct-current generators: (1) permanent magnet field, (2) separately excited field, and (3) self-excited field. The self-excited types are further subdivided according to the method used to connect the armature windings to the field circuit. This can be accomplished by the following connection methods: (1) series, (2) parallel (shunt), or (3) compound. These types of DC generators are shown in figure 4-33.

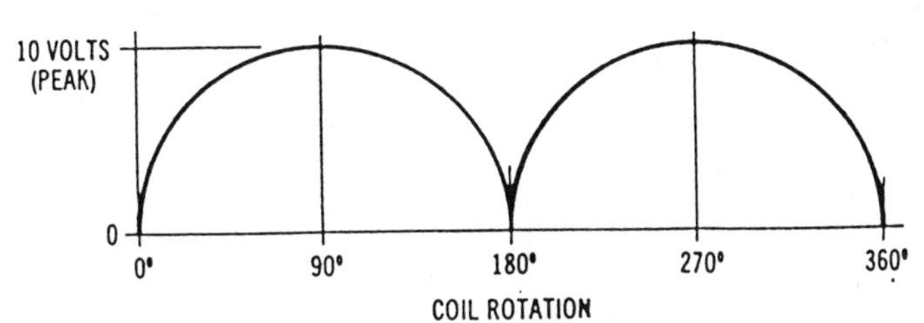

(A) Pulsating dc developed by a simple single-coil generator.

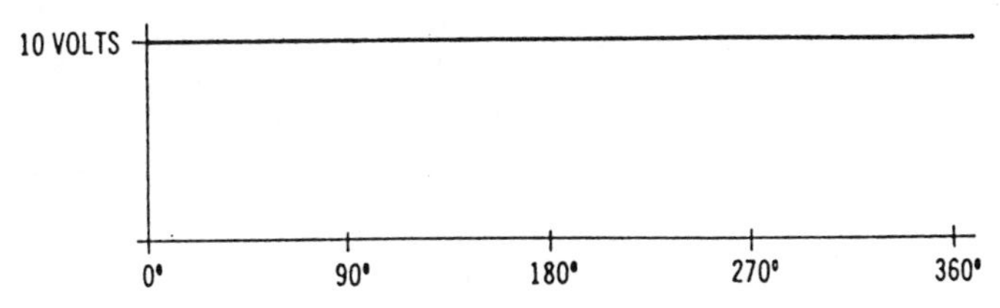

(B) Pure dc developed by a more complex generator using many turns of wire and many commutator segments.

Figure 4-32. Output waveforms of direct-current generators.

Figure 4-33a. Simplified drawing of a permanent-magnet direct-current generator.

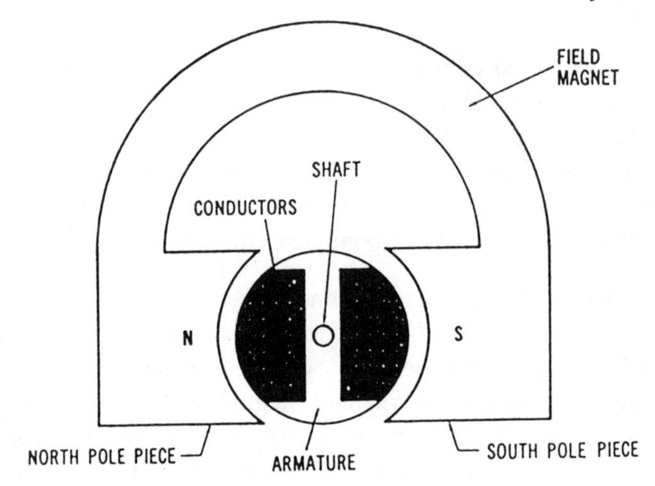

(A) Pictorial diagram.

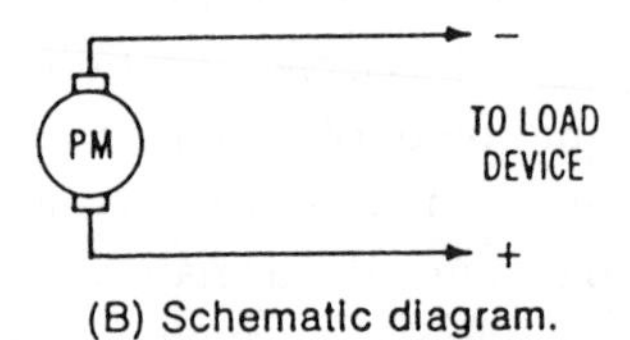

(B) Schematic diagram.

Figure 4-33b. Simplified illustration of a separately excited direct-current generator.

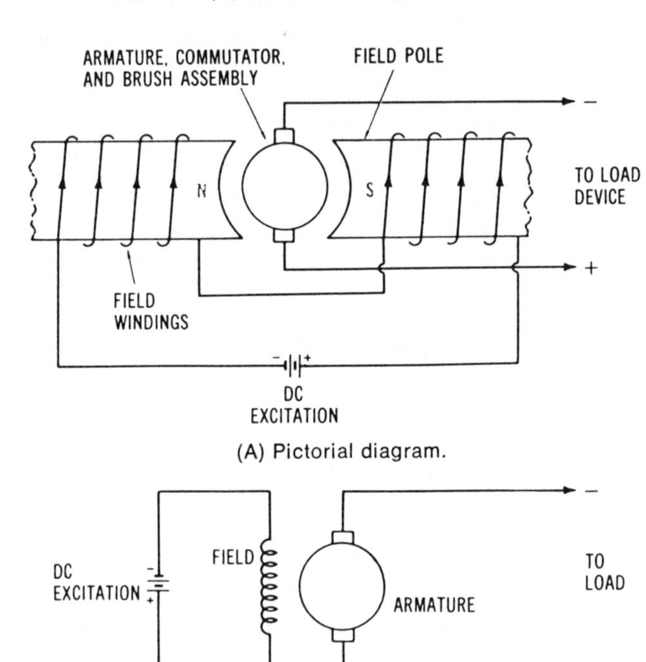

(B) Schematic diagram.

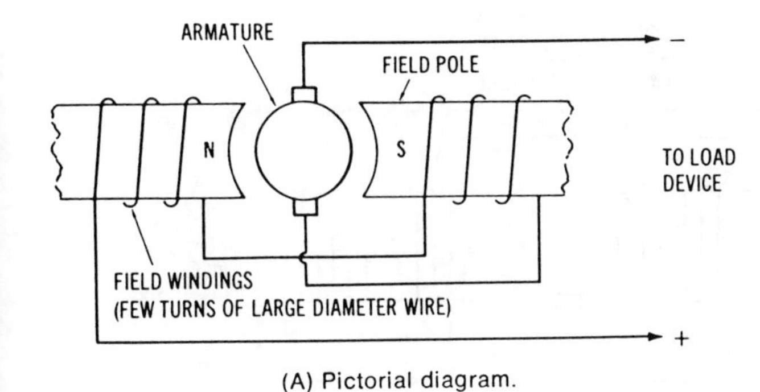

(A) Pictorial diagram.

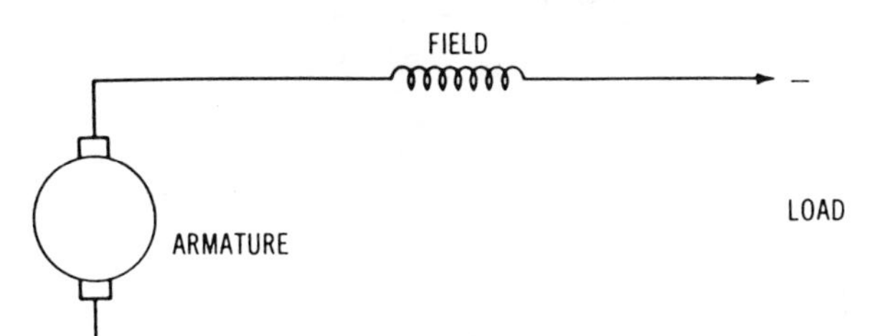

(B) Schematic diagram.

Figure 4-33c. Simplified illustration of a self-excited, series-wound DC generator.

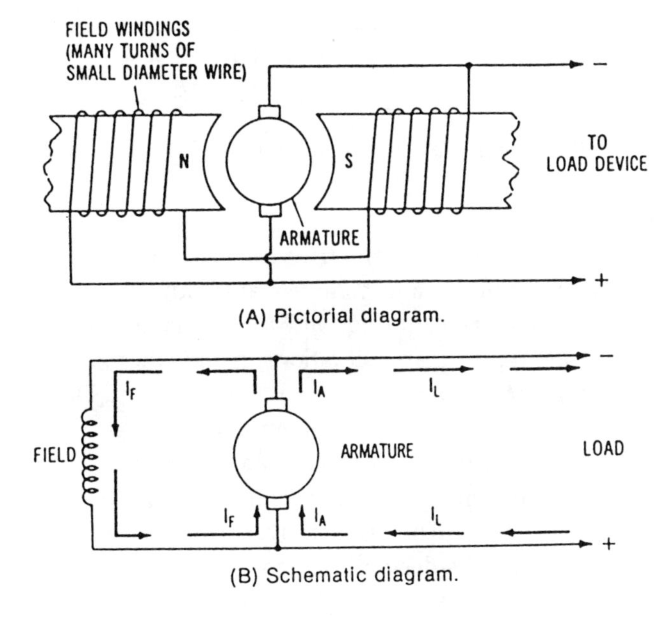

(B) Schematic diagram.

Figure 4-33d. Simplified illustration of a self-excited, shunt-wound DC generator.

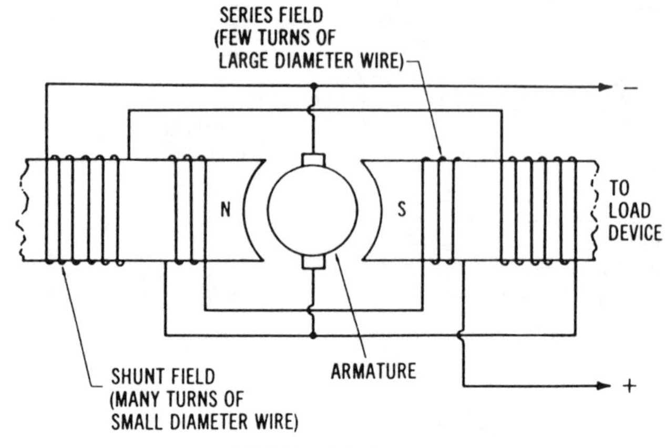

(A) Pictorial diagram.

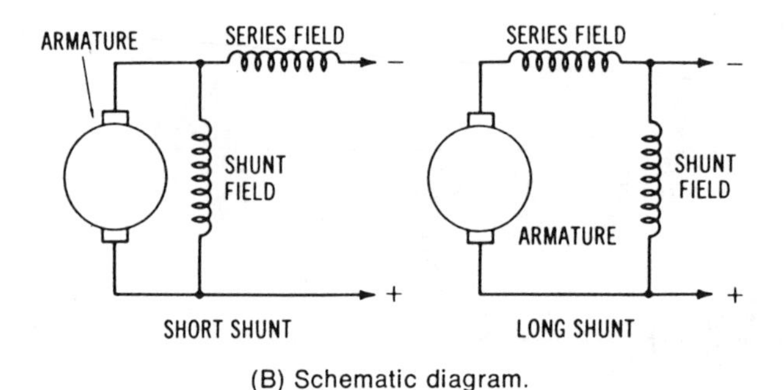

(B) Schematic diagram.

Figure 4-33e. Simplified illustration of a compound-wound DC generator.

DC Generator Characteristics

Direct-current generators are used primarily for operation with mobile equipment. In industrial plants and commercial buildings, they are used for standby power, battery charging, and for specialized DC operations such as electroplating. In many situations, rectification systems that convert alternating current to direct current have replaced DC generators since they are cheaper to operate and maintain.

Direct-current generators supply power to a load circuit by converting the mechanical energy of some prime mover, such as a gasoline or diesel engine, into electrical energy. The prime mover must rotate at a definite speed in order to produce the desired voltage.

Rating of DC Generators—The output of a DC generator is usually rated in kilowatts, which is the electrical power capacity of a machine.

Other ratings, which are specified by the manufacturer on the nameplate of the machine, are current capacity, output voltage, speed, and temperature. Direct-current generators are made in a wide range of physical sizes and with various electrical characteristics.

DC Generator Applications—Direct-current generator use has declined rapidly since the development of the low-cost silicon rectifier. However, there are still certain applications where DC generators are used. These applications include railroad power systems, synchronous motor-generator units, power systems for large earth-moving equipment, and DC motor-drive units for precise equipment control. Some portable generator units are shown in Figure 4-34.

Direct-current Conversion Systems

Most of the electrical power produced is 60-Hz three-phase alternating current. However, the use of direct current is necessary for many applications. For instance, DC motors have more desirable speed-control and torque characteristics than AC motors. We have already discussed two methods of supplying direct current—batteries and DC generators. Batteries and DC generators have been used as direct-current power sources for many years. In many cases today, where large amounts of direct-current power are required, it is more economical to use a system for converting alternating current to direct current. Direct-current load devices may be powered by systems, called either *rectifiers* or *converters,* which change alternating current into a suitable form of direct current.

Rectification Systems

The simplest system for converting alternating current to direct current is single-phase rectification. A single-phase rectifier changes alternating current to pulsating direct current. The most common and economical method is the use of low-cost silicon semiconductor diodes, as shown in Figure 4-35. A three-phase full wave rectifier circuit is shown in Figure 4-36. This type of rectifier circuit is popular for many industrial applications. Six rectifier diodes are required for operation of the circuit. The anodes of D4, D5, and D6 are connected together at point A, while the cathodes of D1, D2, and D3 are connected together at point B. The load device is connected across these two points. The three-phase AC lines are connected to the anode-cathode junctions of D1 and D4, D2 and D5, and D3 and D6. This circuit does not require the neutral line of the three-phase source; therefore, a delta-connected source could be used.

Figure 4-34. Portable generator units: (a) Courtesy Coleman Powermate, Inc.; (b) Courtesy Ag-Tronic.

The resulting DC output voltage of the three-phase, full-wave rectifier circuit is shown in Figure 4-36b. A commercial three-phase rectification system is shown in Figure 4-37.

Rotary Converters

Another method which has been used to convert alternating current to direct current is by use of a *rotor converter*. Rotating AC-to-DC converters are seldom used today. However, motor-driven generator units, such as shown in Figure 4-38, can also be used to convert direct current to alternating current. This system is called an *inverter*. When

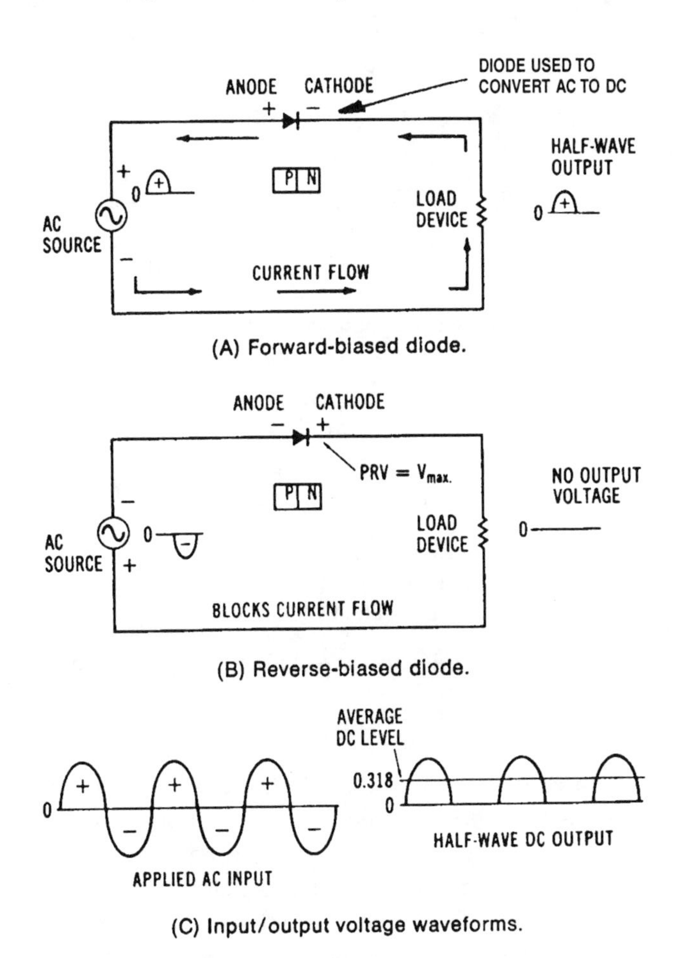

Figure 4-35a. Single-phase half-wave rectification.

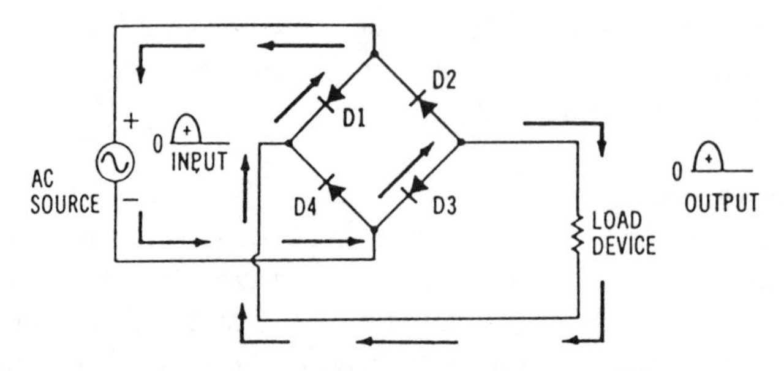

(A) Diodes D1 and D3 are forward biased.

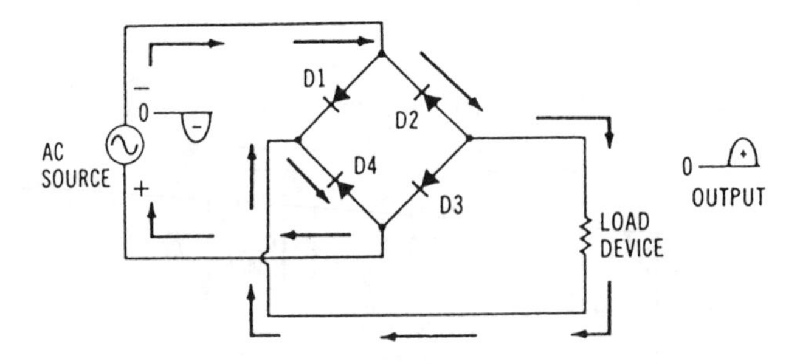

(B) Diodes D2 and D4 are forward biased.

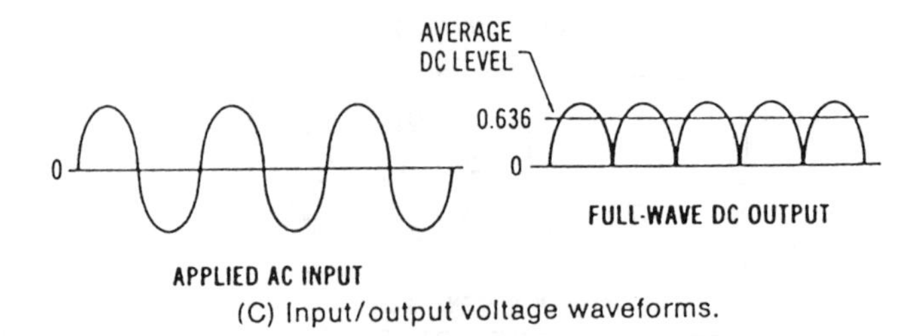

(C) Input/output voltage waveforms.

Figure 4-35b. Single-phase full-wave bridge rectification.

operated as a converter to produce DC, the machine is run off an AC line. The AC is transferred to the machine windings through slip rings and converted to DC by a split-ring commutator located on the same shaft. The amount of DC voltage output is determined by the magnitude of the AC voltage applied to the machine. Converters may be designed as two units, with motor and generator shafts coupled together, or as one unit housing both the motor and generator.

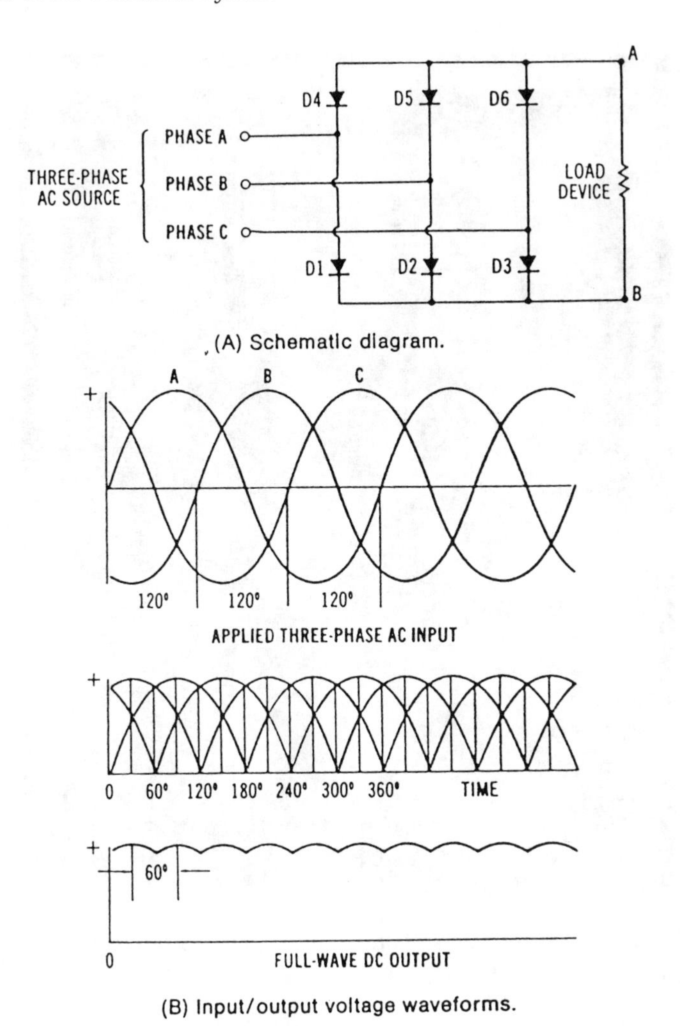

Figure 4-36. Three-phase full-wave rectification circuit.

Figure 4-37. Three-phase rectification system—converts three-phase AC to DC (Courtesy Kinetics Industries).

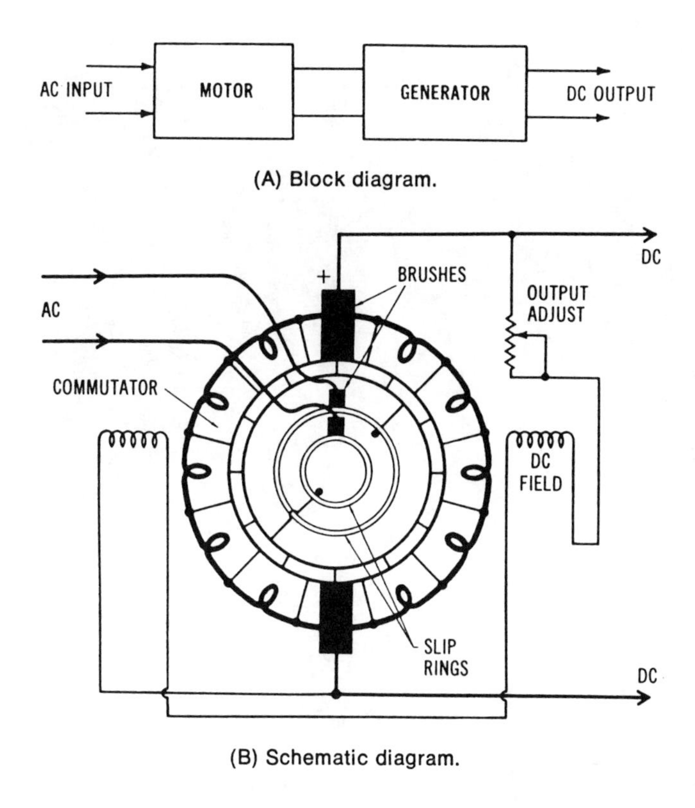

Figure 4-38. Rotary AC-to-DC converter.

Chapter 5

Electrical Distribution System Fundamentals

Electrical distribution systems are an essential part of the electrical power system. In order to transfer electrical power from an alternating-current or a direct-current source to the place where it will be used, some type of distribution network must be utilized. However, the method used to distribute power from where it is produced to where it is used can be quite simple. For example, a battery could be connected directly to a motor, with only a set of wires to connect them together.

More complex power distribution systems are used, however, to transfer electrical power from the power plant to industries, homes, and commercial buildings. Distribution systems usually employ such equipment as transformers, circuit breakers, and protective devices. The essential parts of electrical power distribution systems will be discussed in this text.

POWER SYSTEM BASICS

An electrical power system supplies energy to our homes and industrial buildings. Figure 5-1 shows a sketch of a simple electrical power system. The source of energy may be derived from coal, oil, natural gas, atomic fuel, or moving water. This type of energy is needed to produce mechanical energy, which in turn develops the rotary motion of a turbine. Three-phase alternators are then rotated by the turbine to produce alternating-current electrical energy.

Notice in Figure 5-1 that transformers are located at the power plant and at the industrial site where the power is delivered in this example. The purpose of the transformers at the power is to increase the voltage to a higher level for long distance transmission. Three-phase

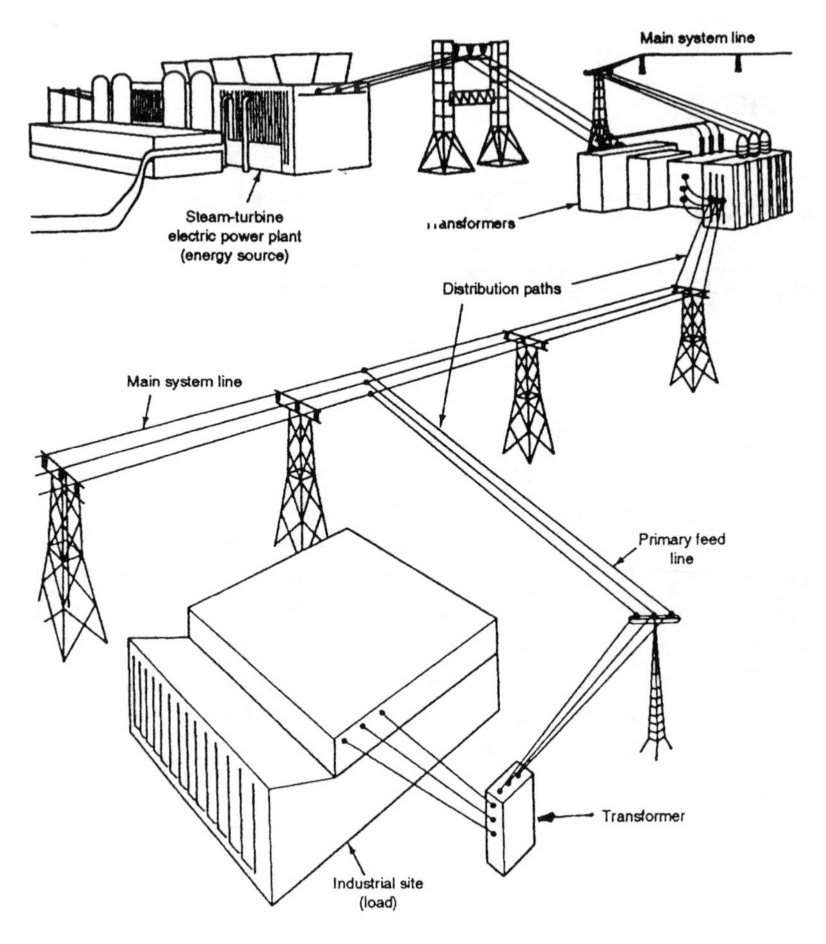

Figure 5-1. Simple electrical power system.

generators at power plants typically produce 13,800 volts (13.8 kV). This voltage may be increased, for example, to 138,000 volts (138 kV) for transmission over distances of 10 to 50 miles. This reduces the current level by a factor of 10 and thus reduces *line loss*. The distribution paths shown in Figure 5-1 typically extend over long distances to an electrical substation or transformers at large consumers like industrial or commercial buildings. The example shown has a transformer at an industrial site used to reduce the voltage level to meet the electrical power requirements of the industry. Also notice in Figure 5-1 that the high-voltage main system lines extend beyond the site shown to serve other residential, commercial and industrial customers.

OVERVIEW OF ELECTRICAL POWER DISTRIBUTION

The distribution of electrical power in the United States is normally in the form of three-phase 60-Hz alternating current. This power, of course, can be manipulated or changed in many ways by the use of electrical circuitry. For instance, a rectification system is capable of converting the 60-Hz alternating current into a direct current, as was discussed in Chapter 4. Also, single-phase power is generally used for lighting and small appliances, such as those in the home or residential environment. However, where a large amount of electrical power is required, three-phase power is more economical.

The distribution of electrical power involves a system of interconnected power transmission. These transmission lines originate at the electrical power-generating stations located throughout the United States. The ultimate purpose of these power transmission and distribution systems is to supply the electrical power necessary for industrial, residential, and commercial use. From the point of view of the system, we may say that the overall electrical power system delivers power from the source to the load which is connected to it. A typical electrical power distribution system is shown in Figure 5-2.

Industries use almost 50% of all the electrical power produced, so three-phase power is distributed directly to most large industries. Electrical *substations*, such as the one shown in Figure 5-3, use massive transformers and associated equipment, such as oil-filled circuit breakers, high-voltage conductors, and huge strings of insulators, in distributing power. From these substations, power is distributed to industrial sites to energize industrial machinery, to residential homes, and to commercial users of electrical power.

POWER TRANSMISSION AND DISTRIBUTION

Power transmission and distribution systems are used to interconnect electrical power production systems and to provide a means of delivering electrical power from the generating station to its point of utilization. Most electrical power systems cast of the Rocky Mountains are interconnected with one another in a parallel circuit arrangement. These interconnections of power production systems are monitored and controlled, in most cases, by a computerized control center. Such control

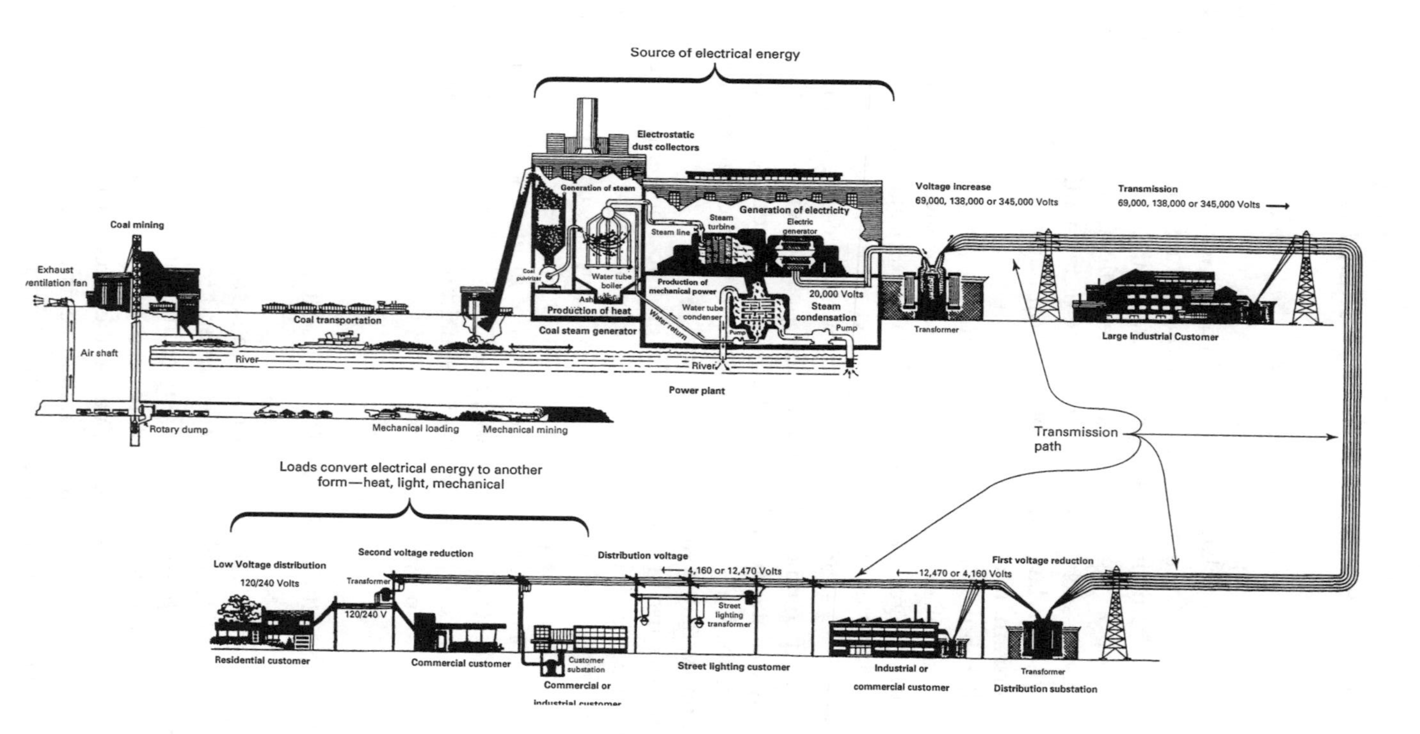

Figure 5-2. A typical electrical power distribution system (Courtesy Kentucky Utilities Co.).

Discussion of Figure 5-2 (Courtesy of Kentucky Utilities Company)

Kentucky Utilities Company manufactures and distributes a product—electricity. Its factories are known as power plants. It transports its product to the customer through miles of wires.

Two elements differentiate electricity from other products: 1) it is weightless and invisible; 2) while most other products can be produced and stored until used, electricity is manufactured, delivered and used instantly, when customers demand it.

The company's system must be large enough to make and deliver the most electricity customers will use at any one time, even though maximum demand occurs only a few hours each year. Equipment and skilled employees must always be ready to meet these fluctuating demands.

In Kentucky, with its wealth of coal, the principal source of heat energy which is changed into electrical energy is coal. Water is used in relatively small amounts to provide steam, and in vast quantities for cooling purposes. It is provided either by large running streams or by recirculation through cooling towers.

At the generating plant, coal is crushed as fine as talcum powder and then blown into huge furnaces where it burns like a gas. En route to the chimneys, flame and hot gases pass over miles of steel tubing in the boilers. Water in the tubing flashes into superheated steam at 1,000 degrees F.

The steam, at 850 to 2400 pounds per square inch pressure, turns the turbine. The principle is similar to that of a windmill, with steam replacing wind.

Nozzles force the steam against a series of blades, spending more energy as it speeds along the length of the turbine, until most of its energy has been used. Additional impetus is given the steam by the partial vacuum in the condenser into which it is finally discharged. There it becomes water again by passing over hundreds of tubes through which cooling water circulates. It is then pumped back to the boiler to be turned into steam once more. The cooling water is returned to the river or to the cooling tower.

The steam-driven turbine, turning at 3600 revolutions per minute, has a common shaft with the electric generator, which makes electricity by turning a huge magnet inside a coil of wire.

Alternating current is generated at 13,800 to 22,000 volts. Yet, it is necessary to increase this to between 69,000 and 345,000 volts since higher voltage is more economical to deliver. The "step up" transformer that does this is a rectangular iron core on which several times as many turns of copper wire are wrapped around the outgoing "leg" as around the incoming one to raise the voltage. Many cores are in oil-filled cases for insulation and to carry off the heat generated. Steel towers or large H-frame wood structures support the many miles of wires which carry the high voltage electricity across the country.

At various points on the transmission system, lines branch off to communities or to large industrial and commercial loads. At these points, the voltage is reduced by transformers to 12,470 or 4,160 volts. These voltages extend through most residential areas, and even to some industrial and large commercial customers. However, the small transformers you see on poles along streets reduce the voltage further to 120/240 to feed electricity directly into homes, offices and stores to be used in countless work and time-saving ways.

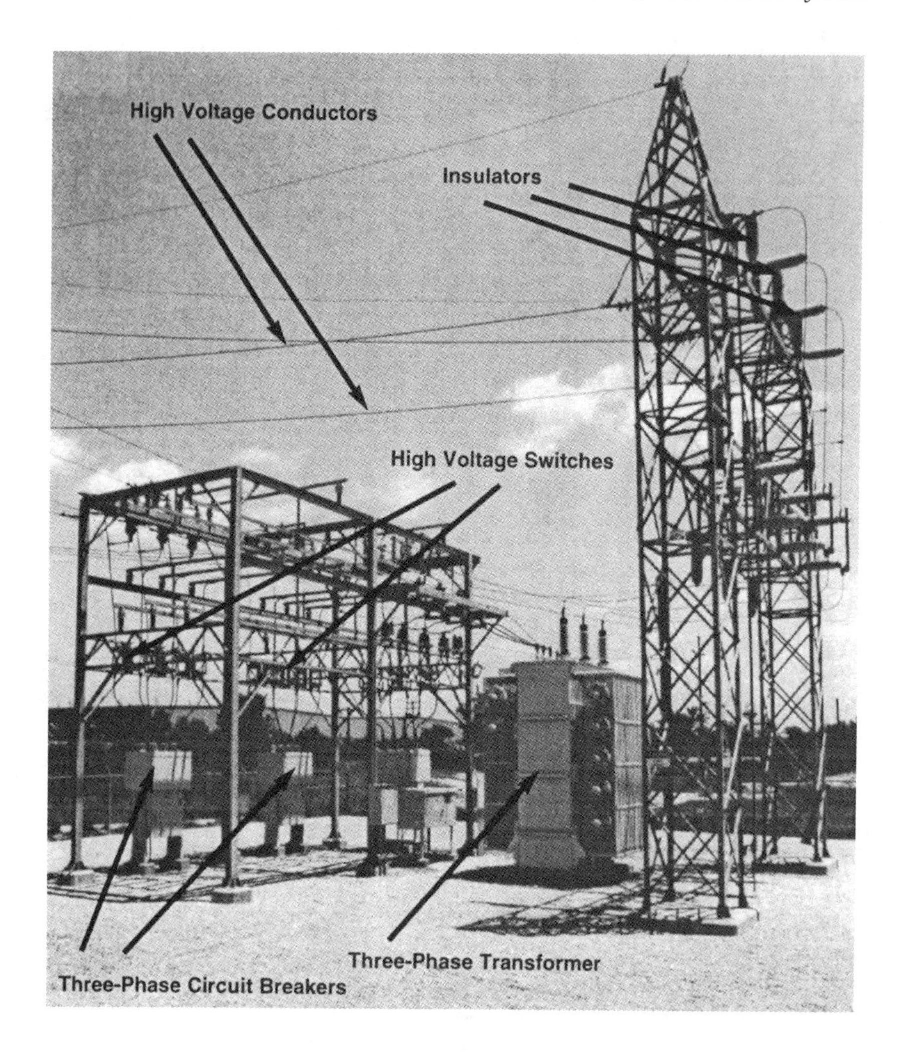
High Voltage Conductors
Insulators
High Voltage Switches
Three-Phase Transformer
Three-Phase Circuit Breakers

Figure 5-3.

centers provide a means of data collection and recording, system monitoring, frequency control, and signaling. Computers have become an important means of assuring the efficient operation of electrical power systems.

The transmission of electrical power requires many long interconnected power lines to carry the electrical current from where it is produced to where it is used (see Figure 5-4). However, overhead power transmission lines require much planning to assure the best use of our land. The location of overhead transmission lines is limited by zoning

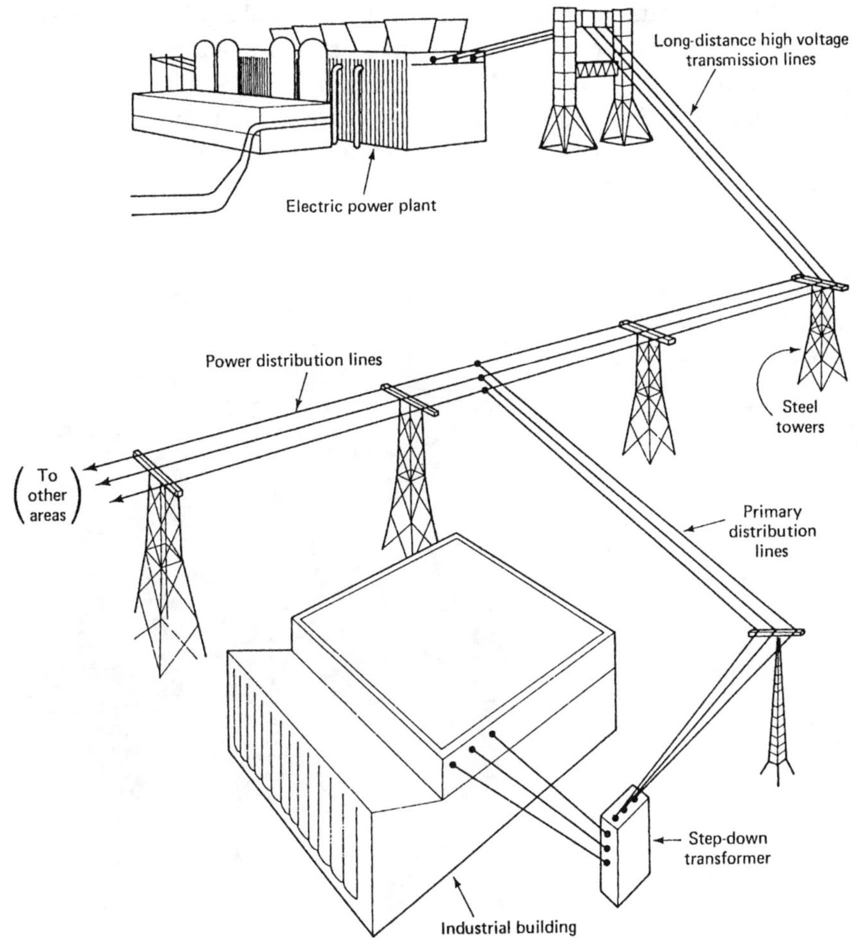

Figure 5-4. Overhead power lines are used for long-distance high-voltage transmission of electrical power from power plants to the areas where it is used.

laws and by populated areas, highways, railroads, and waterways, as well as other topographical and environmental factors. Transmission lines may be supported by steel towers or wood or concrete poles. The type of support is determined by the terrain to be crossed and the size of conductors. Usually steel towers are used for transmission and wood poles for distribution.

Today an increased importance is being placed upon environmental and aesthetic factors. Power transmission lines (see Figure 5-5) ordinarily operate at voltage levels from 12 kilovolts to 500 kilovolts of alternating current. Common transmission-line voltages are in the range of 50 to 150 kilovolts of alternating current. High-voltage direct-current overhead transmission lines may become economical although they are not being used extensively at the present time.

Another option is ultra high-voltage transmission lines which use higher AC transmission voltages. Also, underground transmission methods for urban and suburban areas must be considered since right-of-way for overhead transmission lines is limited. Alternating-current overhead

Figure 5-5. Electrical power transmission lines (Courtesy of Lapp Insulator, Leroy, N.Y.).

transmission voltages have increased to levels in the range of 765 kV, with experimentation dealing with voltages of over 1000 kV. One advantage of overhead cables is their ability to dissipate heat. The use of cryogenic cable may some day bring about a solution to heat dissipation problems in conductors.

UNDERGROUND DISTRIBUTION

The use of underground cable is ordinarily confined to the short lengths required in congested urban areas. The cost of underground cable is much more than for aerial cable. To improve underground cable power-handling capability, research is being done in forced-cooling techniques, such as circulating-oil and with compressed-gas insulation. Another possible method is the use of cryogenic cables or superconductors which operate at extremely low temperatures and have a large power-handling capability.

The original electrical distribution system developed by Thomas Edison was an underground direct current (DC) system. However, expansion of electrical distribution has been primarily accomplished by very economical overhead high voltage alternating current (AC) systems. Limitations to underground systems are primarily economic. The need for duct and manhole construction and conductors with special insulation cause the initial installation cost of underground systems to be very expensive. The primary advantage of underground systems is aesthetic, since conductors, towers, poles and other equipment can be very unsightly.

Overhead power distribution in heavily populated areas can be impractical due to congestion of conductors and associated equipment. Often maintenance is also restricted due to the volume of vehicle traffic. Also the possibility of vehicles striking poles and disrupting the power system is a problem. Underground systems are not as susceptible to problems due to storms and lightning strikes. Greater emphasis on the environment and advances in insulation effectiveness have made underground systems more economical. However, in rural areas and areas of low population density, overhead distribution systems are much more economical.

Placing electrical conductors and equipment underground, as shown in Figure 5-6, presents certain problems which are not common to overhead distribution systems. Distribution equipment in heavily populated downtown areas is usually placed in ducts, manholes and vaults. In less populated areas, conductors are usually buried directly in the ground, with transformers housed in metal enclosures at ground level or partially or completely buried. A consideration with underground systems is assuring that problems can be corrected without causing inconvenience to the public. Cables are placed in ducts and spliced together in manholes or service boxes. Transformers are installed in manholes or vaults. Manholes are often installed with other utilities such as water, gas, sewer, or telephone.

HIGH-VOLTAGE DIRECT-CURRENT (HVDC) TRANSMISSION

An alternative to transmitting AC voltages for long distances is high-voltage direct-current (HVDC) power transmission. The HVDC is suitable for long-distance overhead power lines or for underground power lines. Direct-current power lines are capable of delivering more power per conductor than an equivalent AC power line. Due to its fewer power losses, HVDC is even more desirable for underground distribution.

The primary disadvantage of HVDC is the cost of the necessary AC-to-DC conversion equipment. There are, however, some HVDC systems in operation in the United States. At present, HVDC systems hay, been designed for transmitting voltages in the range of 600 kV. The key to the future development of HVDC systems may be the production of solid-state power conversion systems with higher voltage and current ratings. With a continued developmental effort, HVDC might eventually play a more significant role in future electrical power-transmission systems.

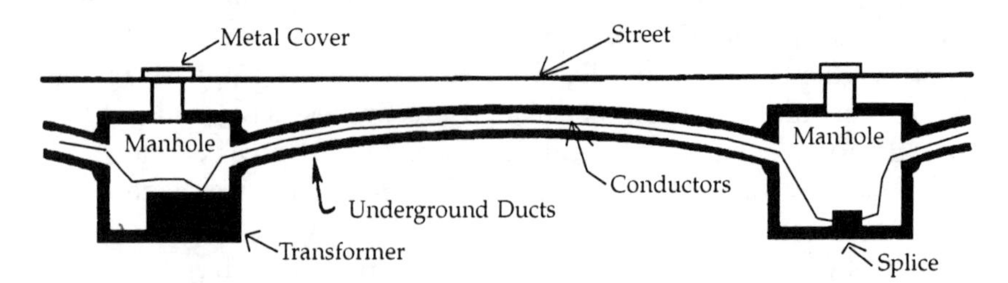

Figure 5-6. Underground electrical distribution.

CRYOGENIC CABLE

There are some problems involved in installing an overhead electrical power distribution system, particularly in urban areas. One of these problems is obtaining a right-of-way through heavily populated areas for the overhead cable. The difficulty is caused primarily by the unattractive appearance of the lines and the potential danger of the high voltage. The problems associated with overhead transmission lines have led to the development of cryogenic cable for underground power distribution.

Cryogenic cables are not considered to be superconductive, but they do have greater electrical conductivity at very low temperatures. These cables, which are still in the developmental stage, will use a metallic conductor cooled to the temperature of liquid nitrogen. One advantage of cryogenic cable over conventional cable is that it will have a lower line loss (I × R loss) due to its greater conductive characteristics.

One design of a cryogenic cable is shown in Figure 5-7. This design involves the use of three separate cables, each having a hollow center for cooling purposes. The conductive portion of the cables is stranded aluminum. The aluminum conductors are wrapped with an insulating material which contains liquid nitrogen. Cryogenic cable has considerable potential in any future development of electrical distribution systems.

PARALLEL OPERATION OF POWER SYSTEMS

Electrical power distribution systems are operated in a parallel circuit arrangement. By adding more power sources (generators) in parallel, a greater load demand or current requirement can be met. On a smaller scale, this is like connecting two or more batteries in parallel to provide greater current capacity. Two parallel-connected three-phase alternators are depicted in Figure 5-8. Most power plants have more than one alternator connected to any single set of power lines inside the plant. These power lines or "bus" lines are usually large copper bar conductors which can carry very high amounts of current. At low-load demand times, only one alternator would be connected to the bus lines.

Figure 5-9 expands the concept of parallel-connected systems. An illustration of two power plants joined together through a distribution substation is shown. The two power plants might be located 100 miles

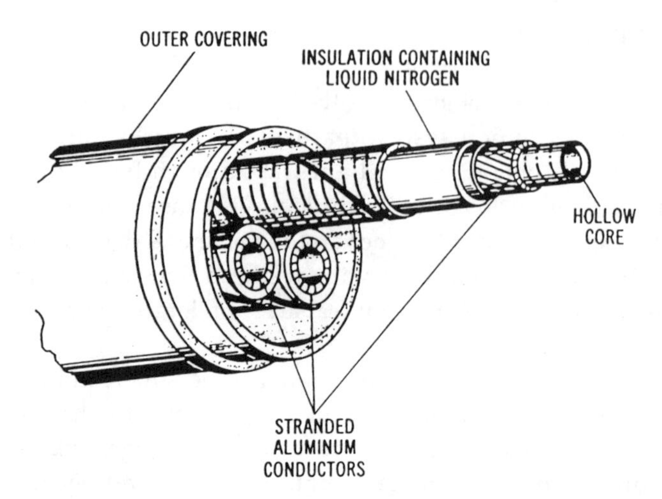

Figure 5-7. Construction of a cryogenic cable.

apart, yet they are connected in parallel to supply power to a specified region. If, for some reason (such as repairs on one alternator), the output of one power plant is reduced, the other power plant is still available to supply power to the requesting localities. It is also possible for power plant No. 1 to supply part of the load requirement ordinarily supplied by power plant No. 2, or vice versa. These regional distribution systems of parallel-connected power sources provide automatic compensation

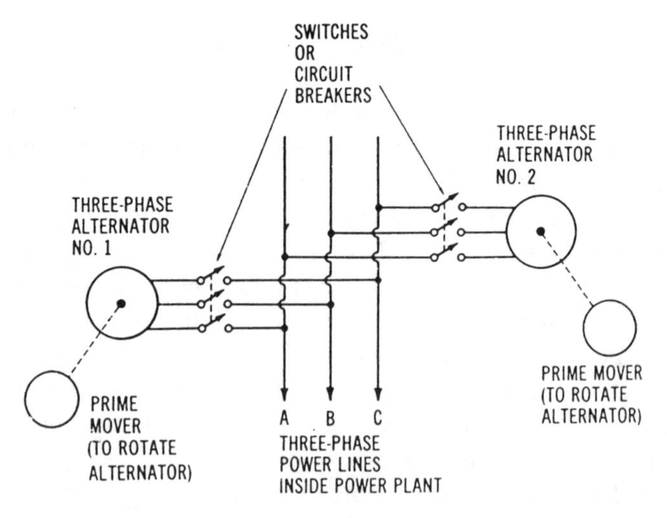

Figure 5-8. Parallel connection of two three-phase alternators.

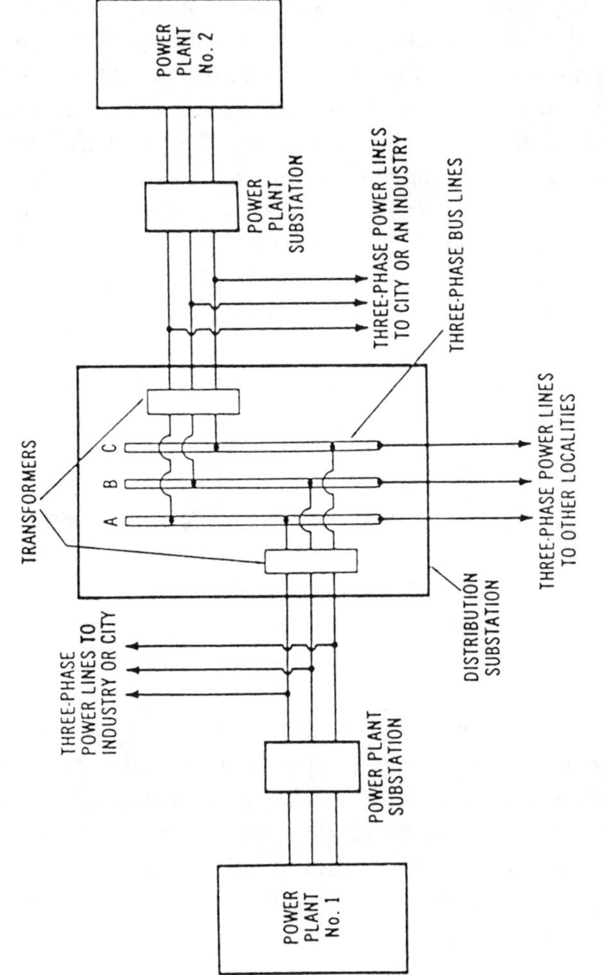

Figure 5-9. Joining two power plants in parallel as part of a regional power system.

for any increased load demand in any area.

The major problem of parallel-connected electrical distribution systems occurs when excessive load demands are encountered by several power systems in a single region. If all of the power plants in one area are operating near their peak power-output capacity, there is no back-up capability. The equipment-protection system for each power plant and, also, for each alternator in the power plant is designed to disconnect from the system when its maximum power limits are reached. When the power demand on one part of the system becomes too excessive, the protective equipment will disconnect that part of the system. This places an even greater load on the remaining parts of the system. The excessive load now could cause other parts of the system to disconnect. This cycle could continue until the entire system is inoperative. This is what occurs when "blackouts" of power systems take place. No electrical power can be supplied to any part of the system until most of the power plants are put back in operation. The process of putting the output of a power plant back "on-line," when the system is down during power outages, can be a long and difficult procedure.

RADIAL, RING, AND NETWORK DISTRIBUTION SYSTEMS

There are three general classifications of electrical power distribution systems. These are the *radial, ring,* and *network* systems shown in Figures 5-10 through 5-12. Radial systems are the simplest type since the power comes from one power source. A generating system supplies power from the substation through radial lines which are extended to the various areas of a community (Figure 5-10). Radial systems are the least reliable in terms of continuous service since there is no back-up distribution system connected to the single power source. If any power line opens, one or more loads are interrupted. There is more likelihood of power outages. However, the radial system is the least expensive. This system is used in remote areas where other distribution systems are not economically feasible.

Ring distribution systems (Figure 5-11) are used in heavily populated areas. The distribution lines encircle the service area. Power is delivered from one or more power sources into substations near the service area. The power is then distributed from the substations through the radial power lines. When a power line is opened, no interruption to other

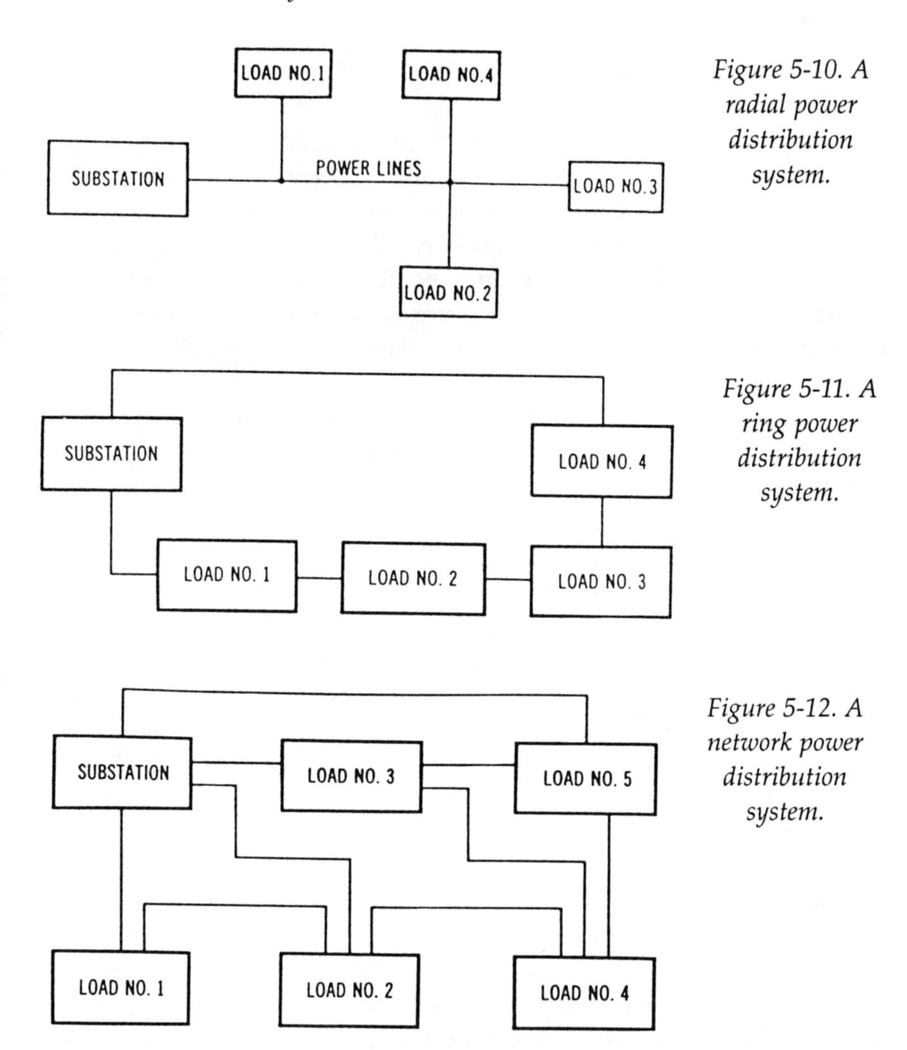

Figure 5-10. A radial power distribution system.

Figure 5-11. A ring power distribution system.

Figure 5-12. A network power distribution system.

loads occurs. The ring system provides a more continuous service than the radial system. Due to additional power lines and a greater circuit complexity, the ring system is more expensive.

Network distribution systems (Figure 5-12) are a combination of the radial and ring systems. They usually result when one of the other systems is expanded. Most of the distribution systems in the United States are network systems. This system is more complex but it provides very reliable service to consumers. With a network system, each load is fed by two or more circuits.

TRANSFORMERS FOR POWER DISTRIBUTION

The heart of an electrical power distribution system is an electrical device known as a transformer. This device is capable of controlling massive amounts of power for efficient distribution. Transformers are also used for many other applications. A knowledge of transformer operation is essential for understanding electrical distribution systems. The distribution of alternating-current power is dependent upon the use of transformers at many points along the route of the power distribution system.

It is economically feasible to transmit electrical power over long distances at high voltages since less current is required at high voltages and, therefore, *line loss* (I^2R) is reduced significantly. A typical high-voltage transmission line may extend a distance of 50-100 miles from the generating station to the first substation. These high-voltage power transmission lines typically operate at 200,000 to 500,000 volts by using step-up transformers to increase the voltage produced by the AC generators at-the-power station. Various substations are encountered along the power distribution system, where transformers are used to reduce the high transmission voltages to a voltage level such as 480 volts which is suitable for industrial motor loads, or to 120/240 volts for residential use.

Transformers provide a means of converting an alternating-current voltage from one value to another. The basic construction of a transformer is illustrated in Figure 5-13. Notice that the transformer shown consists of two sets of windings which are not physically connected. The only connection between the primary and secondary windings is the magnetic coupling effect known as *mutual induction,* which takes place when the circuit is energized by an alternating-current voltage. The laminated iron core plays an important role in transferring magnetic flux from the primary winding to the secondary winding. Single-phase transformers of various types are shown in Figure 5-14.

Transformer Operation

If an AC current which is constantly changing in value flows in the primary winding, the *magnetic field* produced around the primary winding will be transferred to the secondary winding. Thus, an *induced voltage* is developed across the secondary winding. In this way, electrical energy can be transferred from the source (primary-winding circuit) to a load (secondary-winding circuit).

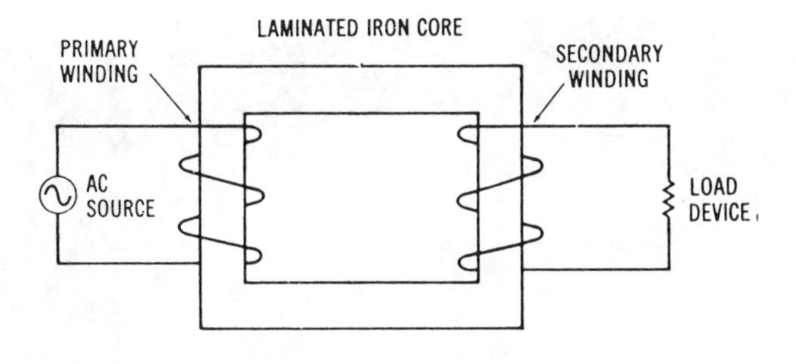

(A) Pictorial diagram.

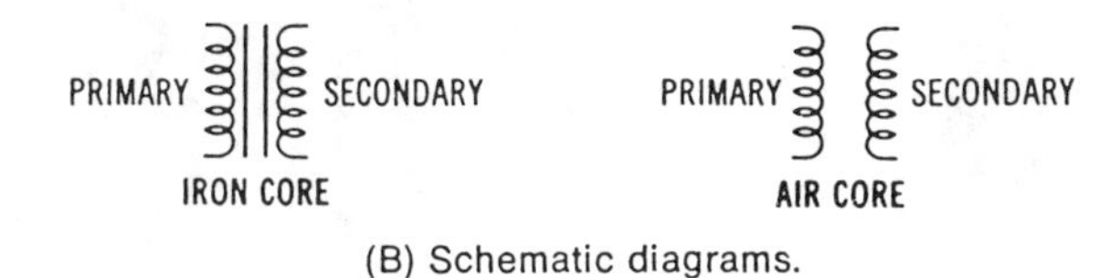

(B) Schematic diagrams.

Figure 5-13. Basic transformer construction.

Figure 5-14a. Single-phase transformers used for electrical power distribution.

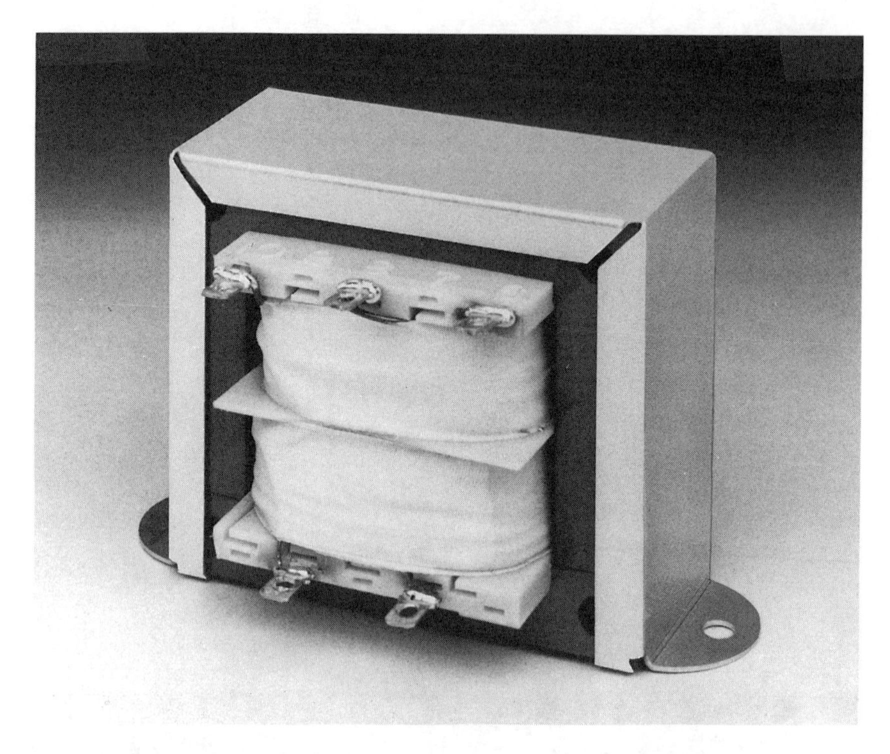

Figure 5-14b. Single-phase transformer with center-tapped secondary winding (top terminals) (Courtesy Signal Transformer).

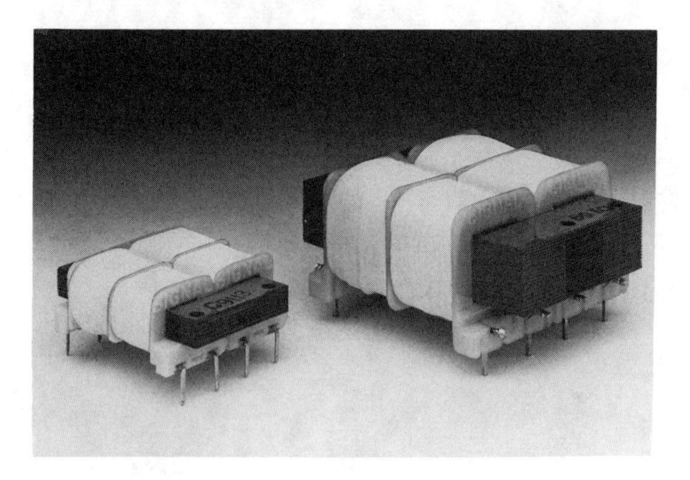

Figure 5-14c. Small transformers used in electronic equipment (Courtesy Signal Transformer).

The efficient transfer of energy from the primary to the secondary windings of a transformer depends on the coupling of the magnetic field between these two windings. Ideally, all magnetic lines of force developed around the primary winding would be transferred by magnetic coupling to the secondary winding. However, a certain amount of magnetic loss takes place as some lines of force escape to the surrounding air.

Transformer Core Construction

In order to decrease the amount of magnetic losses, transformer windings are wound around iron cores. Iron cores concentrate the magnetic lines of force so that better coupling between the primary and secondary windings is accomplished. Two types of transformer cores are illustrated in Figure 5-15. These cores are made of laminated iron to reduce undesirable eddy currents which are induced into the core material. These eddy currents cause power losses. The diagram of Figure 5-15a shows a closed-core transformer construction. The transformer windings of the closed-core type are placed along the outside of the metal core. Figure 5-15b shows the shell-core type of construction. The shell-core construction method produces better magnetic coupling since the transformer windings are surrounded by metal on both sides. Note that the primary and secondary windings of both types are placed adjacent to one another for better magnetic coupling.

TRANSFORMER EFFICIENCY AND LOSSES

Transformers are very efficient electrical devices. A typical efficiency rating for a transformer would be around 98%. Efficiency of electrical equipment is expressed as:

$$\text{Efficiency (\%)} = \frac{P_{out}}{P_{in}} \times 100$$

where,

P_{out} is the power output in watts,

P_{in} is the power input in watts.

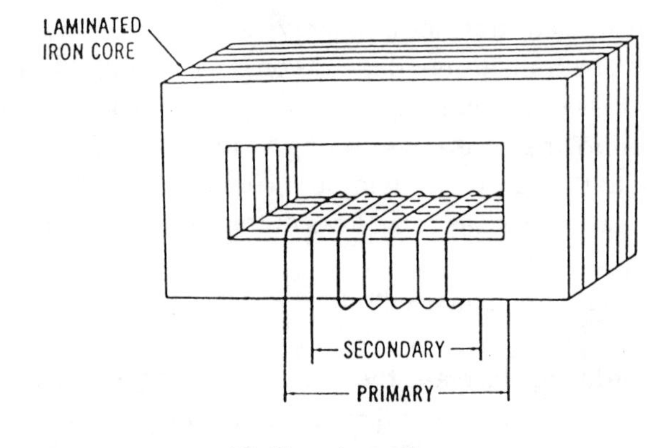

(A) Closed-core type.

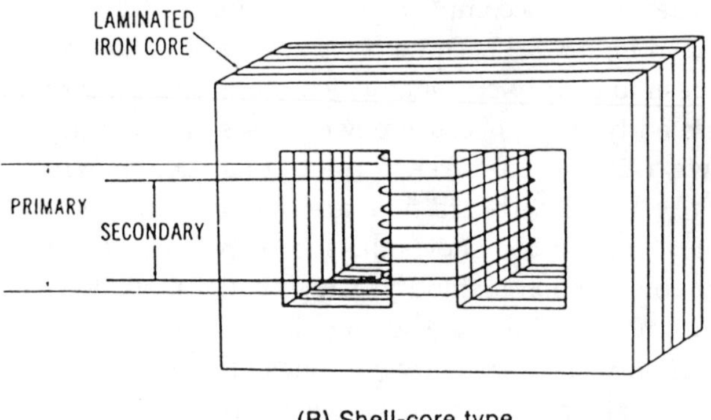

(B) Shell-core type.

Figure 5-15. Types of iron-core transformers.

Sample Problem:

Given: A transformer circuit has a power output of 2.5 kW and a power input of 2,550 W.

Find: The efficiency of the transformer

Solution:

$$\%\ \text{Eff} = \frac{2{,}500\ \text{W}}{2{,}550\ \text{W}} \times 100$$

$$\%\ \text{Eff} = 98\%$$

The losses that reduce efficiency, in addition to flux leakage, are copper and iron losses. Copper loss is the I_2R loss of the windings, while iron losses are those caused by the metallic core material. The insulated laminations of the iron core help to reduce iron losses.

STEP-UP AND STEP-DOWN TRANSFORMERS

Transformers are functionally classified as step-up or step-down types. These types are illustrated in Figure 5-16. The step-up transformer of Figure 5-16a has fewer turns of wire on the primary than on the secondary. If the primary winding has 50 turns of wire and the secondary has 500 turns, a turns ratio of 1:10 is developed. Therefore, if 12 volts AC is applied to the primary from the source, 10 times that voltage, or 120 volts AC, will be transferred to the secondary load (assuming no losses).

The example of Figure 5-16b is a step-down transformer. The step-down transformer has more turns of wire on the primary than on the secondary. The primary winding of the example has 200 turns while the secondary winding has 100 turns, or a 2:1 ratio. If 120 volts AC is applied to the primary from the source, then one-half that-amount, or 60 volts AC, will be transferred to the secondary load.

TRANSFORMER VOLTAGE AND CURRENT RELATIONSHIPS

From the preceding examples, a direct relationship is shown between the primary and secondary turns and the voltages across each winding. This relationship may be expressed as:

$$\frac{V_p}{V_s} = \frac{N_p}{N_s}$$

where,

V_p is the voltage across the primary winding,
V_s is the voltage across the secondary winding,
N_p is the number of turns in the primary winding,
N_s is the number of turns in the secondary winding.

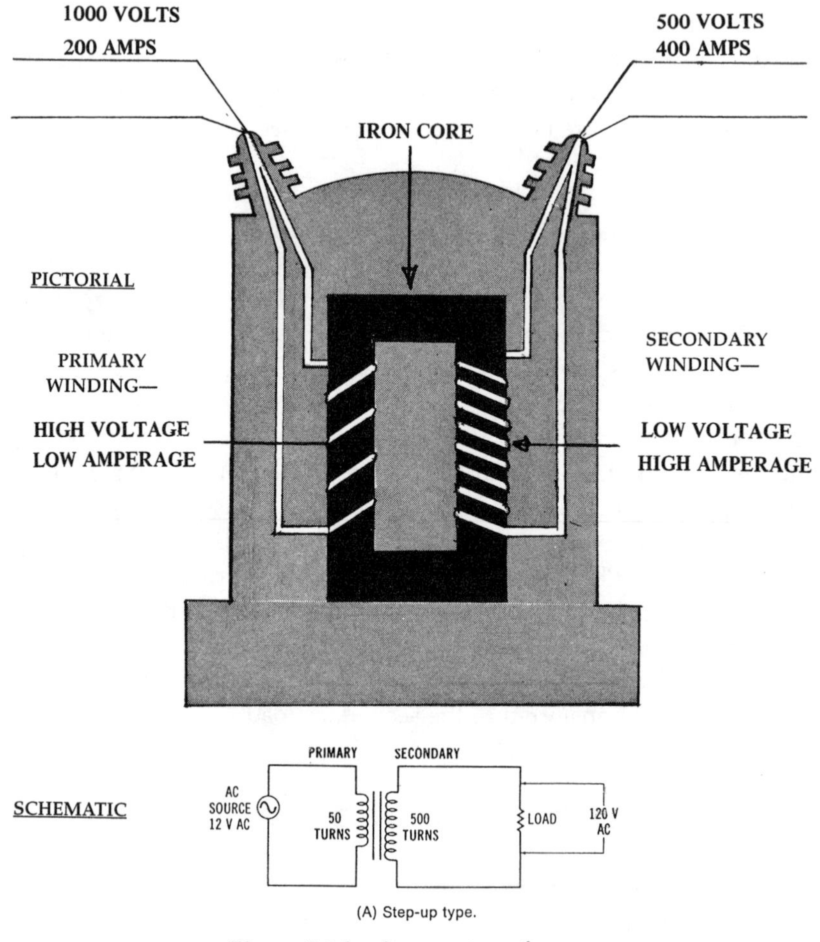

(A) Step-up type.

Figure 5-16a. Step-up transformer.

Sample Problem:

Given: A transformer circuit has the following values:

V_p = 240 volts
N_p = 1000 turns
V_s = 120 volts

Find: The number of secondary turns of wire required to accomplish this step-down of voltage

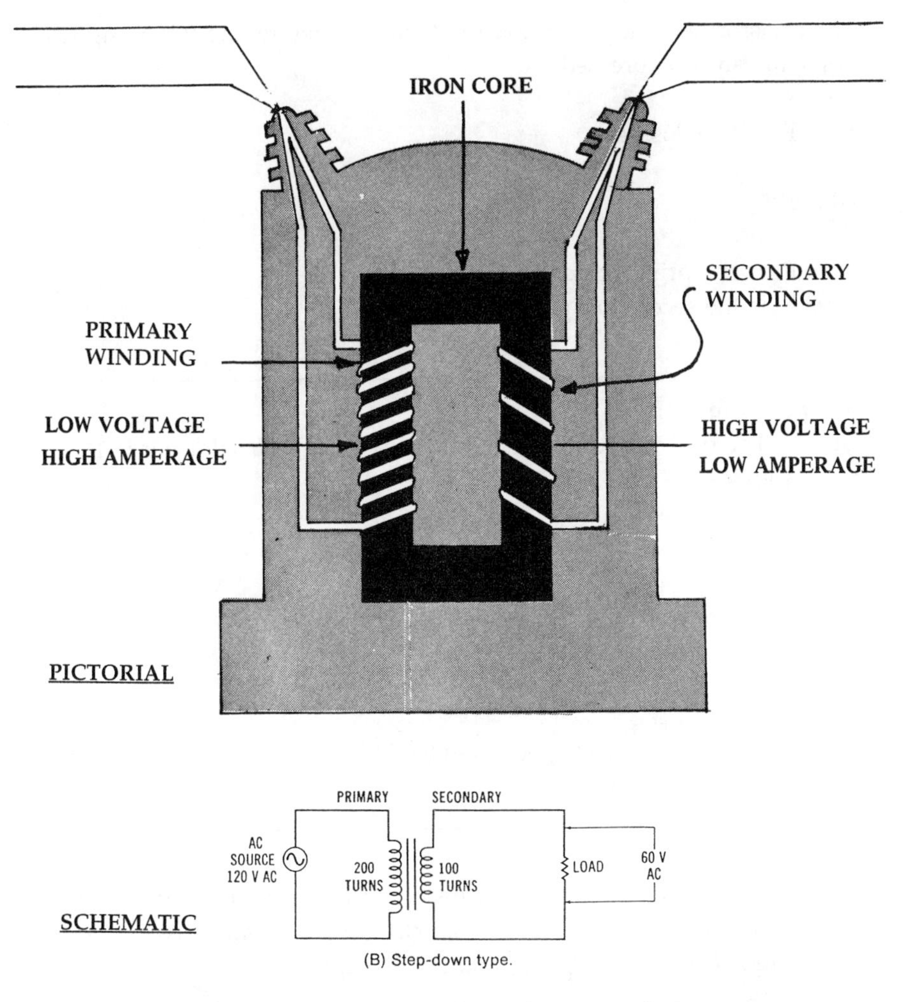

(B) Step-down type.

Figure 5-16b. Step-up and step-down transformer.

Solution:

$$\frac{V_p}{V_s} = \frac{N_p}{N_s}$$

$$\frac{240\text{ V}}{120\text{ V}} = \frac{1000}{N_s}$$

$N_s = 500$ Turns

The transformer is a power-control device; therefore, the following relationship can be expressed:

$$P_p = P_s + \text{losses}$$

where,

P_p is the primary power,
P_s is the secondary power.

Sample Problem:
Given: The power output (secondary power) of a transformer is 15 kW and its losses are as follows:
iron losses - 200 W
copper loss - 350 W

Find: The power input (primary power) required

Solution:

$$P_p = P_s \text{ losses}$$
$$= 15{,}000 \text{ W} + (200 \text{ W} + 350 \text{ W})$$
$$P_p = 15{,}550 \text{ Watts}$$

The losses are those that ordinarily occur in a transformer. In transformer theory, an ideal device is usually assumed, and losses are not considered. Thus, since $P_p = P_s$ and $P = V \times I$, then

$$V_p \times I_p = V_s \times I_s$$

where,
I_p is the primary current,
I_s the secondary current.

Therefore, if the voltage across the secondary is stepped up to twice the voltage across the primary, then the secondary current will be stepped down to one-half the primary current. The current relationship of a transformer is thus expressed as:

$$\frac{I_p}{I_s} = \frac{N_s}{N_p}.$$

Note that whereas the voltage-turns ratio is a *direct* relationship, the current-turns ratio is an *inverse* relationship.

Sample Problem:

Given: An ideal transformer circuit has the following values:

$V_p = 600$ V
$V_s = 2400$ V
$I_s = 80$ A

Find: The primary current drawn by the step-up transformer

Solution:

$$\frac{I_p}{I_s} = \frac{V_s}{V_p}$$

$$\frac{I_p}{80\text{ A}} = \frac{2400\text{ V}}{600\text{ V}}$$

$I_p = 3{,}200$ Amps

MULTIPLE SECONDARY TRANSFORMERS

It is also possible to construct a transformer which has multiple secondary windings, as shown in Figure 5-17. This transformer is connected to a 12-volt AC source which produces the primary magnetic flux. The secondary has two step-down windings and one step-up winding. Between points 1 and 2, a voltage of 5-volts AC could be supplied. Between point 5 and point 6, 30-volts AC may be obtained, and between points 3 and 4, a voltage of 360-volts AC can be supplied to a load. This type of transformer is used for the power supply of various types of electronic equipment and instruments.

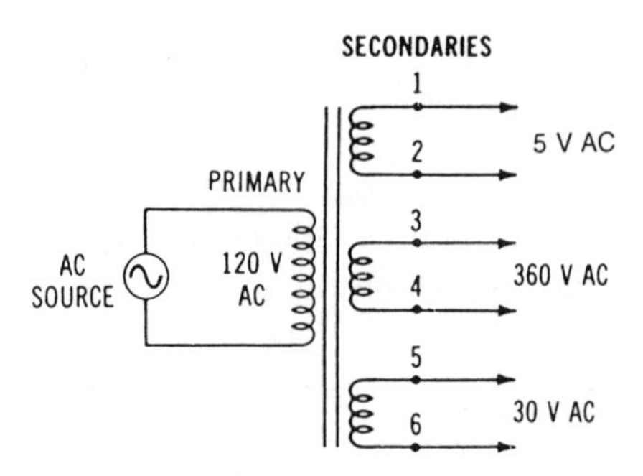

Figure 5-17. Transformer with multiple secondary windings.

Autotransformers

Another specialized type of transformer is the auto-transformer, shown in Figure 5-18. The autotransformer has only one winding, with a common connection between the primary and secondary. The principle of operation of the autotransformer is similar to other transformers. Both the step-up and step-down types are shown in Figure 5-18. The control device shown in Figure 5-19 a variable autotransformer in which the winding tap may be adjusted along the entire length of the winding to provide a variable AC voltage to a load.

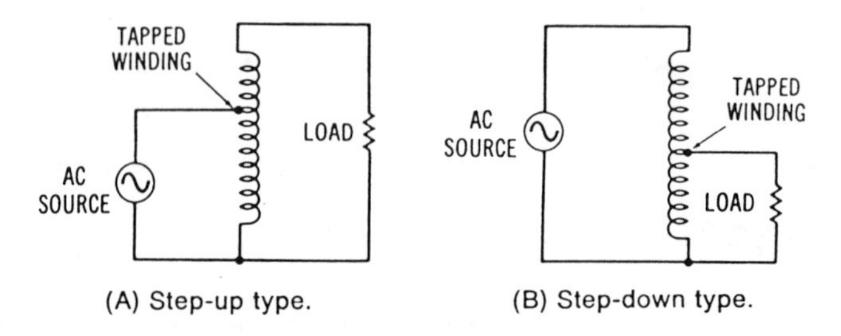

Figure 5-18. Autotransformers.

CURRENT TRANSFORMER

Current transformers are often used to reduce a large value of line current to a smaller value for measurement or control purposes. These transformers are used to measure the current magnitude of high-current systems. Since most metering systems respond linearly to current

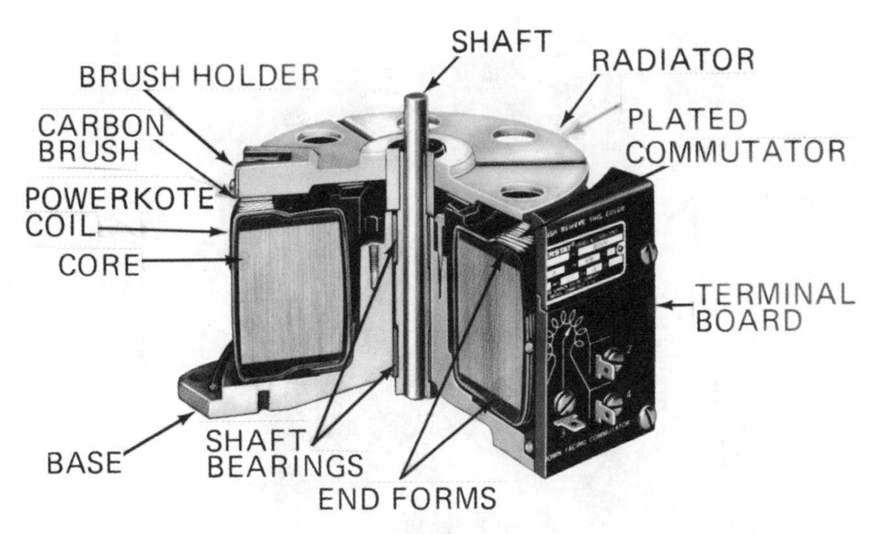

Figure 5-19. Cutaway view of a variable autotransformer (Courtesy Superior Electric Co.).

changes, the current transformer principle could also be used to measure quantities other than current in high-power systems. A single-phase and a three-phase current transformer are shown in Figure 5-20.

TRANSFORMER POLARITY AND RATINGS

Power distribution transformers usually have polarity markings so that their windings may be connected in parallel to increase their current capacity. The standard markings are H_1, H_2, H_3, etc., for the high-voltage windings, and X_1, X_2, X_3, etc., for the low-voltage windings (see Figure 5-21). Many power transformers have two similar primary windings and two similar secondary windings to make them adaptable to different voltage requirements simply by changing from a series to a parallel connection. The voltage combinations available from this type of transformer are shown in Figure 5-22.

The ratings of power transformers are very important. Usually transformers are rated in kilovoltamperes: (kVA). A kilowatt rating is not used since it would be misleading due to the various power-factor ratings of industrial loads. Other power-transformer ratings usually include frequency, rated voltage of each winding, and an impedance rating.

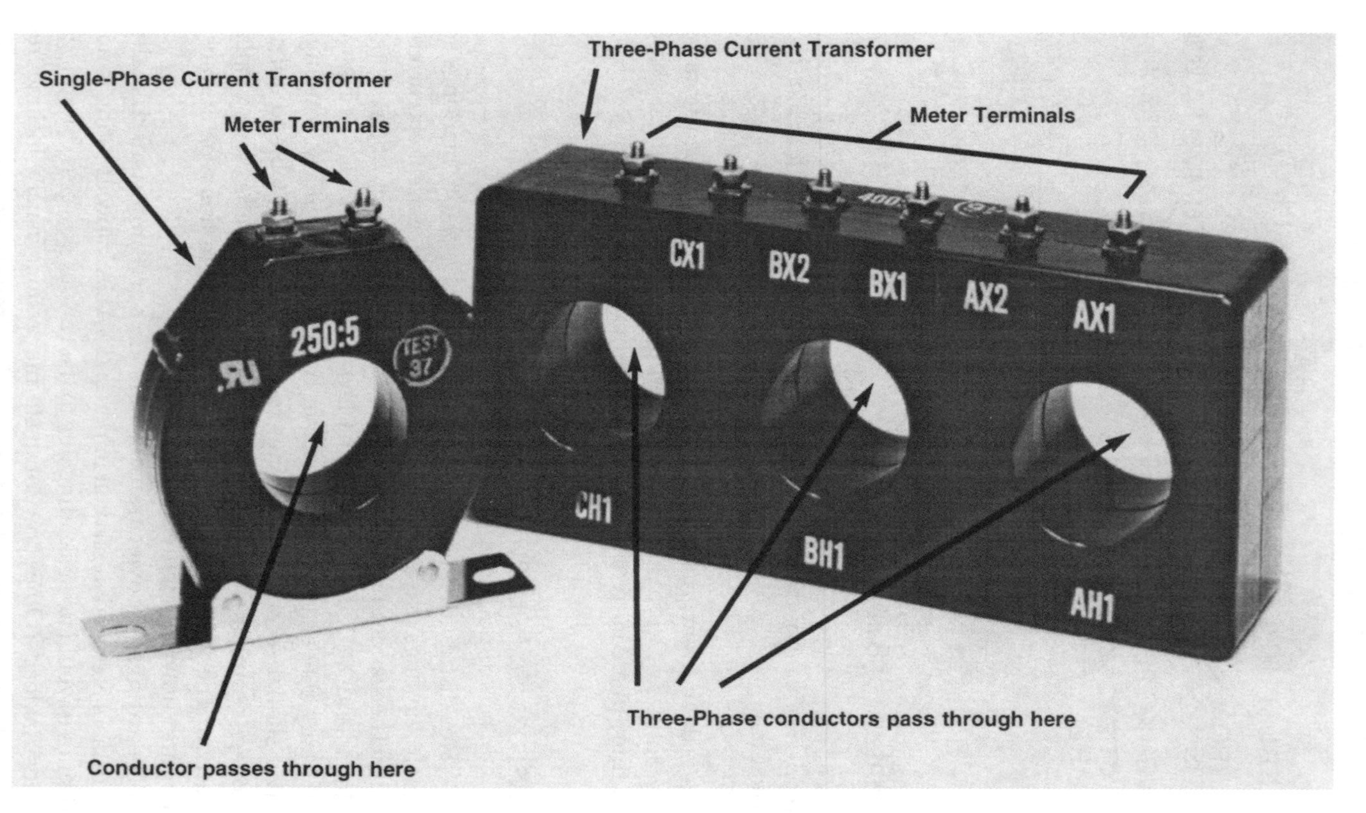
Single-Phase Current Transformer
Meter Terminals
250:5
Three-Phase Current Transformer
Meter Terminals
CX1
BX2
BX1
AX2
AX1
CH1
BH1
AH1
Three-Phase conductors pass through here
Conductor passes through here

Figure 5-20.

Figure 5-21. Transformer polarity marking and connections.

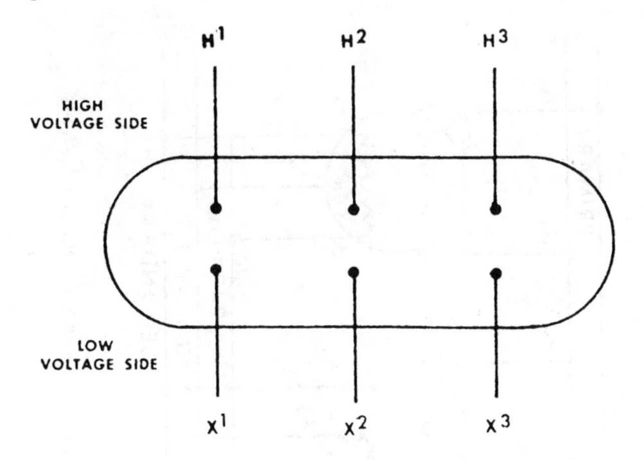

Figure 5-21a. Lead markings of three-phase transformers.

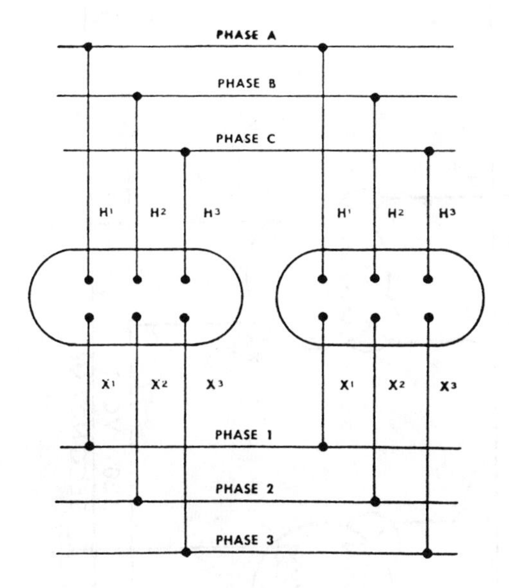

Figure 5-21b. Connecting three-phase transformers in parallel.

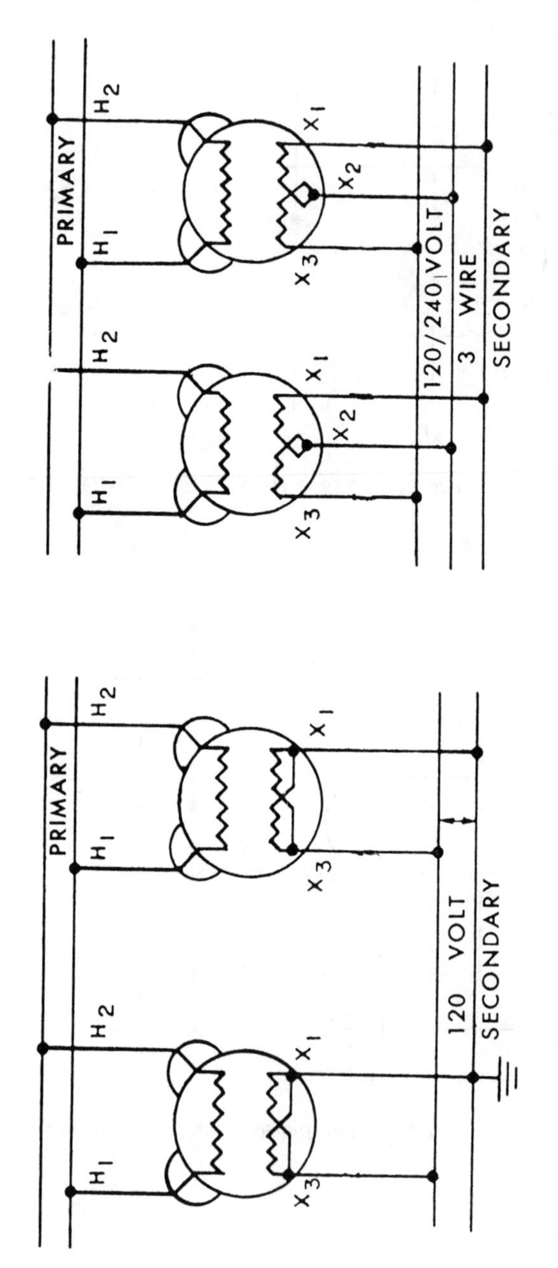

Figure 5-21c. Two single-phase transformers in parallel in two-wire and three-wire secondary systems.

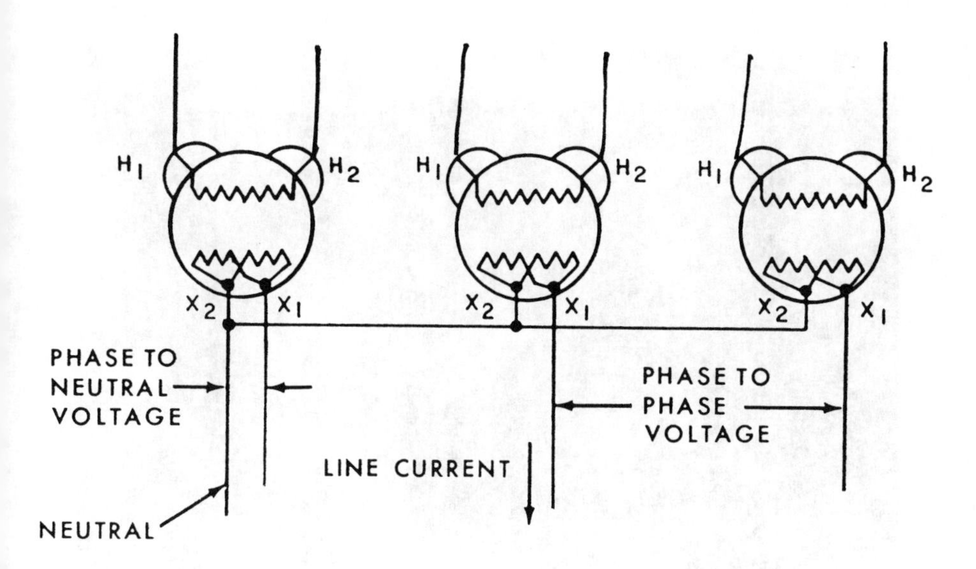

Figure 5-21d. Y-connection.

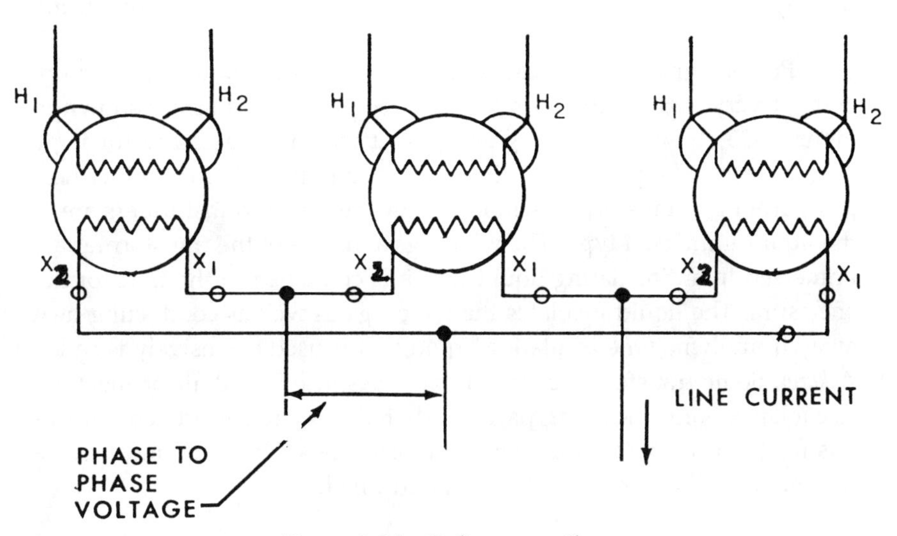

Figure 5-21e. Delta connection.

Figure 5-21f. Three-phase pad-mounted transformer with primary connections (H_1, H_2, H_3) on the left and secondary connections (X_1, X_2, X_3) on the right.

Power transformers located along a power distribution system operate at very high temperatures. The three-phase transformer shown in Figure 5-23 shows the cooling equipment which is necessary for large transformers. The purpose of the cooling equipment is to conduct heat away from the transformer windings. Several power transformers are of the liquid-immersed type. The windings and core of the transformer are immersed in an insulating liquid which is contained in the transformer enclosure. The liquid insulates the windings as well as conducting heat away from them. One insulating liquid that is used extensively is called *Askarel.* Some transformers, called dry types, use forced air or inert gas as coolants. Some locations, particularly indoors, are considered hazardous for the use of liquid-immersed transformers. However, most transformers rated at over 500 kVA are liquid filled.

TRANSFORMER MALFUNCTIONS

Transformer malfunctions result when a circuit problem causes the insulation to break down. Insulation breakdown permits electrical arcs

Figure 5-22. Some transformer connection methods for various voltage combinations.

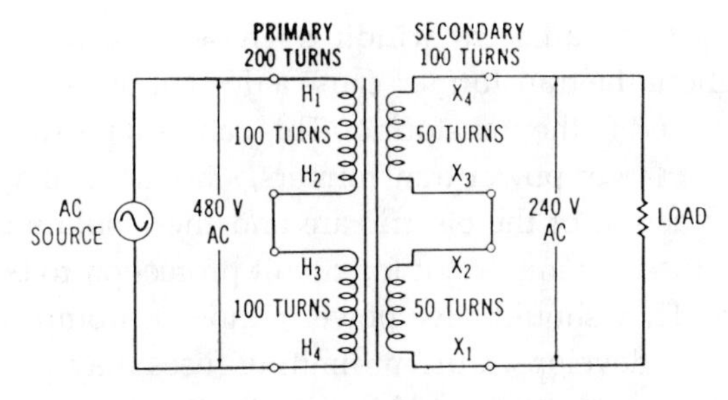

(A) Series-connected primary and secondary.

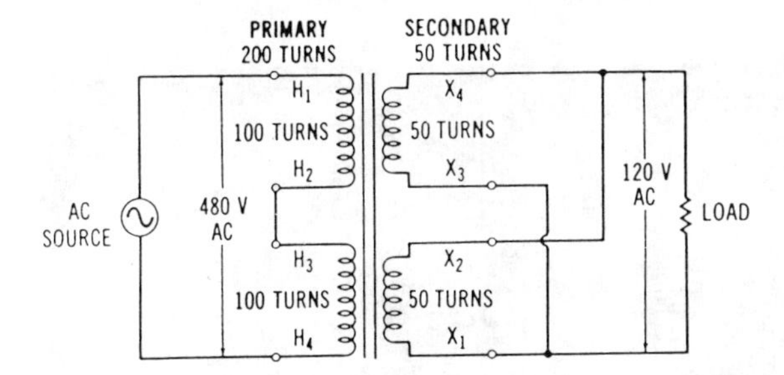

(B) Series-connected primary; parallel-connected secondary.

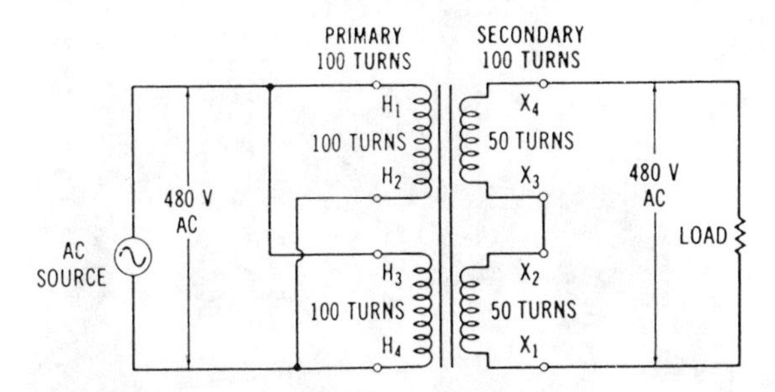

(C) Parallel-connected primary; series-connected secondary.

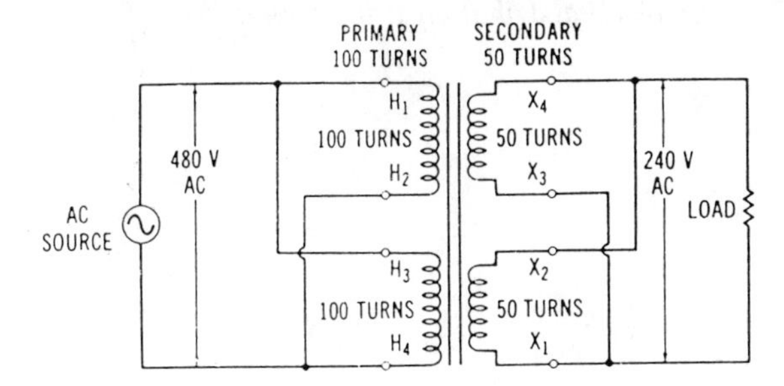

(D) Parallel-connected primary and secondary.

Some transformer connection methods for various voltage combinations.

to flow from one winding to an adjacent winding. These arcs, which may be developed throughout the transformer, cause a decomposition of the paper or oil insulation used in the transformer. This can be a particularly hazardous problem for larger power transformers, since a gas may be produced due to the reaction of the electric are and the insulating material. For this reason, it is very important for circuit protection to be provided for transformers. They should have power removed promptly whenever some type of fault develops. Current-limiting fuses may also be used to respond rapidly to any circuit malfunction. Protective equipment will be discussed in Chapter 7.

Figure 5-23. Three-phase distribution transformers.

Chapter 6

Conductors and Insulators in Electrical Distribution Systems

The portions of the electrical distribution system which carry current are known as conductors. Conductors may be in the form of solid or stranded wires, cable assemblies, or large metallic bus-bar systems. A conductor may have insulation or, in some cases, be bare metal. Chapter 6 will explore some of the characteristics of electrical conductors and insulators.

CONDUCTOR CHARACTERISTICS

Round conductors are measured by using an American Wire Gage (AWG) (see Figure 6-1). The sizes range from No. 36 (smallest) to No. 0000 (largest), with 40 sizes within this range. The cross-sectional area of a conductor doubles with each increase of three sizes and the diameter doubles with every six sizes. The area of conductors is measured in circular mils (cmil).

Almost all conductors are made of either copper or aluminum. Both of these metals possess the necessary flexibility, current-carrying ability, and economical cost to act as efficient conductors. Copper is a better conductor; however, aluminum is 30% lighter in weight. Therefore, aluminum conductors are used when weight is a factor for consideration in conductor selection. One specialized overhead power-line conductor is the aluminum-conductor steel-reinforced (ACSR) type used for long-distance electrical power transmission. This type of conductor has stranded aluminum wires.

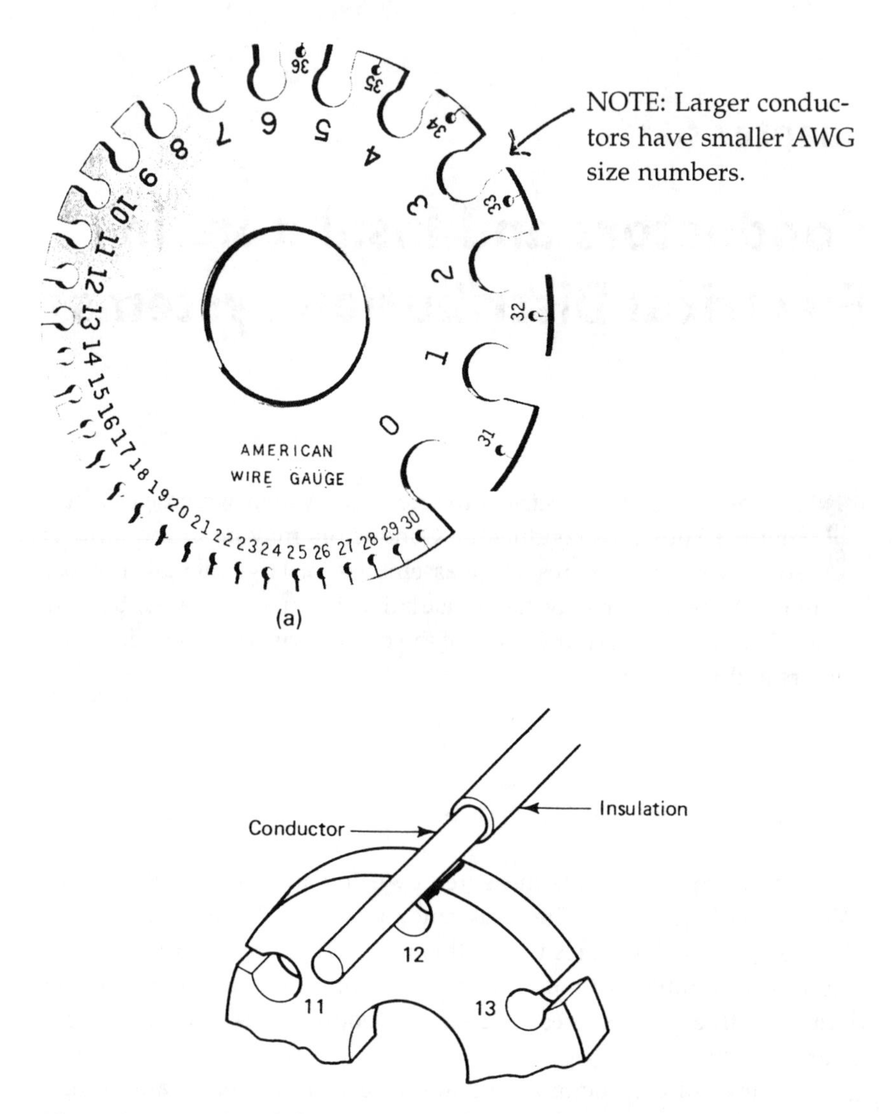

Figure 6-1. American Wire Gauge. (a) The American Wire Gage (AWG) which is used to measure conductor size; (b) Using the AWG.

SIZE			WEIGHT (FEET PER POUND)
AWG or MCM #	Relative Size	DIAMETER (INCHES)	
40	TOO SMALL TO SHOW ACCURATELY	.0031	33,410.
36		.0050	13,210.
30		.0102	3,287.
24		.0201	817.6
18		.0403	203.4
16		.0508	127.9
14		.0640	80.44
12		.0808	50.59
10		.1018	31.82
8		.1284	20.01
6		.184	12.58
4		.232	7.91
2		292	4.97
1		.332	3.94
1/0		.373	3.13
2/0		.419	2.48
3/0		.470	1.97
4/0		.528	1.56
350 MCM		.681	0.925

Figure 6-1c. Relative sizes and weight of copper conductor.

CONDUCTOR TYPES

Copper is still the most widely used conductor material, both for solid and for stranded electrical wire. The availability of a variety of thermosetting and thermoplastic insulating materials offers great flexibility in meeting the requirements for most conductor applications. The operating temperature ranges for various types of insulation are given in Table 6-1.

Table 6-1. Operating temperature ranges of various types of insulation.

Type of Insulation	Temperature Range (°C)
Neoprene	–30° to 90°
Teflon	–70° to 200°
Polyethylene	–60° to 80°
Rubber	–40° to 75°
Vinyl–20° to 80°	
Polypropylene	–20° to 105°

Copper has a combination of various properties, such as malleability, good strength, and high electrical and thermal conductivity. It also has the capability of being alloyed or coated with other metals. Copper may be plated with silver to produce a better solderability and, also, a conductor that has better high-frequency characteristics. This is due to the high conductivity of the silver and the "skin effect" present at higher frequencies.

Where little vibration and no flexing are required of a wire or cable, single-strand conductors may be used. The advantage of a single-strand conductor is its lower cost compared to that of equivalent types of stranded wire. Wire and cable with solid conductors may be used as interconnection wires for electrical instruments and similar equipment. Stranded conductors (see Figure 6-2) are used to provide more flexibility. They, also, have a longer usage life than do solid conductors. If a solid conductor were cut during its installation by wire strippers, it would probably break after being bent a few times. However, stranded wire would not break in this situation. Wires having from 26 to 41 strands may be used where much flexibility is needed while wires with from (65 to 105 strands may be used for special purposes.

Flat or round braided conductors are occasionally used for certain applications where they are usually better suited than are solid or stranded cables. These conductors are seldom insulated since this would hinder their flexibility.

Various types of special purpose conductors which are used in distribution wiring for industrial and commercial facilities are shown in Figure 6-3.

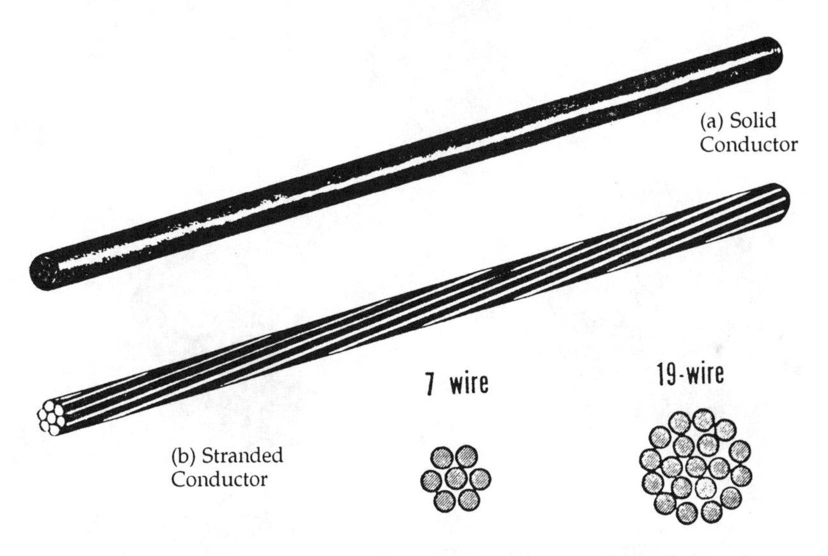

Figure 6-2. Stranded conductors; (a) solid conductor; (b) stranded conductor; (c) cross-section of stranded conductors.

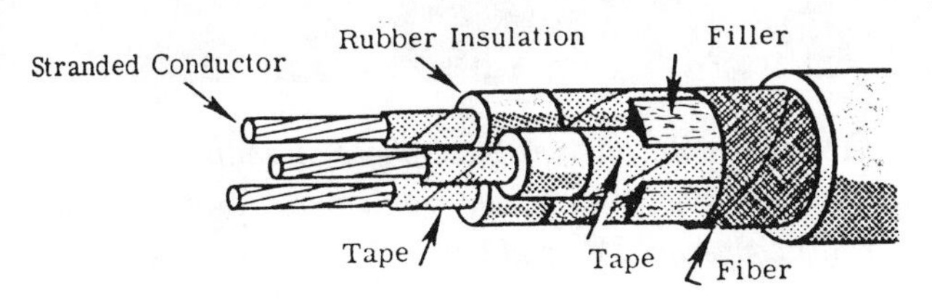

Figure 6-2d. Single stranded conductor cable.

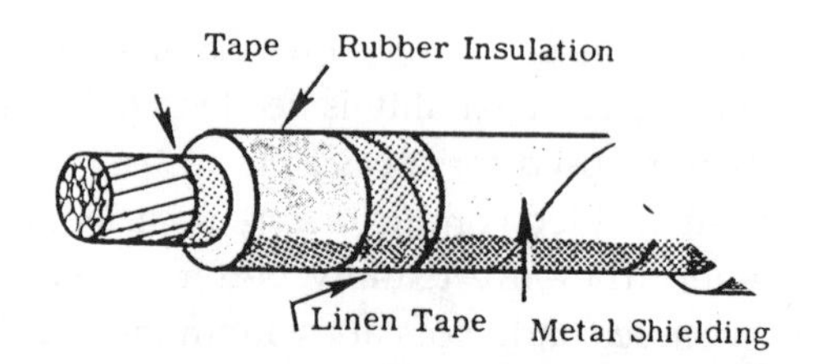

Figure 6-2e. Three conductor cable.

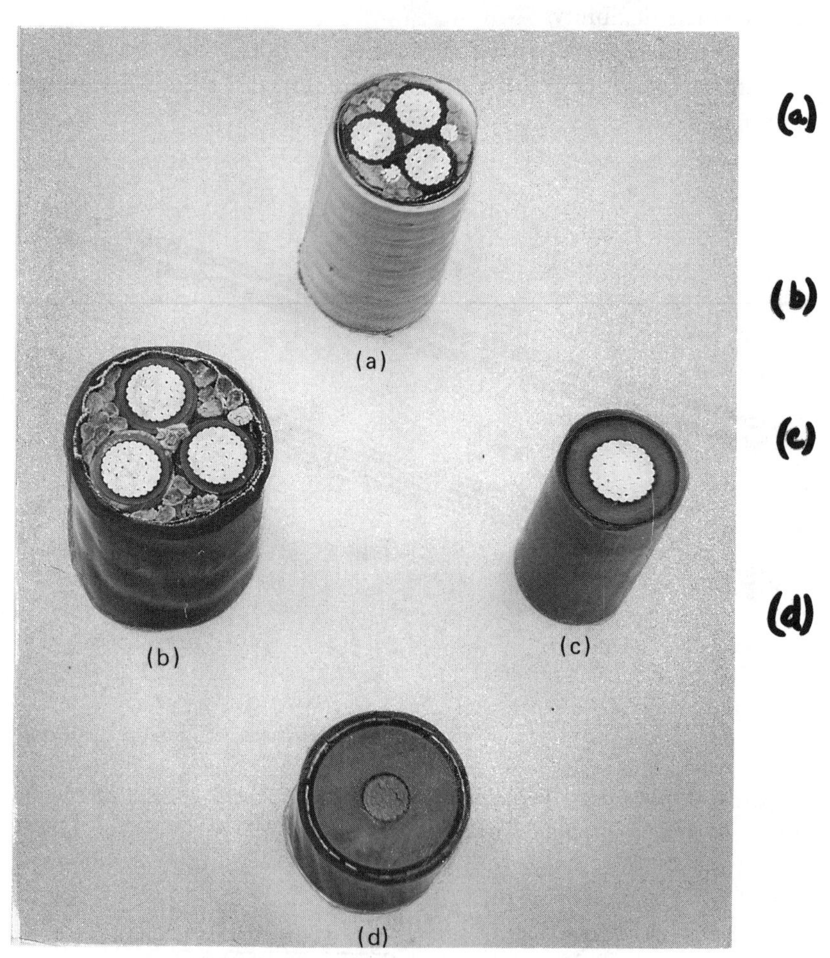

Figure 6-3. Electrical conductors (a) three-phase aluminum conductor group, with grounds; (b) three-phase aluminum conductors; (c) aluminum high-current conductor; (d) copper 69-kV conductor

Figure 6-3. Electrical conductors (Cont'd): ; (e, left) industrial control cable; (f, below) control wire assembly (Courtesy of Special Purpose Conductors); (e & f Courtesy Clifford of Vermont, Inc.).

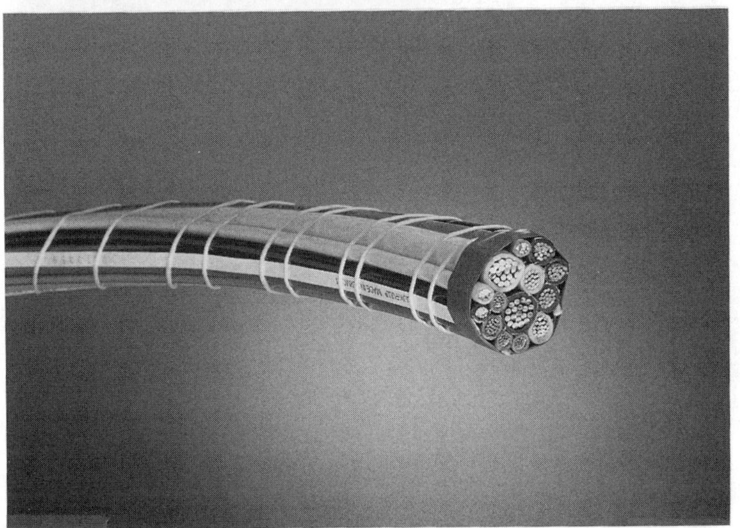

CONDUCTOR AREA

The unit of measurement for conductors is the *circular mil* (cmil) since most conductors are round. One mil is equal to 0.001 inch (0.0254 mm); thus, one circular mil is equal to a circle whose diameter is 0.001 inch. The cross-sectional area of a conductor in circular mils) is equal to its diameter D (in mils) squared, or cmil = D^2 (see Figure 6-4). For example, if a conductor is 1/4-inch (6.35 mm) in diameter, its circular mil area can be found as follows. The decimal equivalent of 1/4 inch is 0.250 inch which equals 250 mils. Inserting this value into the formula for the cross-sectional area of a conductor gives you:

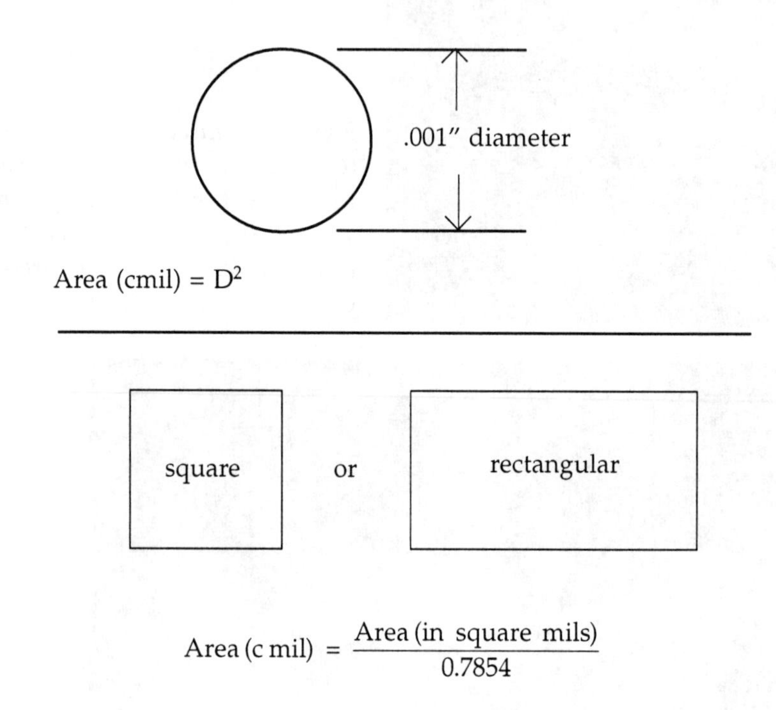

Figure 6-4. Conductor area.

$$\begin{aligned} \text{Area} &= D^2 \text{ (in mils)} \\ &= (250)^2 \\ &= 62{,}500 \text{ cmils} \end{aligned}$$

If the conductor is not round in shape, its area may still be found by applying the following formula:

$$\text{Area (in cmil)} = \frac{\text{Area (in square mils)}}{0.7854}$$

This formula allows us to convert the dimensions of a conductor to square mils and, then, to an equivalent value in circular mils. For example, if a conductor is 1/2 inch × 3/4 inch (12.7 mm × 19.05 mm), the circular mil area may be found using the following method. Again, convert the fractional inches into decimal values and, then, into equivalent mil values. Thus, 1/2 inch equals 0.5 inch which equals 500 mils, and 3/4 inch is 0.75 inch which equals 750 mils. Using the formula for square mils and substituting, you have:

$$\text{Area} = \frac{\text{Area (in square mils)}}{0.7854}$$

$$= \frac{500 \text{ mils} \times 750 \text{ mils}}{0.7854}$$

$$= \frac{375{,}000 \text{ mils}^2}{0.7854}$$

$$= 477{,}463.7 \text{ cmils}$$

We can also use the circular mil area of a conductor, which may be found in any standard conductor table, to find the diameter of a conductor. If a table shows that the circular mil area of a conductor is equal to 16,510 cmils (a No. 8 AWG conductor), its diameter is found by the following method. Since

$$D^2 = \text{cmil},$$

then,

$$D = \sqrt{\text{cmil}}$$

$$= \sqrt{16{,}510}$$

$$= 128.5 \text{ mils}$$

$$= 0.1285 \text{ inch.}$$

RESISTANCE OF CONDUCTORS

The *resistance* of a conductor expresses the amount of opposition it will offer to the flow of electrical current. The unit of measurement for resistance is the ohm (Ω). The *resistivity* (ρ) of a conductor is the resistance for a specified cross-sectional area and length. This measurement is given in circular mil-feet (cmil-ft).: The resistivity of a conductor changes with the temperature, so resistivity is usually specified at a temperature of 20° Celsius. The resistivity for some common types of conductors is listed in Table 6-2.

Table 6-2. Resistivity of common conductors.

Conductor	Resistivity in ohms per cmil-ft
Silver	9.8
Copper	10.4
Aluminum.	17.0
Tungsten	33.0
Nickel	50.0
Iron	60.0

We can use Table 6-2 to calculate the resistance of any size conductor. We know that resistance increases as the length increases and decreases as the cross-sectional area increases. The following method can be used to find the resistance of 500 feet (152.4 meters) of aluminum conductor that is 1/4 inch (6.35 mm) in diameter. According to Table 6-2 aluminum has a resistivity of 17 ohms. The diameter (D) equals 1/4 inch which equals 0.250 inch which is the equivalent of 250 mils. Using the formula and substituting, we have:

$$\text{Resistance} = \frac{\text{Resistivity} \times \text{Length (in feet)}}{\text{Diameter}^2 \text{ (in mils)}}$$

$$= \frac{17 \times 500}{(250)^2}$$

$$= \frac{8{,}500}{62{,}500}$$

$$= 0.136 \text{ ohm}$$

CONDUCTOR SIZES

Table 6-3 lists the sizes of copper and aluminum electrical conductors. The American Wire Gage (AWG) is the standard used to measure the diameter of conductors. The sizes range from No. 40 AWG which is the smallest to No. 0000 AWG. Sizes larger than No. 0000 AWG are expressed in thousand circular mil (MCM) units.

Table 6-3. Sizes of copper and aluminum conductors.

Size (AWG or MCM)	Area (cmil)	Number of Wires	Diameter of Each Wire (in.)	DC Resistance; (Ω/1000 ft) 25°C	
				Copper	Aluminum
18	1,620	1	0.0403	6.51	10.7
16	2,580	1	0.0508	4.10	6.72
14	4,110	1	0.0641	2.57	4.22
12	6,530	1	0.0808	1.62	2.66
10	10,380	1	0.1019	1.018	1.67
8	16,510	1	0.1285	0.6404	1.05
6	26,240	7	0.0612	0.410	0.674
4	41,740	7	0.0772	0.259	0.424
3	52,620	7	0.0867	0.205	0.336
2	66,360	7	0.0974	0.162	0.266
1	83,690	19	0.0664	0.129	0.211
0	105,600	19	0.0745	0.102	0.168
00	133,100	19	0.0837	0.0811	0.133
000	167,800	19	0.0940	0.0642	0.105
0000	211,600	19	0.1055	0.0509	0.0836
250	250,000	37	0.0822	0.0431	0.0708
300	300,000	37	0.0900	0.0360	0.0590
350	350,000	37	0.0973	0.0308	0.0505
400	400,000	37	0.1040	0.0270	0.0442
500	500,000	37	0.1162	0.0216	0.0354
600	600,000	61	0.0992	0.0180	0.0295
700	700,000	61	0.1071	0.0154	0.0253
750	750,000	61	0.1109	0.0144	0.0236
800	800,000	61	0.1145	0.0135	0.0221
900	900,000	61	0.1215	0.0120	0.0197
1000	1,000,000	61	0.1280	0.0108	0.0177

Note, in Table 6-3 that as the AWG size number becomes smaller, the conductor is larger. Sizes up to No. 8 AWG are solid conductors, while larger wires have from 7 to 61 strands. Table 6-3 also lists the DC resistance (in ohms per 1000 feet) of the copper and aluminum conductors. These values are used to determine conductor voltage drop in power distribution systems.

AMPACITY OF CONDUCTORS

A measure of the ability of a conductor to carry electrical current is called *ampacity.* All metal materials will conduct electrical current to some extent; however, copper and aluminum are the two most desirable types used. Copper is the most used metal since it is the better conductor of the two and is physically stronger. However, aluminum is usually used where weight is a factor, such as for long-distance overhead power lines. The weight of copper is almost three times that of a similar volume of aluminum; however, the resistance of aluminum is over 150% that of copper. The ampacity of an aluminum conductor is, therefore, less than a similar size copper conductor. In a wiring design, aluminum conductors, that are one size larger than the copper conductors necessary to carry a specific amount of current, are ordinarily used to allow for this difference (see Table 6-3).

The ampacity of conductors depends upon several factors such as the type of material, cross-sectional area, and the type of area in which they are installed. Conductors in the open or in "free air" dissipate heat much more rapidly than they do if they are enclosed in a metal raceway or plastic cable. When several conductors are contained within the same enclosure, heat dissipation is a greater problem.

THE NATIONAL ELECTRIC CODE (NEC)©

The *National Electrical Code* (NEC) is a very important document to understand. All electrical distribution equipment and wiring must conform to the NEC standards. The NEC is not difficult to use. The user should become familiar with the comprehensive index contained in the NEC and the organization of the various sections. For instance, if you wished to review the standards related to "system grounding," you should look in the index and locate this term. The index will refer you to the appropriate sections (articles) the NEC which discuss "system grounding." A table of contents of the current NEC is shown below. This listing provides an overview of the organization of the NEC.

National Electric Code (NEC)

Table of Contents

344 Underplaster Extensions
345 Intermediate Metal Conduit
346 Rigid Metal Conduit
347 Rigid Nonmetallic Conduit
348 Electrical Metallic Tubing
349 Flexible Metallic Tubing
350 Flexible Metal Conduit
351 Liquidtight Flexible Metal Conduit
352 Surface Raceways
353 Multioutlet Assembly
354 Underfloor Raceways
356 Cellular Metal Floor Raceways
358 Cellular Concrete Floor Raceways
362 Wireways
363 Flat Cable Assemblies
364 Busways
365 Cablebus
366 Electrical Floor Assemblies,
370 Outlet, Switch and Junction Boxes, and Fittings
373 Cabinets and Cotout Boxes
374 Auxiliary Gutters
380 Switches
384 Switchboards and Panelboards

Chapter 4 - Equipment for General Use
400 Flexible Cords and Cables
402 Fixture Wires
410 Lighting Fixtures, Lampholders, Lamps, Receptacles and Rosettes
422 Appliances
424 Fixed Electric Space Heating Equipment
426 Fixed Outdoor Electric De-Icing and Snow-Melting Equipment
427 Fixed Electric Heating Equipment for Pipelines and Vessels
430 Motors, Motor Circuits and Controllers
440 Air-Conditioning and Refrigerating Equipment
445 Generators
450 Transformers and Transformer Vaults
460 Capacitors
470 Resistors and Reactors
480 Storage Batteries

Chapter 5 - Special Occupancies

500 Hazardous (Classified) Locations
501 Hazardous Locations - Class I Installations
502 Hazardous Locations - Class II Installations
503 Hazardous Locations - Class III Installations
510 Hazardous (Classified) Locations - Specific
511 Commercial Garages, Repair and Storage
513 Aircraft Hangars
514 Gasoline Dispensing and Service Stations
515 Build-Storage Plants
516 Finishing. Processes
517 Health Care Facilities
518 Places of Assembly
520 Theaters and Similar Locations
530 Motion Picture and Television Studios and Similar Locations
540 Motion-Picture Projectors
545 Manufactured Building
547 Agricultural Buildings
550 Mobile Homes and Mobile Home Parks
551 Recreational Vehicles and Recreational Vehicle Parks
555 Marinas and Boatyards

Chapter 6 - Special Equipment

600 Electric Signs and Outline Lighting
610 Cranes and Hoists
620 Elevators, Dumbwaiters, Escalators, and Moving Walks
630 Electric Welders
640 Sound-Recording and Similar Equipment
645 Data Processing Systems
650 Organs
660 X-ray Equipment
665 Induction and Dielectric Heating Equipment
668 Electrolytic Cells
670 Metal Working Machine Tools
675 Electrically Driven or Controlled Irrigation Machines
680 Swimming Pools, Fountains and Similar Installations

Chapter 7 - Special Conditions

700 Emergency Systems

710 Over 600 Volts, Nominal - General
720 Circuits and Equipment Operating at Less than 50 Volts
725 Class I, Class II, and Class III Remote-Control, Signaling, and Power Limited Circuits
750 Stand-By Power Generation Systems
760 Fire Protective Signaling Systems

Chapter 8 - Communications Systems
800 Communication Circuits
810 Radio and Television Equipment
820 Community Antenna Television and Radio Distribution Systems

Chapter 9 - Tables and Examples
Tables
Examples

Index

AMPACITY TABLES

Tables 6-4 through 6-6 are used for conductor ampacity calculations for electrical wiring design. These tables are simplified versions of those given in the National Electrical Code (NEC). Table 6-4 is used to determine conductor ampacity when a single conductor is mounted in free air. Table 6-5 is used to find the ampacity of conductors when not more than three are mounted in a raceway or cable. These two tables are based on ambient temperatures of 30° Celsius (86° Fahrenheit). Table 6-6 lists the correction factors which are used for temperatures over 30° C.

As an example, we will find the ampacity of three No. 10 copper conductors with RHW insulation which are mounted in a raceway. They will be located in a foundry area where temperatures reach 50°C. The ampacity for No. 10 RHW copper wire is 30 amperes (from Table 6-5). The correction factor for an RHW-insulated conductor at 50°C ambient temperature is 0.75 (Table 6-6). Therefore,

$$30 \text{ amperes} \times 0.75 = 22.5 \text{ amperes.}$$

Figure 6-4. Allowable ampacities of single conductors in free air.

	Copper		Aluminum	
Wire Size	With R, T, TW Insulation	With RH, RHW, TH, THW Insulation	With With R, T, TW Insulation	RH, RHW, TH, THW Insulation
14	20	20		
12	25	25	20	20
10	40	40	30	30
8	55	65	45	55
6	80	95	60	75
4	105	125	80	100
3	120	145	95	115
2	140	170	110	135
1	165	195	130	155
0	195	230	150	180
00	225	265	175	210
000	260	310	200	240
0000	300	360	230	280
250	340	405	265	315
300	375	445	290-	350
350	420	505	330	395
400	455	545	355	425
500	515	620	405	485
600	575	690	455	545
700	630	755	500	595
750	655	785	515	620
800	680	815	535	645
900	730	870	580	700
1000	780	935	625	750

Table 6-5. Ampacities of conductors in a raceway or cable (3 or less).

	Copper		Aluminum	
Wire Size	With R, T, TW Insulation	With RH, RHW, TH, THW Insulation	With With R, T, TW Insulation	RH, RHW, TH, THW Insulation
14	15	15		
12	20	20	15	15
10	30	30	25	25
8	40	45	30	40
6	55	65	40	50
4	70	85	55	65
3	80	100	65	75
2	95	115	75	90
1	110	130	85	100
0	125	150	100	120
00	145	175	115	135
000	165	200	130	155
0000	195	230	155	180
250	215	255	170	205
300	240	285	190	230
350	260	310	210	250
400	280	335	225	270
500	320	380	260	310
600	355	420	285	340
700	385	460	310	375
750	400	475	320	385
800	410	490	330	395
900	435	S20	355	425
1000	455	545	375	445

Table 6-6. Correction factors for temperatures above 30°C.

Ambient Temperature		Conductor Correction Factor	
C°	F°	R, T, TW	RH, RHW, TH, THW
40	104	0.82	0.88
45	113	0.71	0.82
50	122	0.58	0.75
55	131	0.41	0.67
60	140	—	0.58
70	158	—	0.35

USE OF INSULATION IN POWER DISTRIBUTION SYSTEMS

Synthetic insulation for wire and cable is classified into two broad categories—thermosetting and thermoplastic. The mixtures of materials within each of these categories are so varied as to make the available number of insulations almost unlimited. Most insulation is composed of compounds made of synthetic rubber polymers (thermosetting) and from synthetic materials (thermoplastics). These synthetic materials are combined to provide specific physical and electrical properties.

Thermosetting materials are characterized by their ability to be stretched compressed or deformed within reasonable limits under mechanical strain and then return to their original shape when the stress is removed.

Thermoplastic insulation materials are best known for their excellent electrical characteristics and relatively low cost. These materials are popular since much thinner insulation thicknesses may be used to obtain good electrical properties. Particularly at the higher voltages.

There are many types of insulation used today for electrical conductors. Some new materials have been developed that will last for exceptionally long periods of time and will withstand very high operating temperatures. The operating conditions where the conductors are used mainly determine the type of insulation required. For instance system voltage heat and moisture affect the type of insulation required. Insulation must be used which will withstand both the heat of the surrounding atmosphere and that developed by the current flowing through the conductor. Exceptionally large currents will cause excessive heat to be developed in a conductor. Such heat could cause insulation to melt or burn. This is why overcurrent protection is required as a safety factor to prevent fires. The ampacity or current-carrying capacity of a conductor depends upon the type of insulation used. The NEC has developed a system of abbreviations for identifying various types of insulation. Some of the abbreviations are shown in Table 6-7.

Table 6-7. Common abbreviations for types of electrical insulation.

Abbreviation	Type of Insulation
R	Rubber—140°F
RH	Heat-Resistant Rubber—167°F
RHH	Heat-Resistant Rubber—194°F
RHW	Moisture and Heat-Resistant Rubber—167°F
T	Thermoplastic—140°F
THW	Moisture and Heat-Resistant Thermoplastic—167°F
THWN	Moisture and Heat-Resistant Thermoplastic With Nylon 194°F

Chapter 7

Power Distribution Equipment

In order to distribute electrical power, it is necessary to use many types of specialized equipment. The electrical power system consists of such specialized equipment as power transformers, high-voltage fuses and circuit breakers, lightning arresters, power-factor-correcting capacitors, and power-metering systems. Types of specialized power distribution equipment will be discussed in this chapter.

ELECTRICAL SUBSTATIONS

Substations are essential parts of the electrical distribution system. The link between high-voltage transmission lines and low-voltage power distribution systems is the substation. The function of a distribution substation, such as the one shown in Figure 7-1, is to receive electrical power from a high-voltage transmission system and convert it to voltage levels suitable for industrial, commercial, or residential use. The major functional component of a substation is the transformer, whose basic characteristics were discussed in Chapter 5. However, there are many other types of specialized equipment required for the operation of a substation.

The design of typical substations which are used for electrical distribution are shown in Figure 7-2. Some of the symbols used for substation equipment and other parts of the electrical distribution system are shown in Figure 7-3. The symbols are used to show a substation design in Figure 7-4.

HIGH-VOLTAGE FUSES

Since electrical power lines are frequently short circuited, various protective equipment is used to prevent damage to both the power lines and the equipment. This protective equipment must be designed to handle high voltages and currents.

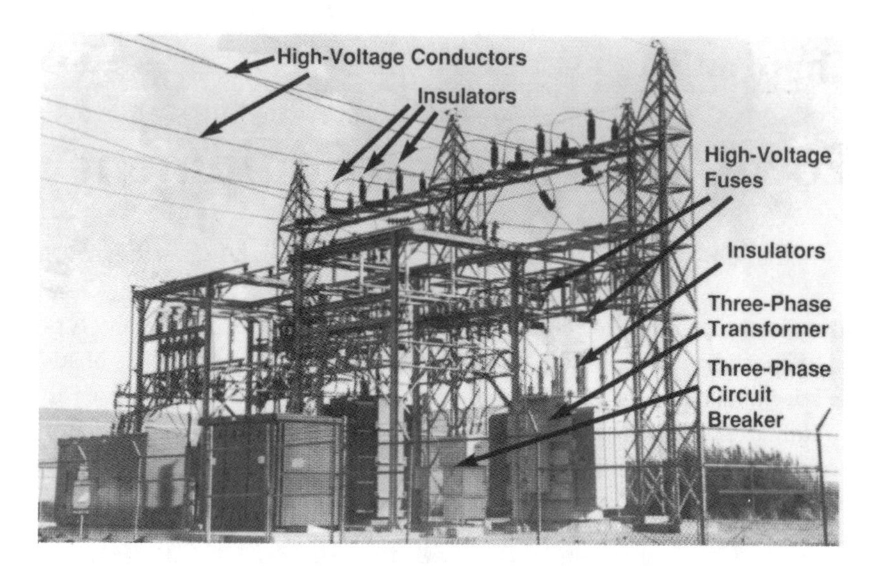

Figure 7-1a. An electrical substation (Courtesy Kentucky Utilities Co.).

Figures 7-1b .

Figures 7-1c.

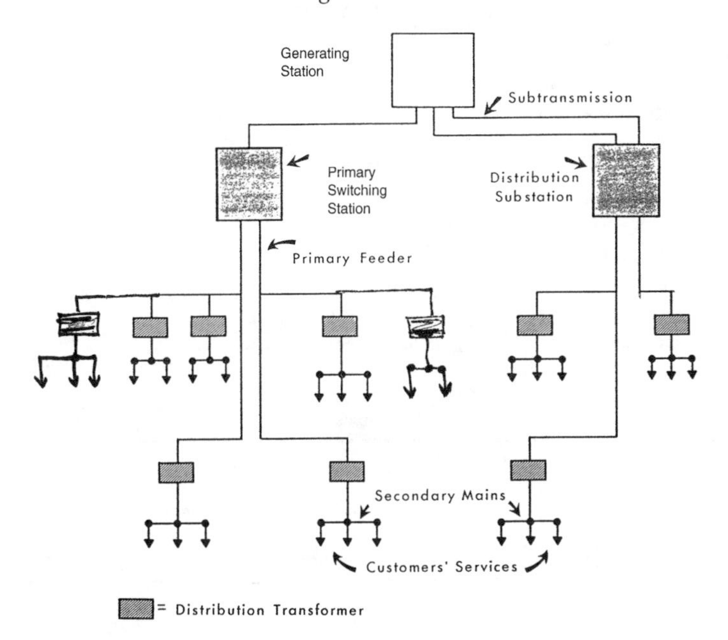

Figure 7-2. Typical substation design using a block diagram.

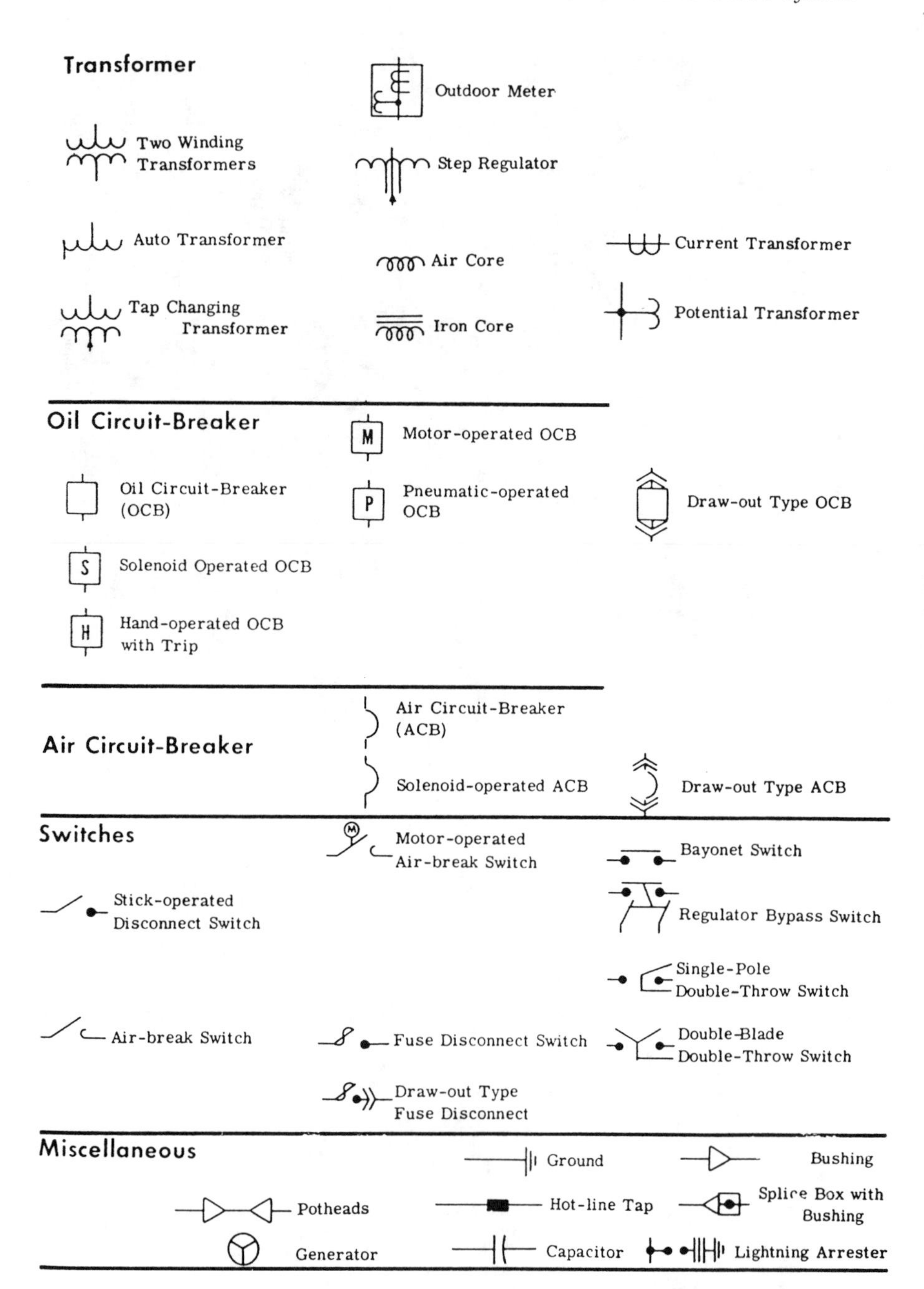

Figure 7-3. Symbols used for substation and associated equipment.

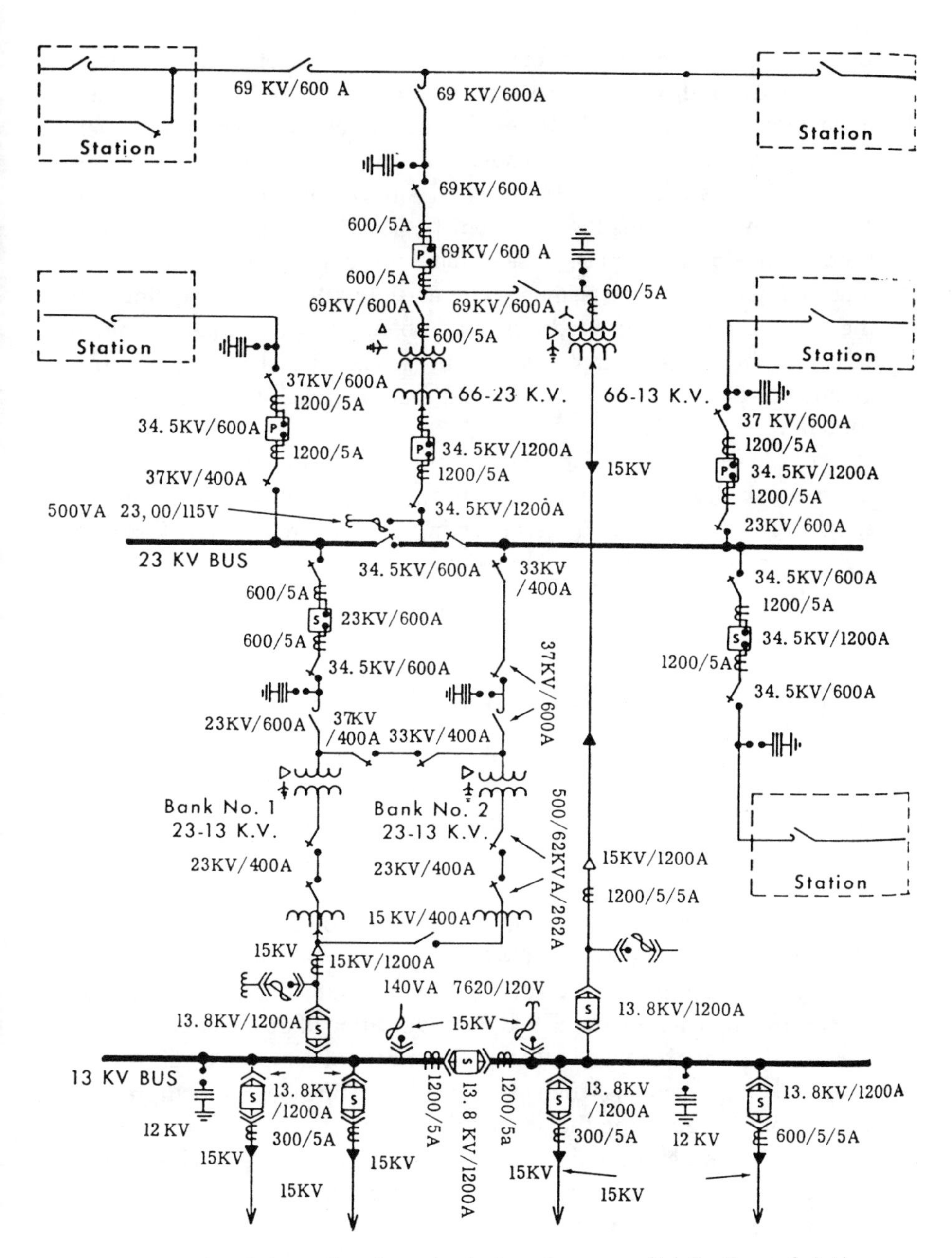

Figure 7-4. Symbols used to show the design of a power distribution substation.

Either fuses or circuit breakers may be used to protect high-voltage power lines. High-voltage fuses (those used for over 600 volts) are made in several ways. An *expulsion-type fuse* has an element which will melt and vaporize when it is overloaded, causing the power line it is connected in series with to open. *Liquid fuses* have a liquid-filled metal enclosure, which contains the fuse element. The liquid acts as an are suppressing medium. When the fuse element melts due to an excessive current in a power line, the element is immersed in the liquid to extinguish the are. This type of fuse reduces the problem of high-voltage arcing. A *solid-material fuse* is similar to a liquid fuse except that the are is extinguished in a chamber filled with solid material. High-voltage fuses are shown in Figure 7-5.

Ordinarily, high-voltage fuses at substations are mounted adjacent to air-break disconnect switches. These switches provide a means of switching power lines and disconnecting them for repair. The fuse and switch enclosure is usually mounted near the overhead power lines at a substation.

HIGH-VOLTAGE CIRCUIT BREAKERS

Circuit breakers which control high voltages are also located at electrical substations. Many outdoor substations use oil-filled circuit breakers, such as shown in Figure 7-6. In-this type of circuit breaker, the contacts are immersed in an insulating oil contained in a metal enclosure. Another type of high-voltage circuit breaker is the magnetic air breaker in which the contacts separate, in the air, when the power line is overloaded. Magnetic blowout coils are used to develop a magnetic field which causes the are, that is produced when the contacts break, to be concentrated into are chutes where it is extinguished. A modification of this type is the compressed-air circuit breaker. In this type, a stream of compressed air is concentrated on the contacts when the power line is opened. The compressed air aids in extinguishing the are which is developed when the contacts open. It should be pointed out that large arcs are present whenever a high-voltage circuit is interrupted. This problem is not encountered to any great extent in low-voltage protective equipment.

There are two major types of high-voltage circuit breakers—oil-filled and oilless. These circuit breakers, are designed to operate on voltages of 1000 volts to over 500,000 volts. Oil-filled circuit breakers are

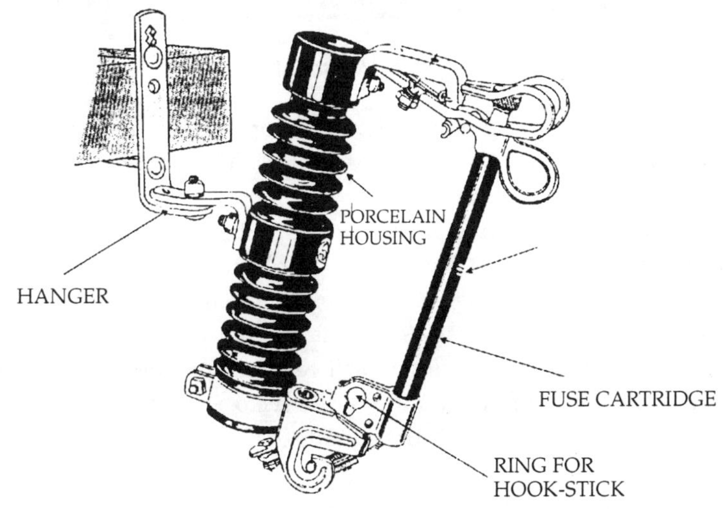

Figure 7-5. High-voltage fuese used for electrical power transmission: (a) Pictorial of fuses at a substation; (b) detail of fuse construction.

Figure 7-6. Oil-filled circuit breakers.

used primarily for outdoor substations, except for very high voltages in the range of 500,000 volts and higher. Oilless circuit breakers are ordinarily used for indoor operation.

Reclosers are protective devices wed in substations to open when a fault current occurs on that part of the distribution. A timing device allows the recloser to automatically reclose a predetermined number of times for short durations to assess the nature of the fault current. If the fault is only temporary, the circuit will remain operational and power will be restored. If the fault continues, the recloser will open and cause power, to remain disconnected from the load.

HIGH-VOLTAGE DISCONNECT SWITCHES

High-voltage disconnect switches are used to disconnect electrical equipment from the power lines which supply the equipment. Ordinarily, disconnect switches are not operated when current is flowing through them. A high-voltage arcing problem would occur if disconnect switches were opened while current was flowing through them. They are opened mainly to isolate equipment from power lines for safety purposes. Most disconnect switches are the "air-break" type which is similar in construction to knife switches. These switches are available for indoor or outdoor use in both manual and motor-operated designs. An air break high-voltage disconnect switch is shown in Figure 7-7.

LIGHTNING ARRESTERS

The purpose of using lightning arresters on power lines is to cause the conduction to ground of excessively high voltages that are caused by lightning strikes or other system-problems. Without lightning arresters, power lines and associated equipment could become inoperable when struck by lightning. Arresters are designed to operate rapidly and repeatedly if necessary. Their response time must be more rapid than the other protective equipment used on power lines.

Lightning arresters must have a rigid connection to ground on one side. The other side of the arrester is connected to a power line. Sometimes, they are connected to transformers or the insides of switchgear. Lightning is a major cause of power-system failures and equipment damage, so lightning arresters have a very important function.

Lightning arresters are also used at outdoor substations. The lightning arrester, such as that shown in Figure 7-8, is used to provide a path to ground for lightning strikes or hits. This path eliminates the flashover between power lines, which causes short circuits. Valve-type lightning

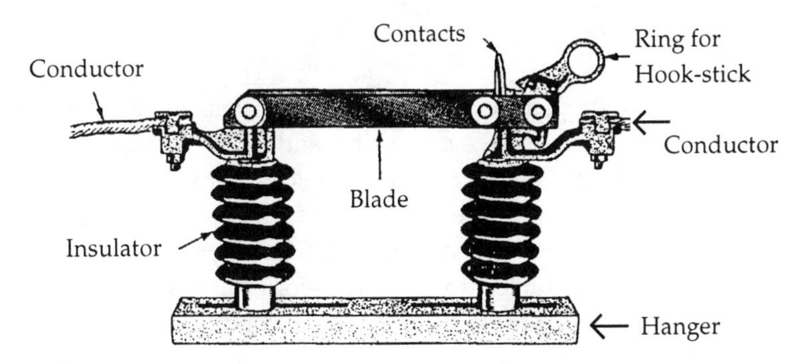

Figure 7-7. High-voltage disconnect switch.

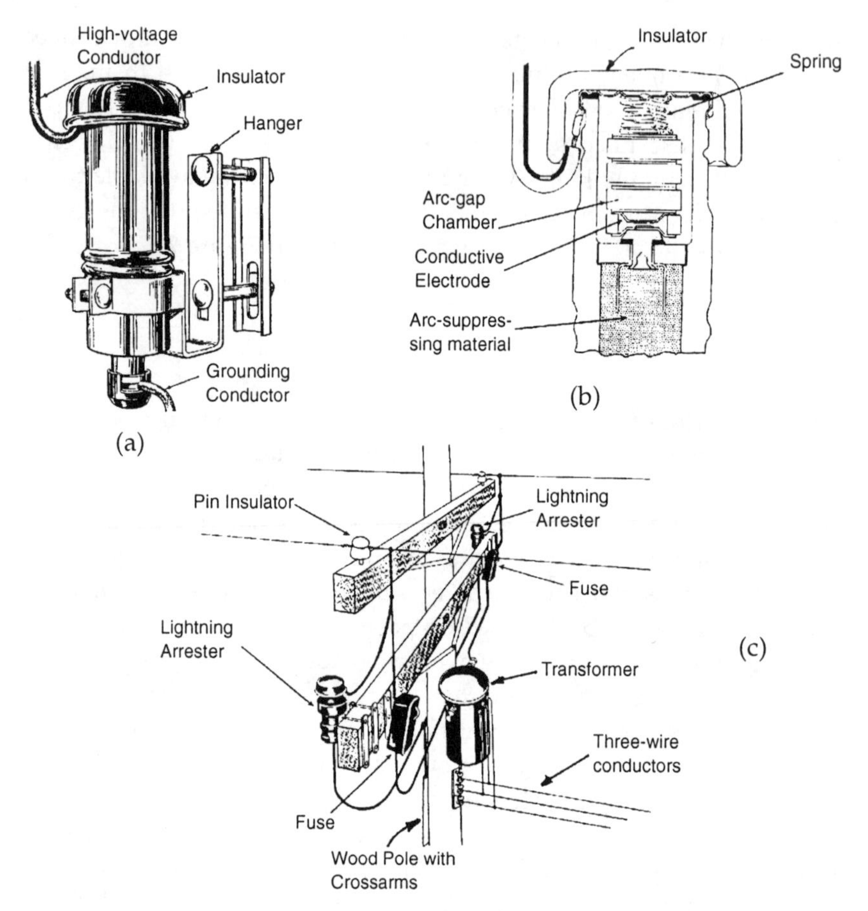

Figure 7-8. Lightning arresters: (a) external detail; (b) internal detail; (c) placement on a wood pole.

arresters are used frequently. They are two-terminal devices in which one terminal is connected to the power line and the other is connected to ground. The path from line to ground is of such high resistance that it is normally open. However, when lightning, which is a very high voltage, strikes a power line, it causes conduction from line to ground. Thus, voltage surges are conducted to ground before flashover between the lines occurs. After the lightning surge has been conducted to ground, the valve assembly then causes the lightning arrester to become nonconductive once more.

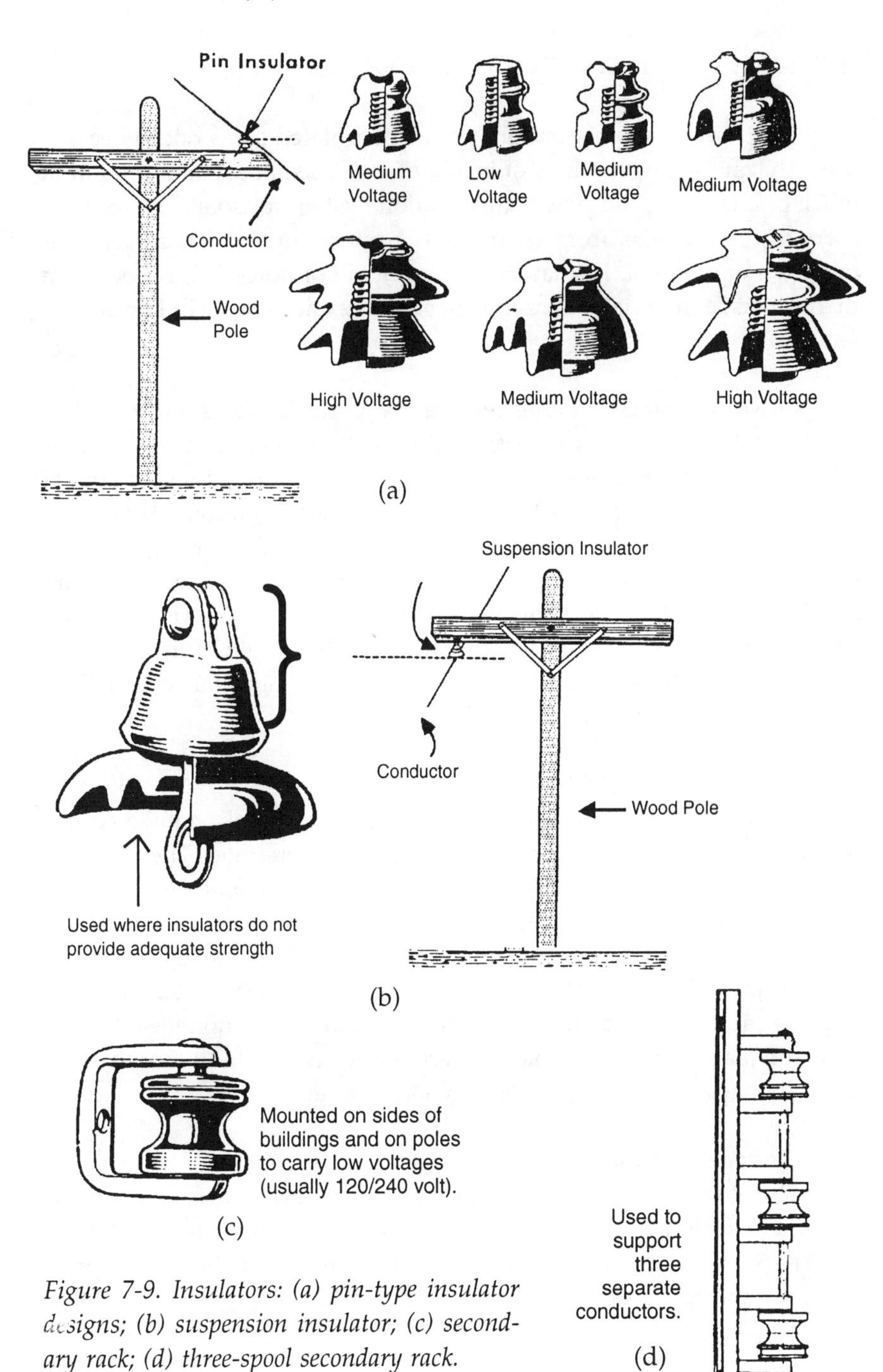

Figure 7-9. Insulators: (a) pin-type insulator designs; (b) suspension insulator; (c) secondary rack; (d) three-spool secondary rack.

HIGH-VOLTAGE INSULATORS AND CONDUCTORS

All power transmission lines must be isolated so as not to become safety hazards. Large strings of insulators are used at substations and at other points along the power distribution system to isolate the current carrying conductors from their steel supports or any other ground-mounted equipment. Insulators may be made of porcelain, rubber, or a thermoplastic material. High-voltage insulators are shown in Figures 7-9 and 7-10.

Power transmission lines require many insulators in order to electrically isolate the power lines from the steel towers and wooden poles which support the lines. Insulators must have enough mechanical strength to support power lines under all weather conditions. They must also have sufficient insulating properties to prevent any arcing between the power lines and their support structures. High-voltage insulators are usually made of porcelain. Insulators are constructed in "strings" which are suspended from steel or wooden towers. The design of these insulators is very important since design affects their capacitance and their ability to withstand weather conditions

Basic Insulation Level (BIL)

Electrical distribution equipment is subject to high voltage surges resulting from lightning and other switching operations. Their insulation must be capable of withstanding these high instantaneous currents. Lighting arresters are installed as close to other distribution equipment as possible to divert voltage surges to ground. Basic insulation level (BIL), especially for circuit breakers and transformers at substations, must be carefully calculated. BIL is the minimum insulation level to provide adequate insulation protection economically and limit the possibility of damage to equipment due to voltage surges.

High-Voltage Conductors

The conductors used for power distribution are, ordinarily, uninsulated aluminum wires or aluminum-conductor steel-reinforced (ACSR) wires for long-distance transmission, and insulated copper wires for shorter distances.

Figure 7-10. Insulators for high-voltage transmission lines: (a) strings of insulators at a power plant hang frxom a steel tower; (b) typical insulator used for isolating high voltage conductors from steel transmission towers.

Substation Location

Distribution substations should be located as close to the load to be served as possible. In addition, future load requirement should be planned accurate. The level of distribution voltage is also a consideration. Generally, the higher the distribution voltage, the farther apart substations may be located. However, they become larger in capacity and in number of customers served as distance apart increases. The decision of substation location must be based upon system reliability and economic factors. Among these factors are the availability of land, estimated operating costs, taxes, local zoning laws, environmental factors and potential public opinion. Also considered is the fact that conductor size increases as the size of the load supplied increases. The primary voltage level affects not only the size of conductors, but also the size of regulation equipment, insulation and other equipment ratings. You should review Figure 7-1 to notice the layout of various types of equipment at a substation.

Voltage Regulators

Voltage regulators are an important part of the power distribution system. They are used to maintain the voltage levels at the proper value as a constant voltage must be maintained in order for the electrical equipment to function properly. For instance, motors do not operate properly when a reduced or an excessive voltage is applied to them. Transformer tap-changers, illustrated in Figure 7-11 may be used as voltage regulators. By either manual or automatic changing the secondary tap, the voltage output can be changed to compensate for changes in the load voltage. As load current increases, line loss ($I \times R$) also increases. Increased line loss causes the secondary voltage (V_s) to decrease. If the secondary tap is initially connected to tap No. 4, the secondary voltage can be boosted by reconnecting to either tap No. 3, No. 2, or No. 1. This can be done automatically with a motor-controlled tap changer. There are various other types of automatic voltage regulators which can be used with electrical power distribution systems, such as shown in Figure 7-12. This transformer has primary taps to deliver 120 volt AC at the secondary with three different primary voltages.

POWER-SYSTEM PROTECTIVE EQUIPMENT

There are many devices that are used to protect electrical distribution systems from damages due to abnormal conditions. For instance, switches, fuses, circuit breakers, lightning arresters, and protective relays are all used for this purpose. Some of these devices automatically disconnect the equipment from power lines before any damage can occur. Other devices sense changes from the normal operation of the system and make the changes necessary to compensate for abnormal circuit conditions. The most common electrical problem which requires protection is short circuits. Other problems include overvoltage undervoltage and changes in frequency. Generally more than one method of protection is used to protect electrical circuits from faulty conditions. The purpose of any type of protective device is to cause a current-carrying conductor to become inoperative when an excessive amount of current flows through it.

Types of Fuses

The simplest type of protective device is a fuse. Fuses are low-cost items and have a fast operating speed. However, in three-phase systems since each hot line must be fused two lines are still operative if only one

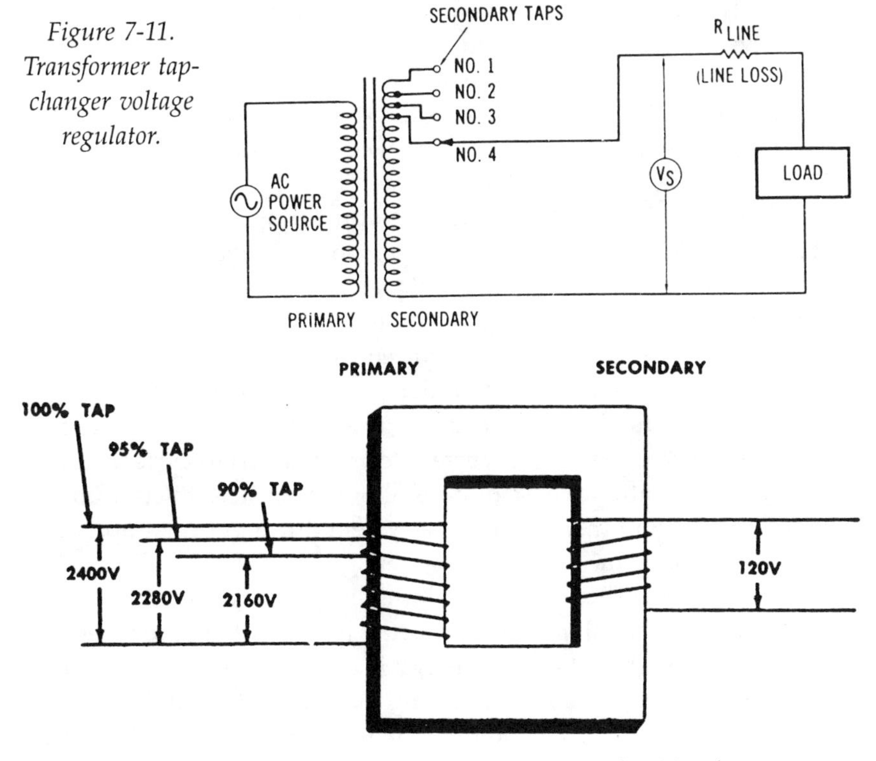

Figure 7-11. Transformer tap-changer voltage regulator.

Figure 7-12. Transformer with primary taps to provide 120-volt output.

fuse burns out. Three-phase motors will continue to run with one phase removed. This condition is undesirable in most instances since motor torque is greatly reduced and overheating may result. Another obvious disadvantage of fuses is that replacements are required. All protective devices including fuses have an operating-characteristic time curve, as shown in Figure 7-13. To prevent any possible damage to equipment, circuit protection should be planned utilizing these curves. They show the response time required when an overload occurs for a protective device to interrupt a circuit.

Plug Fuses—Fuses are used in safety switches and power distribution panels. The plug fuse shown in Figure 7-14, is a common type of fuse. Standard sizes for this fuse are 10, 15, 20, 25, and 30 amperes at voltages of 125 volts or below. These fuses have a zinc or metallic-alloy-fusible element enclosed in a case made of an insulating material. Their most common use is in safety switches and fuse panelboards.

Cartridge Fuses—Cartridge fuses such as shown in Figure 7-15, are commonly used in power distribution systems for voltages up to 600 volts. They have a zinc or alloy-fusible element which is housed in round fiber enclosure. One type has a nonrenewable element while another type such as the one shown in Figure 7-16, has a renewable element. Cartridge fuses may be used to protect high-current circuits since they come in sizes of 60, 100, 200, 400, 600, and 1000 amperes.

Time-delay Fuses—A modification of the plug or cartridge fuse is called a time-delay fuse. This type of fuse, shown in Figure 7-17, is used to delay the circuit-interrupting action. It is useful where momentary high currents exist periodically, such as motor-starting currents. The fuse element melts only when an excessive current is sustained over the time-lag period; thus, sufficient circuit protection is still provided.

Time delay fuses are used to limit current on systems including electric motors which draw higher currents during their start cycle than during normal operation. These devices allow the system to start up at a higher current and then protect the system during normal operation without disrupting the distribution system.

Fuse Metals—The type of metal used in fuses is ordinarily an alloy material or, possibly, aluminum. All metals have resistance, so when current Rows through metal, heat energy is produced. As the current increases, more heat is produced, causing the temperature of the metal to increase. When the melting point of the fuse metal is reached, the fuse will open, causing the circuit to which it is connected to open. Metals

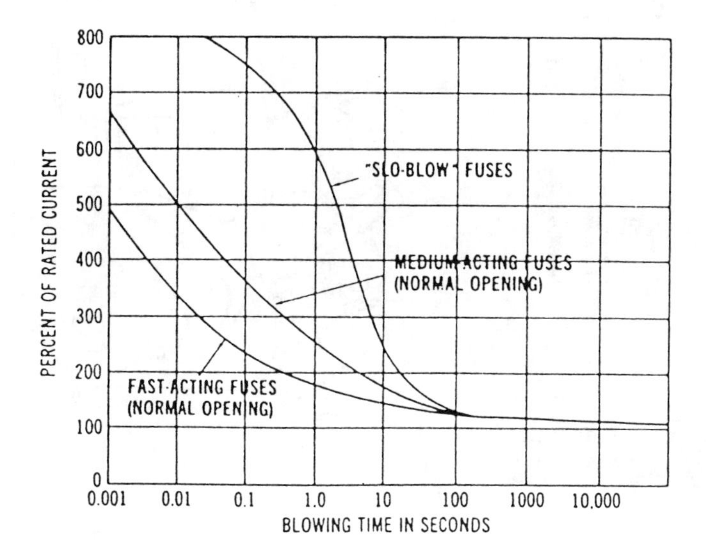

Figure 7-13. Typical operating-characteristic time curves for three types of fuses.

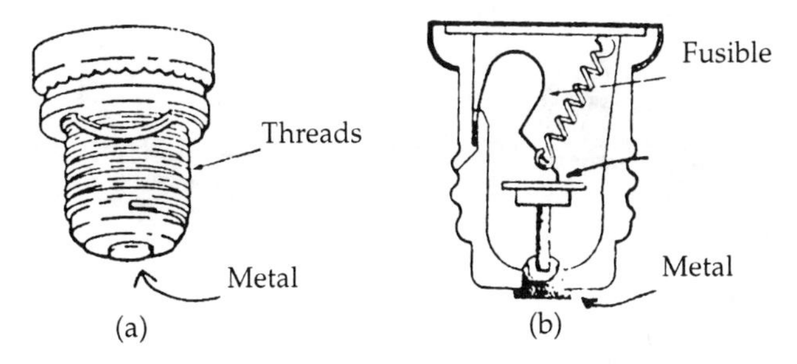

Figure 7-14. Plug fuses: (a) external appearance; (b) internal detail.

which decompose rapidly are used rather than ones which produce small metallic globules when they melt. This reduces the likelihood of any are-over occurring after the fuse metal has melted. The current rating of fuses depends upon the melting temperature of the fuse metal, as well as its shape, size, and the type of enclosure used.

Low-Voltage Circuit Breakers

Circuit breakers are somewhat more sophisticated overload devices than are fuses. Although their function is the same as that of fuses, circuit breakers are much more versatile. In three-phase systems, circuit breakers can open all three hot lines when an overload occurs. They may also be

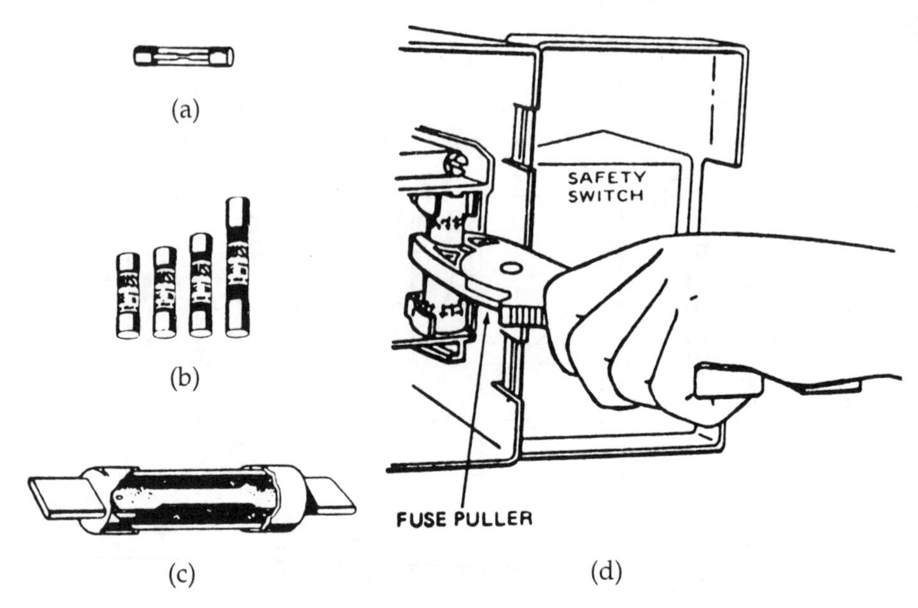

Figure 7-15. Cartridge fuses: (a) glass tube type; (b) various sizes of insulated fiber fuses; (c) knife type used for high current applications; (d) removing a cartridge fuse with a fuse puller—the safe way.

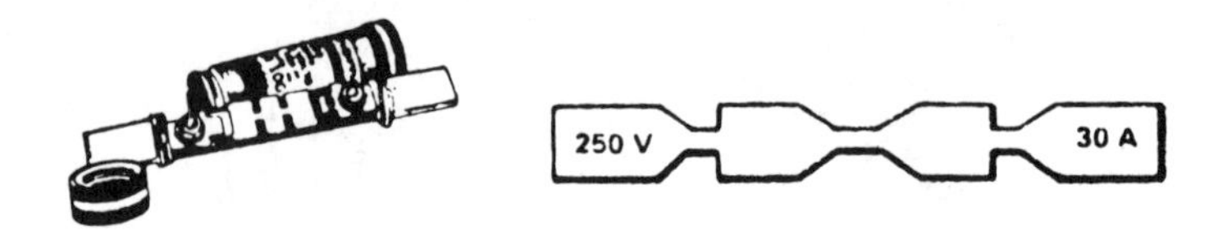

Figure 7-16. Renewable element cartridge fuses: (a) body of fuse with element removed; (b) renewable fuse link.

activated by remote-control relays. Relay systems may cause circuit breakers to open due to changes in frequency, voltage, current, or other circuit variables. Circuit breakers, such as those shown in Figure 7-18, are used in industrial plants and are usually of the low-voltage variety (less than 600 volts). They are not nearly as complex in construction as their high-voltage counterparts. Most low-voltage circuit breakers are housed in molded-plastic cases which mount in metal power distribution panels. Circuit breakers are designed so that they will automatically open when a current occurs which exceeds the rating of the breaker. Ordinarily, the circuit breakers must be reset manually. Most circuit breakers employ either a thermal tripping element or a magnetic trip element. Ratings of circuit breakers extend into current ranges that are as high as 800 to 2000 amperes.

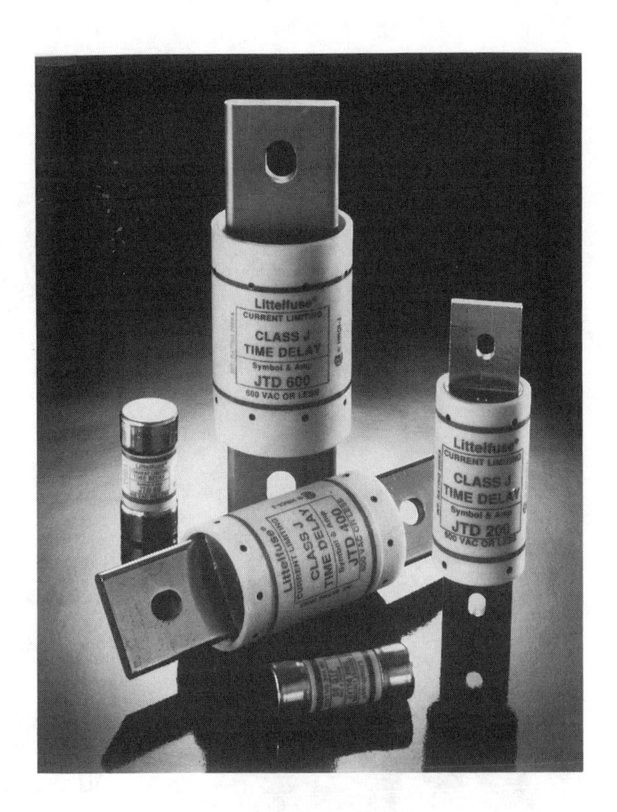

Figure 7-17. Time-delay fuses (Courtesy Littlefuse, Inc.).

Protective Relays

Protective relays provide an accurate and sensitive method of protecting electrical distribution equipment short circuits and other abnormal conditions. Overcurrent relays are used to cause the rapid opening of electrical power lines when the current exceeds a predetermined value. The response time of the relays is very important in protecting the equipment from damage. Some common types of faults which may be protected by relays are line-to-ground short circuits, line-to-line short circuits, double line-to-ground short circuits, and three-phase line short circuits. Each of these conditions is caused by faulty circuit conditions which draw abnormally high current (fault currents) from the power lines.

MOTOR-FAULT CURRENT PROTECTION

Motor-fault currents are excessive currents which occur in motors due to some unnatural malfunction. Since motor-fault currents cannot be withstood for any duration of time, some type of protection must be pro-

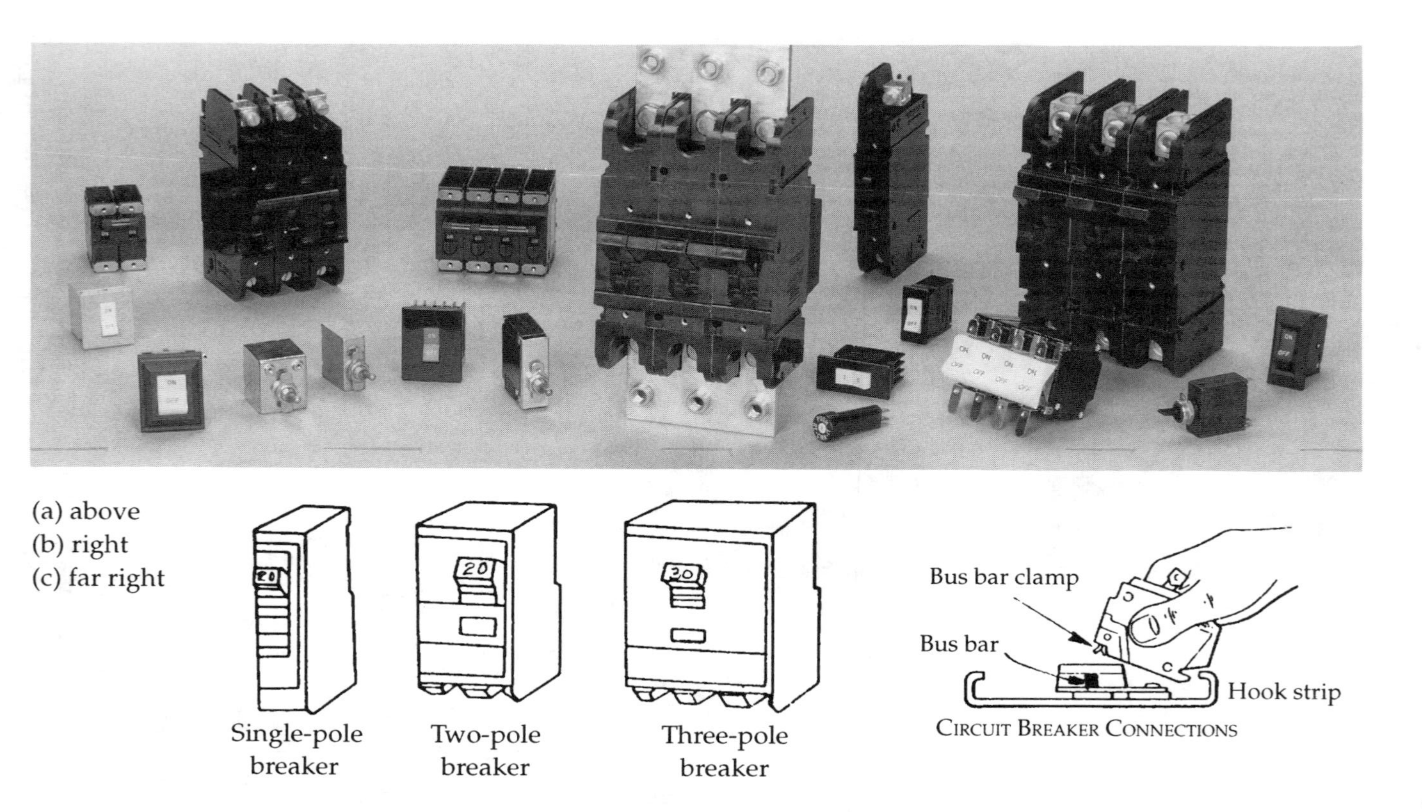

Figure 7-18. Low-voltage circuit breakers: (a) pictorial (Courtesy Heinemann Electric Co.; (b) types of circuit breakers; (c) replacing a breaker.

vided to disconnect the motor from the power distribution system when a fault condition occurs. Such protection may be provided by motor starters, circuit breakers, or fuses. The type of protection used is dependent upon several characteristics of the power distribution system and the motor.

Motor Protective Devices

The distribution of electrical power to motors is a particularly important function of industrial and commercial power distribution systems. Since motors are used in some way with most industrial processes, the efficient distribution of energy to industrial electric motors is very important. Basic functions which motors are expected to perform are starting, stopping, reversing, and speed variation. These functions may be manually or automatically controlled. Various types of protective devices are used to provide for the efficient distribution of power to electric motors.

Overload protection is the most important motor protection function. Such protection should serve the motor, its branch circuit, and associated control equipment. The major cause of motor overload is an excessive mechanical load on the motor which causes it to draw too much current from the power source. A block diagram of a motor protection system is shown in Figure 7-19.

Thermal overload relays are often used as protective devices. Thermal relays may be reset either manually or automatically. One type of thermal overload relay, shown in Figure 7-20, uses a bimetallic heater element. The bimetallic element bends as it is heated by the current flowing through it. When the current reaches the rating of the element, the relay opens the branch circuit. Another type of element is the melting-alloy type. This device has contacts held closed by a ratchet wheel. At the current capacity of the device, the fusible alloy melts causing the ratchet wheel to turn. A spring then causes the device to open the circuit. Thermal motor overload devices are shown in Figure 7-21.

Overheating Protection for Motors

Motors must be protected from excessive overheating. This protection is provided by magnetic or thermal protective devices which are ordinarily within the motor-starter enclosure. Protective relays or circuit breakers can also perform this function. When an operational problem causes the motor to overheat, the protective device is used to automatically disconnect the motor from its power supply.

Undervoltage Protection

Motors do not operate efficiently when less than their rated voltage is applied and some types of motors can be destroyed if they are operated continuously at reduced voltages. Magnetic contactors (see Chapter 11) may be used effectively to protect against undervoltages. A specific level of voltage is required to cause magnetic contactors to operate. If the voltage is reduced below a specified level, the magnetic contactor will open, thus disconnecting the circuit between the power source and the motor and stopping the motor before any damage can be done.

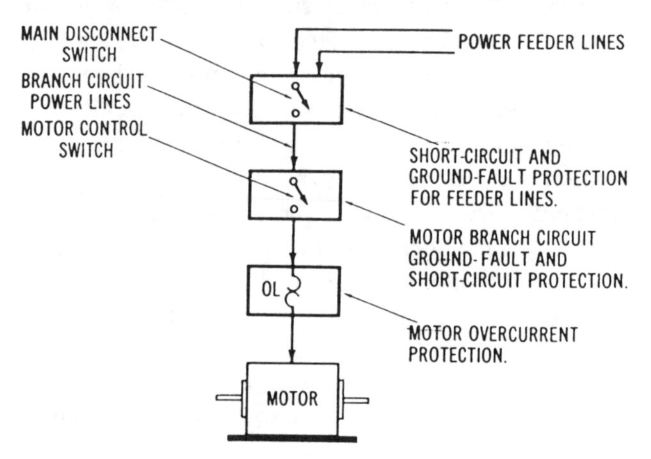

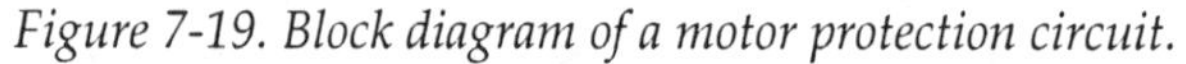
Figure 7-19. Block diagram of a motor protection circuit.

Figure 7-20. Thermal motor overload devices (Courtesy Fumas Electric Co.).

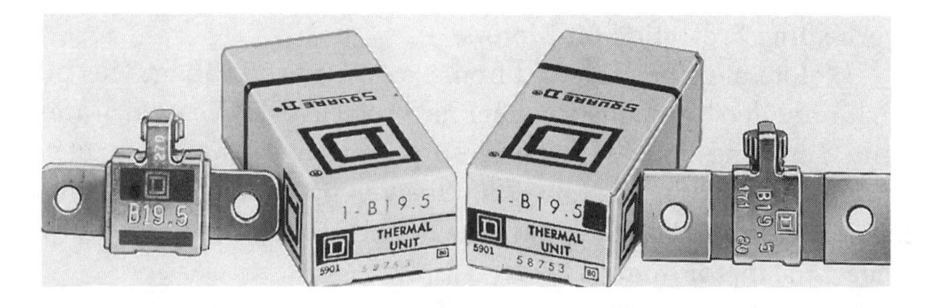

Figure 7-21. Thermal motor overload devices (Courtesy of Square-D Co.).

Chapter 8

Electrical Distribution Inside Buildings

Electrical power is distributed to the location where it is to be used and then distributed within a building by the power distribution system. Various types of circuit breakers and switchgear are employed for power distribution. Another factor involved in power distribution is the distribution of electrical energy to the many types of loads which are connected to the system. This part of the distribution system is concerned with the conductors, feeder systems, branch circuits, grounding methods, and protective and control equipment which is used. The primary emphasis in this chapter will be industrial and commercial power systems.

RACEWAYS

Most electrical distribution to industrial and commercial loads is through wires and cables contained in raceways. These raceways carry the conductors which carry the power to the various equipment throughout a building. Copper conductors: are ordinarily used for indoor power distribution. The physical size of each conductor is dependent upon the current rating of the branch circuit. Raceways may be large metal ducts, such as those shown in Figure 8-1 or rigid metal conduit. These raceways provide a compact and efficient method of routing cables, wires, etc., throughout an industrial complex. A cable tray raceway design for industrial applications is shown in Figure 8-2.

FEEDER LINES AND BRANCH CIRCUITS

The conductors which carry current to the electrical load devices in a building are called *feeders* and *branch circuits.* Feeder lines supply power to

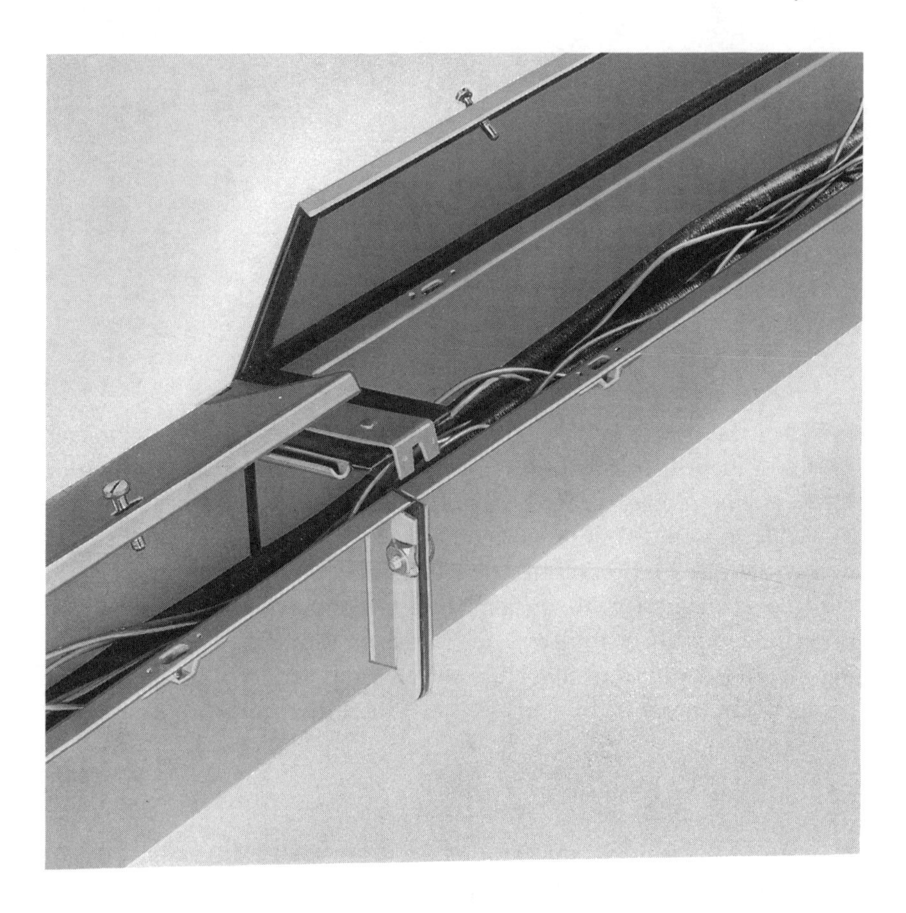

Figure 8-1. Typical raceway. Opened raceway showing conductors (Courtesy Square-D Co.).

branches which are connected to them. Primary feeder lines may be either overhead or underground. Usually, overhead lines are preferred because they permit flexibility for future expansion. Underground systems cost more, but they are much more attractive in appearance. Secondary feeders are connected to the primary feeder lines and supply power to individual sections within the building. Either aluminum or copper feeder lines may be used, depending on the specific power requirements. The distribution is from the feeder lines, through individual protective equipment, to branch circuits which supply the various loads. Each branch circuit has various protective devices according to the needs of that particular branch. The overall feeder-branch system may be a very complex net-

1 Ladder tray straight section.
2 Ventilated trough tray.
3 Splice plate-bolted connector.
4 Horizontal bend (30° to 90°).
5 Tee.
6 Cross.
7 Vertical bends (30° to 90°).
INSIDE VERTICAL = transition from level to upward direction.
OUTSIDE VERTICAL = transition from level to downward direction.
8 Vertical tee.
9 Reducer.
10 Channel tray.
11 Barrier.
12 Cover
13 Tray-to-box connector.
14 Channel-outside bend.

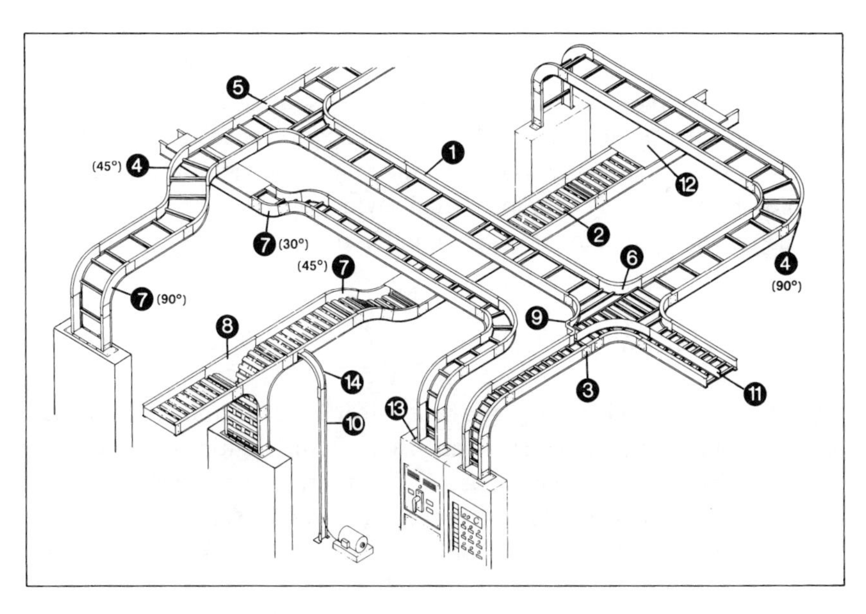

Figure 8-2. Industrial cable tray (Courtesy Chalfant Manufacturing Co.).

work of switching equipment, transformers, conductors, and protective equipment.

SWITCHING EQUIPMENT

In addition to circuit protection, power distribution systems must have equipment which can be used to connect or disconnect the entire system or parts of the system. Various types of switching devices are used to perform this function. A simple type of switch is the *safety switch.* This type of switch is mounted in a metal enclosure and operated by means of an external handle. A single-phase safety switch is shown in Figure 8-3. Safety switches are used only to turn a circuit off or on; however, fuses are often mounted in the same enclosure with the safety switch.

DISTRIBUTION PANELBOARDS

Another type of switch is the kind used in conjunction with a circuit-breaker panelboard. Panelboards are metal cabinets which enclose the main disconnect switch and the branch-circuit protective equipment. Distribution panelboards are usually located between the power feed lines within a building and the branch circuits which are connected to it. A typical distribution panelboard is shown in Figure 8-4.

LOW-VOLTAGE SWITCHGEAR

Metal-enclosed low-voltage switchgear is used in many industrial and commercial buildings. These are used as a distribution control center to house the circuit breakers, bus bars, and terminal connections which are part of the power distribution system. Ordinarily, a combination of switchgear and distribution transformers is placed in adjacent metal enclosures, such as that shown in Figure 8-5. This combination is referred to as a load-center unit substation since it is the central control for several loads. The rating of these load centers is usually 15,000 volts or lower for the high-voltage section and 600 volts or less for the low-voltage section. Load centers provide flexibility in the electrical power distribution design of industrial plants and commercial buildings.

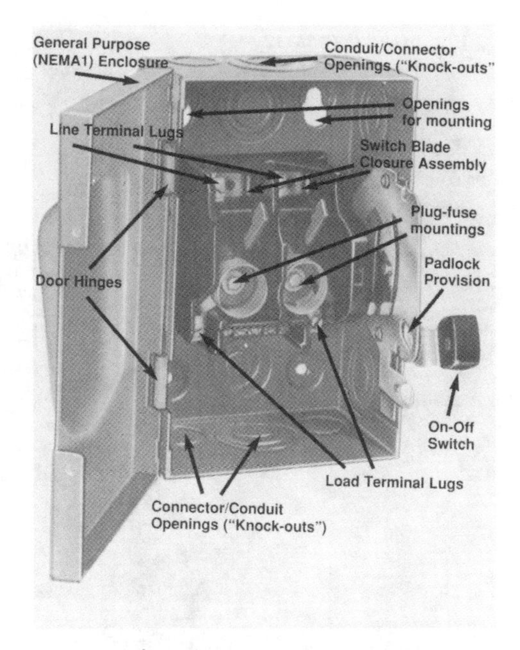

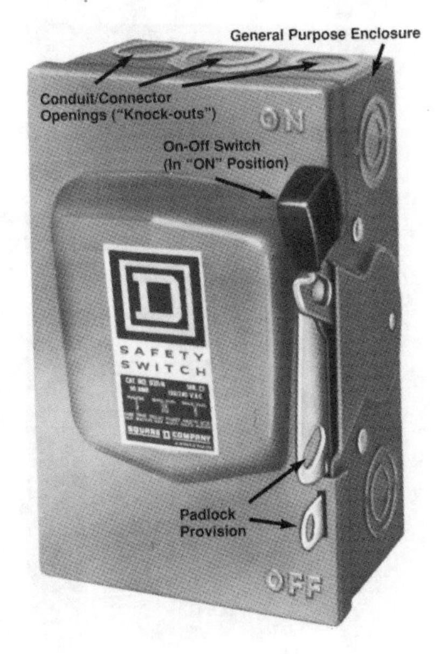

(b) Safety switch (exterior)

Figure 8-3. Safety switch: (a) interior; (b) exterior; (Courtesy of Square-D Co.).

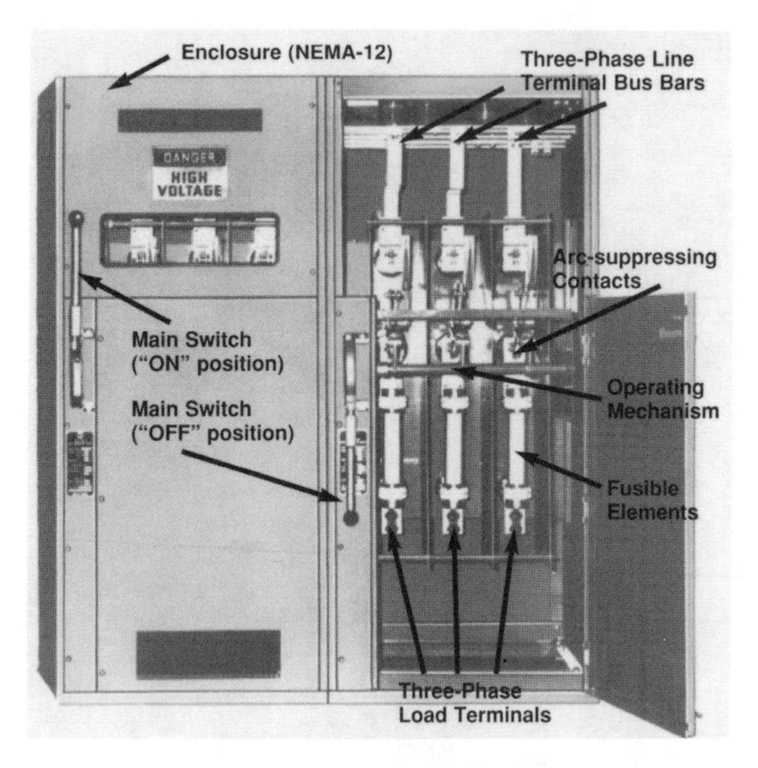

Figure 8-4. Power distribution panelboard (Courtesy Square-D Co.).

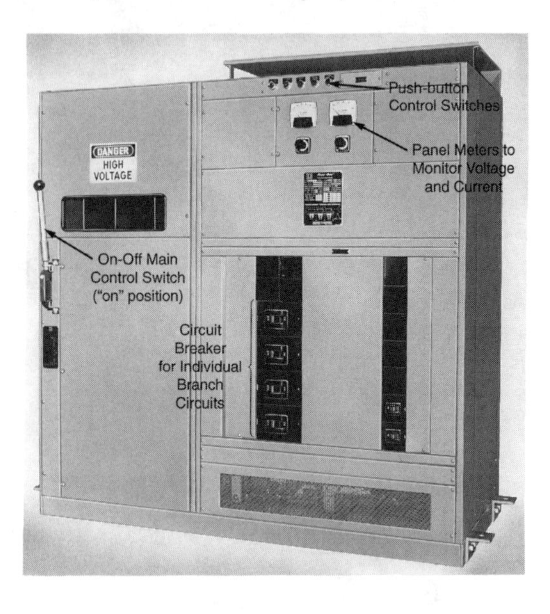

Figure 8-5. Load-center unit substation (Courtesy Square-D Co.).

Metal-enclosed switchgear or metal-clad switchgear is a type of equipment which houses all the necessary control devices for the electrical circuits that are connected to them. The control devices contained inside the switchgear include circuit breakers, disconnect switches, interconnecting cables and buses, transformers, and the necessary measuring instruments. Switchgear is used for indoor and outdoor applications at industrial plants, commercial buildings, and at substations. The voltage ratings of switchgear are usually from 13.8 to 138 kilovolts with 1 megavolt-ampere to 10 megavolt-ampere power ratings.

THE ELECTRICAL SERVICE ENTRANCE

Electrical power is brought from the overhead power lines or from the underground cable into a building by what is called a *service entrance.* A good working knowledge of the *National Electrical Code* (NEC) specifications and definitions is necessary to understand service-entrance equipment. The NEC sets the minimum standards that are necessary for wiring design inside a building.

The type of equipment used for an electrical service entrance of a building may include high-current conductors and insulators, disconnect switches, protective equipment for each load circuit which will be connected to the main power system and the meters needed to measure power, voltage, current, or frequency. It is also necessary to ground the power system at the service-entrance location. This is done by a "grounding electrode" which is a metal rod driven deep into the ground. The grounding conductor is attached securely to this grounding electrode. Then, the grounding conductor is used to make contact with all neutral conductors and safety grounds of the system. Service entrances are shown in Figure 8-6.

SERVICE-ENTRANCE TERMINOLOGY

There are several terms associated with service-entrance equipment. The service-entrance *conductors* are a set of conductors which are brought to a building by the local electrical utility company. These conductors must be capable of carrying all of the electrical current that is to be delivered to the various loads inside the building which are to be supplied by the power

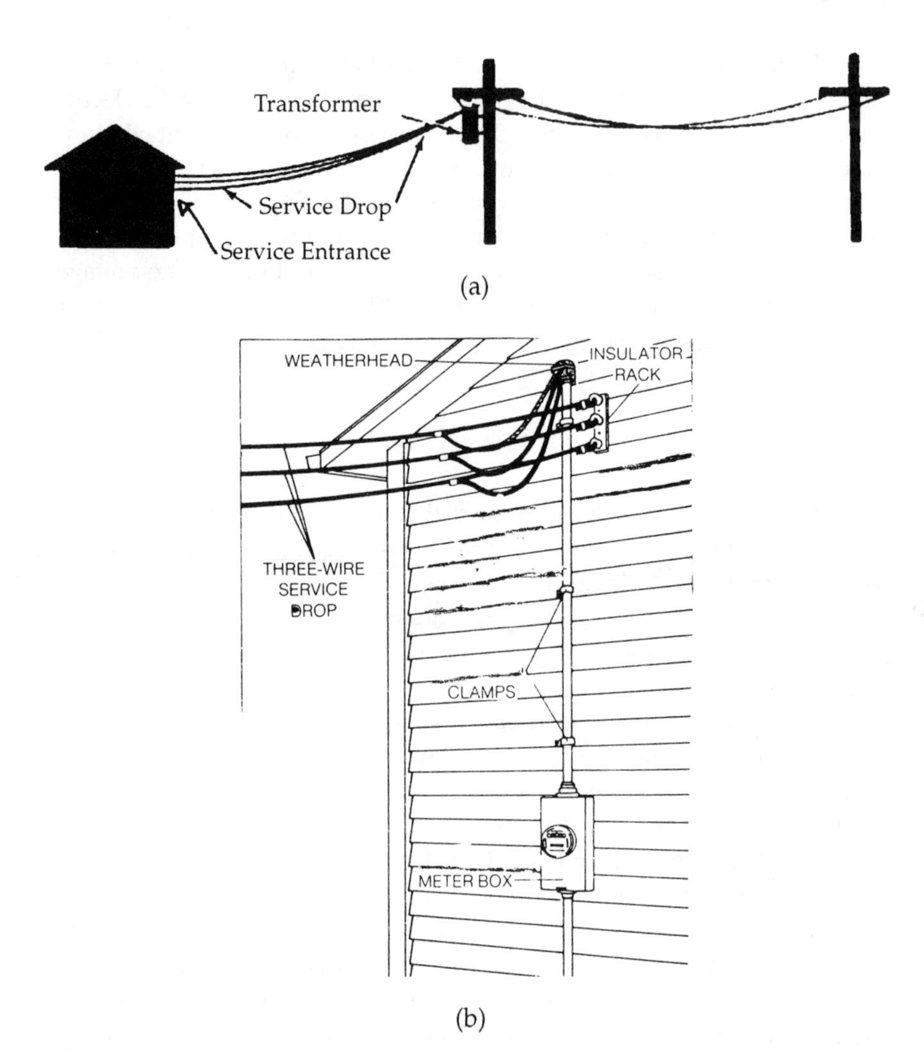

Figure 8-6. Electrical service entrance: (a) conductors from distribution system are connected through a transformer mounted on a pole; (b) the three-wire service drop is connected into a building through the service entrance.

system. Conductors which extend from the service entrance to a power distribution panel or other type of overcurrent protective equipment are called *feeders.* Feeders are power lines which supply branch circuits. A *branch circuit* is defined as conductors which extend beyond the last overcurrent protective equipment of the power system. Usually, each branch circuit delivers electrical power to a small percentage of the total load of the main power system.

In commercial and industrial installations, *switchboards* and *panelboards* are used to supply power to various loads throughout the power system. A switchboard is a large enclosure which has several overcurrent protective devices (fuses or circuit breakers). Each feeder is connected to the proper type of overcurrent device. Often, switchboards contain metering equipment for the power system. Panelboards are smaller than switchboards but are used for a similar purpose. They are enclosures for overcurrent devices for either branch circuits or feeder circuits. A common example of a panelboard is the main power-distribution panel which houses the circuit breakers used for the branch circuits of a home. For more specific definitions of terms, you should refer to the most recent edition of the *National Electrical Code.*

POWER DISTRIBUTION SYSTEM COMPONENTS

Several specialized types of power distribution system components are available today and should be reviewed.

Uninterruptible Power Supply—Figure 8-7 shows an uninterruptible power supply (UPS). An uninterruptible power supply has computer controlled diagnostics and monitoring to provide constant on-line power for today's modern equipment.

Power Filters and Conditioners—Figure 8-8 shows AC power line filters which plug into interior power distribution systems. These filters have power outlets for obtaining filtered AC power. Power conditioners are also used to protect power distribution systems from spikes, surges, or other interference which may be damaging to certain types of equipment.

Floor-mounted Raceways—Floor-mounted raceways are shown in Figure 8-9. Surface raceways and power outlets are used in most commercial and industrial facilities. The designs shown in Figure 8-10 are typical underfloor systems which are available.

Conduit Connectors—Figure 8-11 shows some types of conduit bodies used today. These bodies are used to provide a means of connecting conductors and to allow angular bends in conduit runs throughout a building.

Wire Connectors—A simple but essential component of electrical power distribution systems are wire connectors (sometimes called "wire nuts"). Figure 8-12 shows some typical types of wire connectors for electrical power systems.

Plastic Components—Flexible plastic conduit is shown in Figure 8-13a. This product provides an alternative to electrical metallic tubing (EMT) and rigid conduit in certain distribution systems. Plastic boxes which are compatible with flexible conduit are shown in Figure 8-13b.

Power Outlets—Power outlets have a standard configuration which has been established by the National Electrical Manufacturers Association (NEMA). The specific configuration indicates the voltage, current and phase ratings of the distribution system. NEMA designs are shown in Figure 8-14.

International Power Sources—The international power source shown in Figure 8-15 provides a convenient means of converting North American voltage and frequency (120 volt/60 Hertz) to international voltages and frequencies. Output power is obtained through the appropriate standard international socket. The system shown has adjustable voltages and frequencies for obtaining power to match that of most countries. This allows testing of products manufactured for export to other countries or use of products purchased in other countries without modification of power supplies.

Figure 8-7. Uninterruptible power supplies (Courtesy Best Power Technology) (a): Best Power Technology, Inc. Uninterruptible power supply (UPS) from 250 VA to 15 kVA features new technology with computer-controlled diagnostics and monitoring. Data-Save Software, plus highest efficiency in the industry. Models 2, 3, or 5 kVA shown above (batteries included).

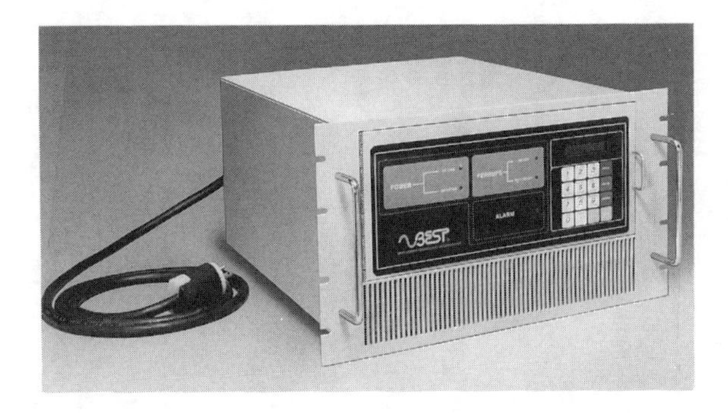

Figure 8-7b.

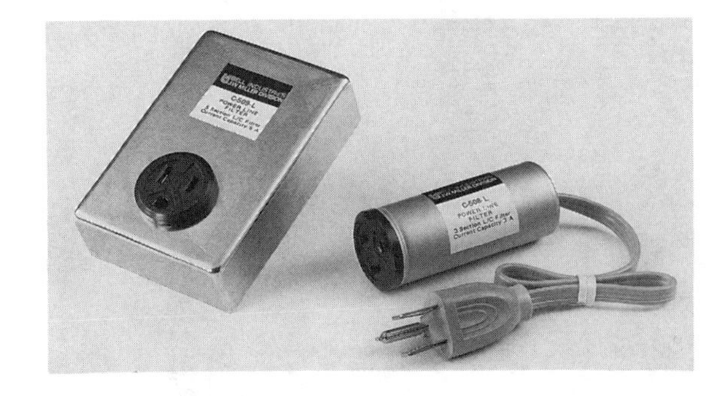

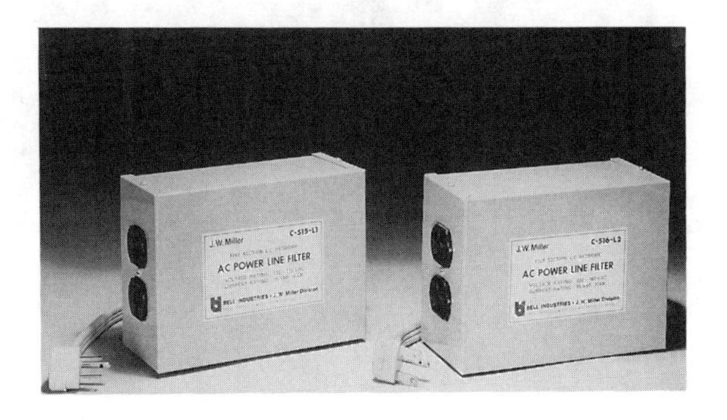

Figure 8-8. AC power line filters (Courtesy J.W. Miller Div./Bell Industries).

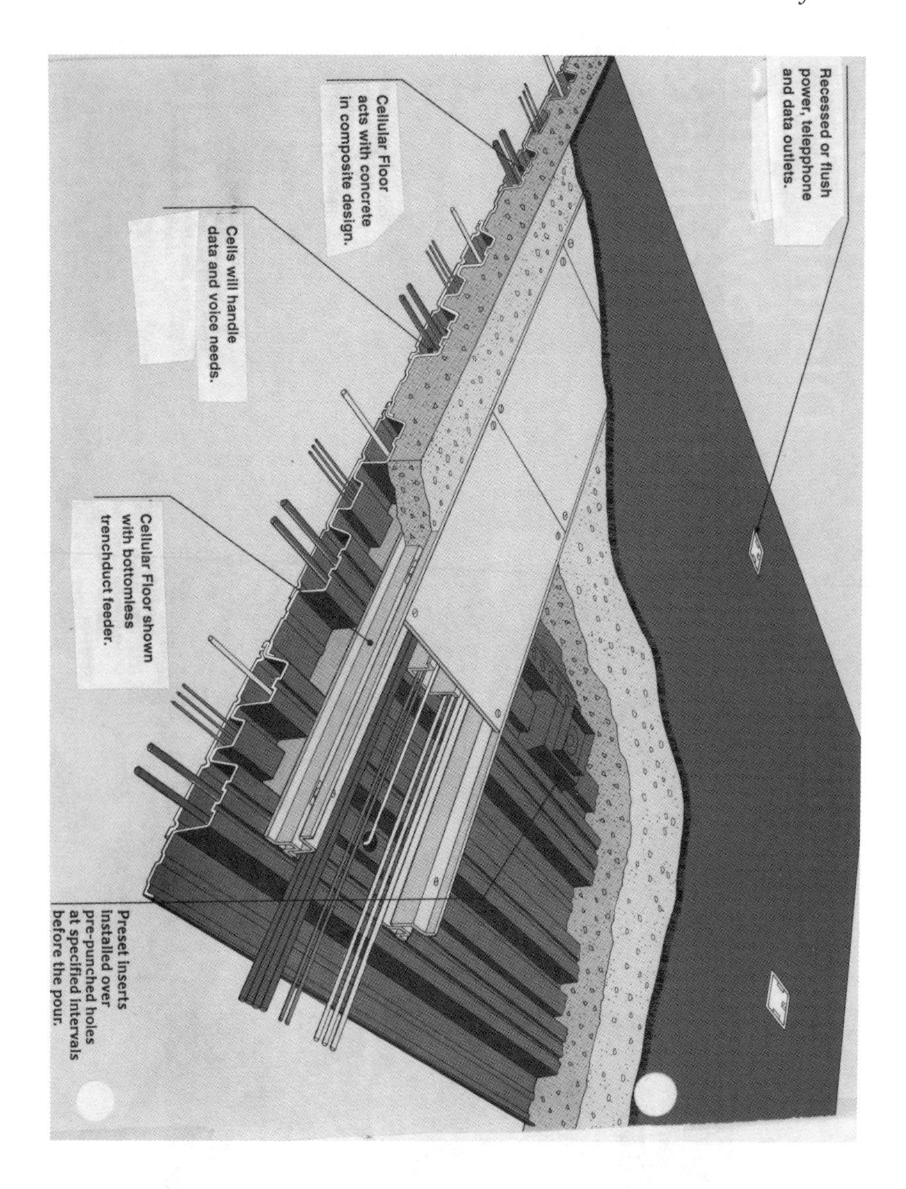

Figure 8-9. Floor-mounted raceway design (Courtesy Walker Co.).

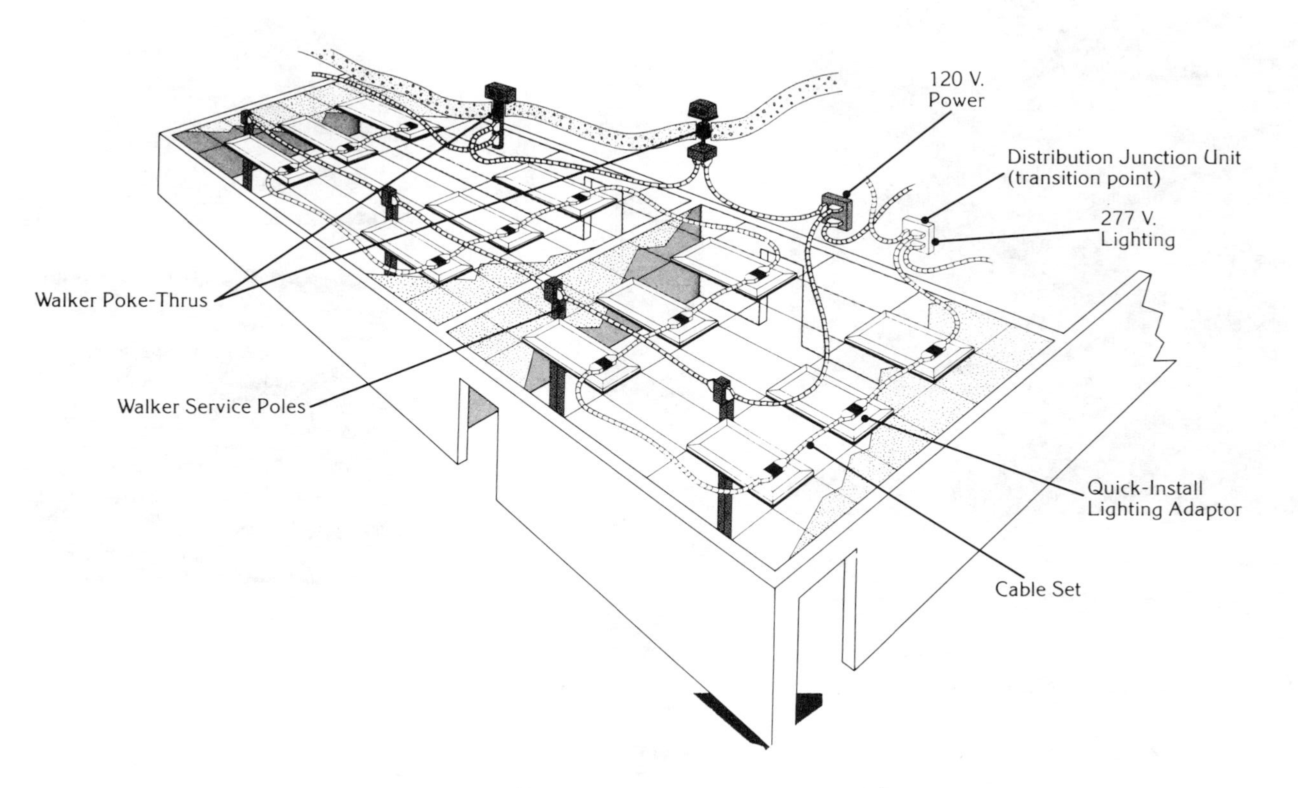

Figure 8-10a. Underfloor raceway systems (Courtesy Walker Co.).

Figure 8-10b (left). Surface raceway (Courtesy Walker Co.).

Figure 8-11 (below). Conduit bodies (Courtesy RACO Inc.).

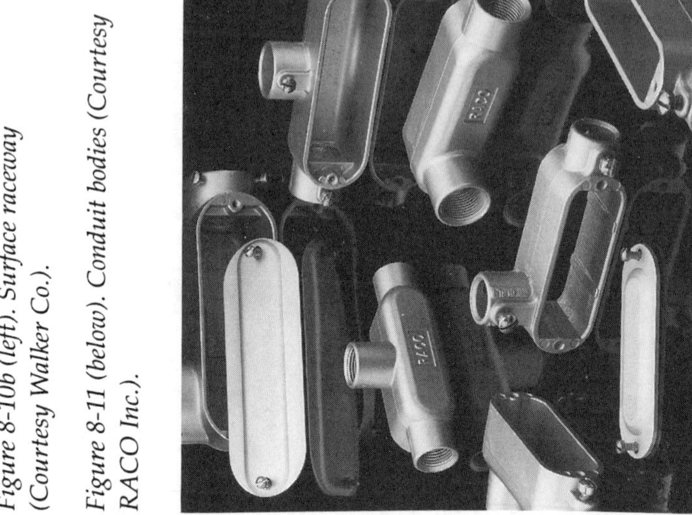

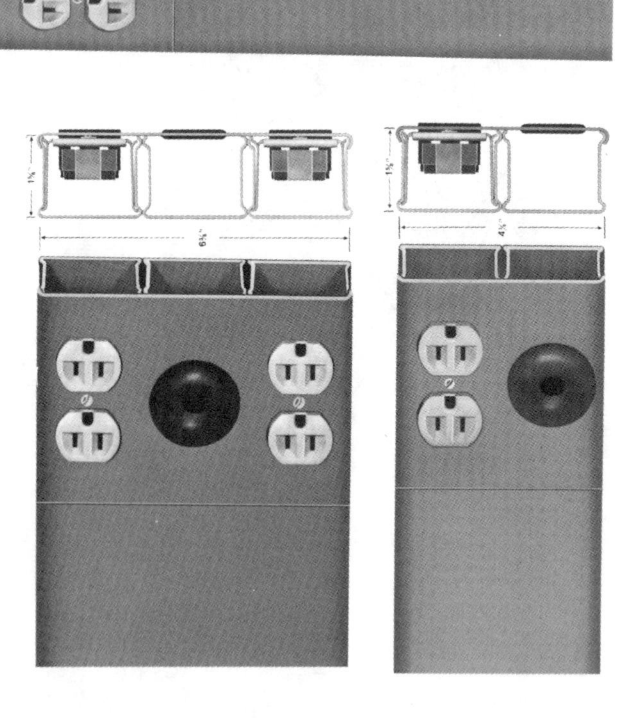

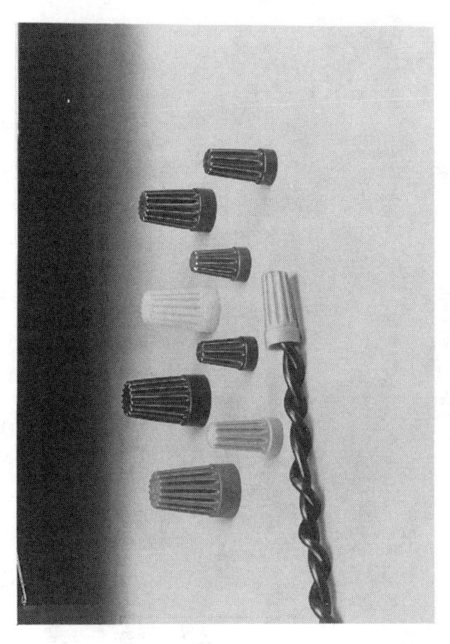

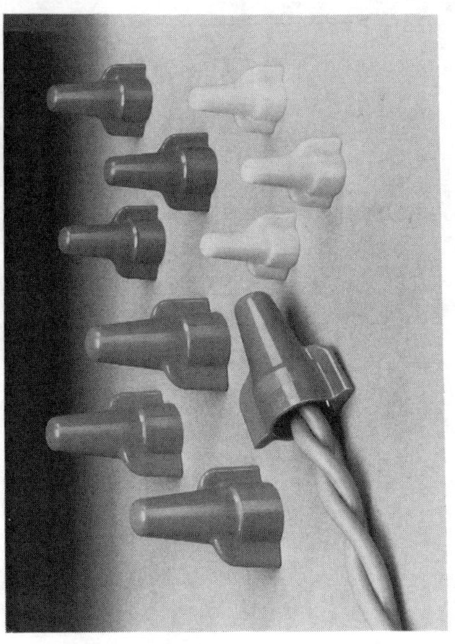

Figure 8-12. Wire connectors: (a-above) plastic "wire nuts"; (b-above right) metal power connectors; (c-right) plastic wire connectors (Courtesy Panduit Co.).

Figure 8-13: Plastic Components used for power distribution: (a) flexible plastic conduit and connectors (Courtesy Carlon Co.); (b) Plastic boxes, cover and connectors.

Figure 8-14. Standard NEMA designs for power outlets (Courtesy Pulizzi Engineering Inc.). NOTE that these NEMA configurations are not to scale. Canadian, United Kingdom, European and other required configurations are also available.

		15 AMP		20 AMP		30 AMP	
		RECEPTACLE	PLUG	RECEPTACLE	PLUG	RECEPTACLE	PLUG
125 VOLT	5	5-15R	5-15P	5-20R	5-20P	5-30R	5-30P
250 VOLT	6	6-15R	6-15P	6-20R	6-20P	6-30R	6-30P
277 VOLT	7	7-15R	7-15P	7-20R	7-20P	7-30R	7-30P
125/250 VOLT	10			10-20R	10-20P	10-30R	10-30P
3 Ø 250 VOLT	11	11-15R	11-15P	11-20R	11-20P	11-30R	11-30P
125/250 VOLT	14	14-15R	14-15P	14-20R	14-20P	14-30R	14-30P
3 Ø 250 VOLT	15	15-15R	15-15P	15-20R	15-20P	15-30R	15-30P
3 Ø Y 120/208 VOLT	18	18-15R	18-15P	18-20R	18-20P	18-30R	18-30P

		15 AMP		20 AMP		30 AMP	
		RECEPTACLE	PLUG	RECEPTACLE	PLUG	RECEPTACLE	PLUG
125 VOLT	L5	L5-15R	L5-15P	L5-20R	L5-20P	L5-30R	L5-30P
250 VOLT	L6	L6-15R	L6-15P	L6-20R	L6-20P	L6-30R	L6-30P
277 VOLT	L7	L7-15R	L7-15P	L7-20R	L7-20P	L7-30R	L7-30P
480 VOLT	L8			L8-20R	L8-20P	L8-30R	L8-30P
600 VOLT	L9			L9-20R	L9-20P	L9-30R	L9-30P
125/250 VOLT	L14			L14-20R	L14-20P	L14-30R	L14-30P
3 Ø 250 VOLT	L15			L15-20R	L15-20P	L15-30R	L15-30P
3 Ø 480 VOLT	L16			L16-20R	L16-20P	L16-30R	L16-30P
3 Ø 600 VOLT	L17					L17-30R	L17-30P
3 Ø 208Y/120 VOLT	L21			L21-20R	L21-20P	L21-30R	L21-30P
3 Ø 480Y/277 VOLT	L22			L22-20R	L22-20P	L22-30R	L22-30P
3 Ø 600Y/347 VOLT	L23			L23-20R	L23-20P	L23-30R	L23-30P

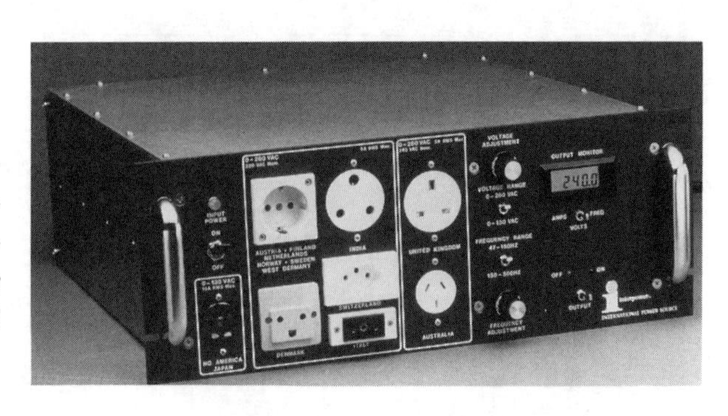

Figure 8-15. International power distribution: (a) international power source (Courtesy Panel Components Corp.).

Country	Rating	Cord Design
Continental Europe	10A 250V	
Australia, New Zealand	10A 250V	
U.K., Ireland	10A 250V	
Denmark	10A 250V	
India	10A 250V	
Israel	10A 250V	
Italy	10A 250V	
No. America	10A 125V	
Switzerland	10A 250V	

Figure 8-15b. Cord designs used for power distribution by several countries (Courtesy Panel Components Corp.).

Figure 8-15c. Voltage-selectable power distribution and control system.

Figure 8-15d. Three-phase to single-phase AC power controller.

Chapter 9

Single-phase and Three-phase Distribution Systems

When electrical power is distributed to its point of utilization, it is normally either in the form of single-phase or three-phase AC voltage. Single-phase AC voltage is distributed into residences and other smaller commercial buildings. Normally, three-phase AC voltage is distributed to industries and larger commercial buildings. Thus, the main types of power distribution systems are residential (single-phase) and industrial or commercial (three-phase).

An important aspect of both single-phase and three-phase distribution systems is grounding. Two grounding methods, system grounding and equipment grounding, will be discussed, along with ground-fault protective equipment.

SINGLE-PHASE SYSTEMS

Most electrical power when produced at the power plants is produced as three-phase AC voltage. Electrical power is also transmitted in the form of three-phase voltage over long-distance power-transmission lines. At its destination, three-phase voltage can be changed into three separate single-phase voltages for distribution into the residential areas.

Although single-phase systems are used mainly for residential power distribution systems, there are some industrial and commercial applications of single-phase systems. Single-phase power distribution usually originates from three-phase power lines so electrical power systems are capable of supplying both three-phase and single-phase loads from the same power lines. Figure 9-1 shows a typical power distribution system from the power station (source) to the various single-phase and three-phase loads which are connected to the system.

Single-phase systems can be of two major types—single-phase two-wire systems or single-phase three wire systems. A single-phase two-wire system is shown in Figure 9-2a (the top diagram). This system uses a 10-kVA transformer whose secondary produces one single-phase voltage, such as 120 or 240 volts. This system has one hot line and one neutral line. In residential distribution systems, several years ago, this was the most used type to provide 120-volt service. However, as appliance power requirements increased, the need for a dual-voltage system was evident.

To meet the demand for more residential power, the single-phase three-wire system is now used. A home-service entrance can be supplied with 120/240-volt energy by the methods shown in Figures 9-2b and 9-2c (center and bottom diagrams). Each of these systems is derived from a three-phase power line. The single-phase three-wire system has two hot lines and a neutral line. The hot lines, whose insulation is usually black and red, are connected to the outer terminals of the transformer secondary windings. The neutral (white insulated wire) is connected to the center tap of the distribution transformer. Thus, from neutral to either hot line, 120 volts for lighting and small-power requirements may be obtained. Across the hot lines, 240 volts is supplied for higher-power requirements. Therefore, the current requirement for large power-consuming equipment is cut in half, since 240 volts rather than 120 volts is used. Either the single-phase two-wire or the single-phase three-wire system can be used to supply single-phase power for industrial or commercial use. However, these single-phase systems are mainly for residential power distribution.

THREE-PHASE SYSTEMS

Since industries and commercial buildings use three-phase power predominantly, they rely upon three-phase distribution systems to supply this power. Large three-phase distribution transformers are usually located at substations adjoining the industrial plants or commercial buildings. Their purpose is to supply the proper AC voltages to meet the necessary load requirements. The AC voltages that are transmitted to the distribution substations are high voltages which must be stepped down by three-phase transformers.

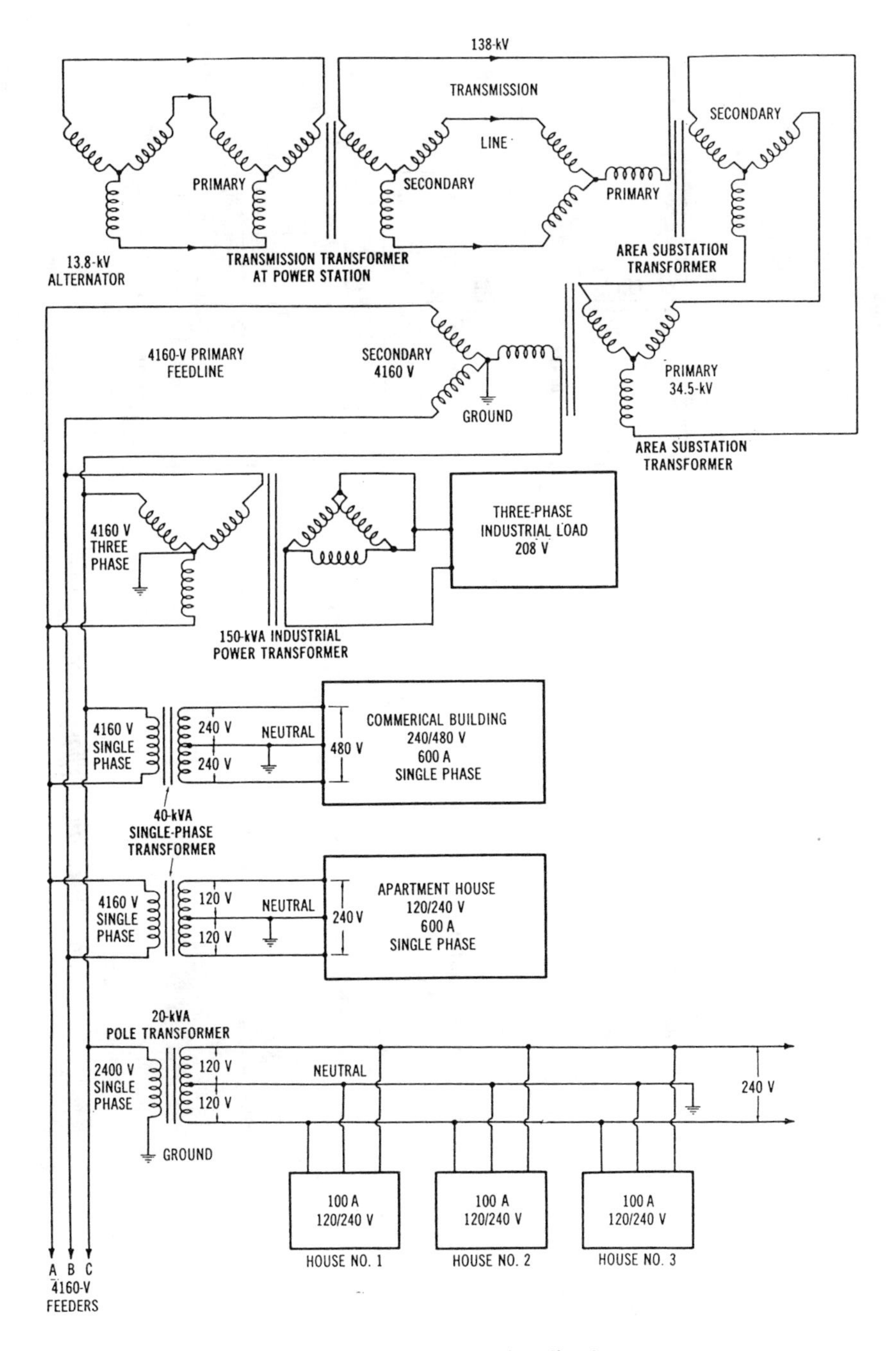

Figure 9-1. A typical power distribution system.

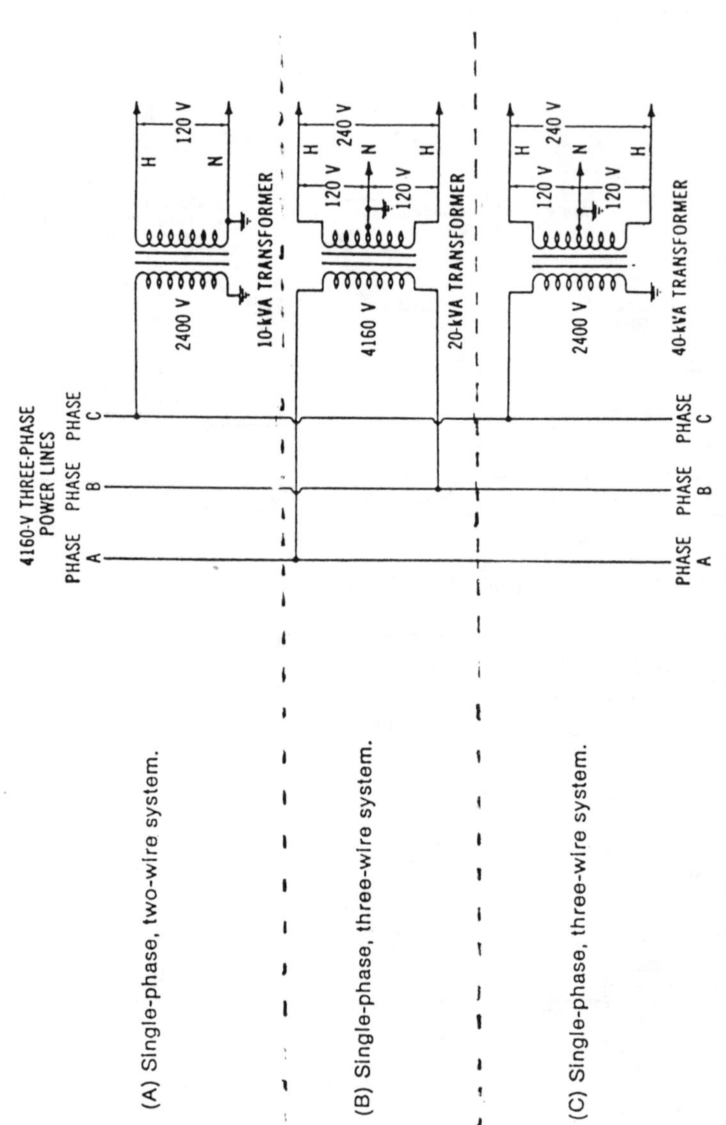

Figure 9-2. Single-phase power distribution systems.

THREE-PHASE TRANSFORMER CONNECTIONS

There are five ways in which the primary and secondary windings of three-phase transformers may be connected. These are the delta-delta, the delta-wye, the wye-wye, the wye-delta, and the open-delta connections. These basic methods are illustrated schematically in Figure 10-3. The delta-delta connection (Figure 9-4) is used for some lower voltage applications. The delta-wye method (Figure 9-5) is commonly used for stepping up voltages since an inherent step-up factor of 1.73 times is obtained due to the voltage characteristic of the wye-connected secondary. The wye-wye connection of Figure 9-6 is ordinarily not used, while the wye-delta method (Figure 9-7) may be used advantageously to step voltages down. The open-delta connection (Figure 9-8) is used if one transformer winding becomes damaged or is taken out of service. The transformer will still deliver three-phase power, but at a lower current and power capacity. This connection may also be desirable when the full capacity of three transformers is not needed until a later time. Two identical single-phase transformers can be used to supply power to the load until, at a later time, the third transformer is needed to meet increased load requirements.

TYPES OF THREE-PHASE SYSTEMS

Three-phase power distribution systems which supply industrial and commercial buildings are classified according to the number of phases and number of wires required. These systems, shown in Figure 9-9 are: (1) three-phase three-wire system; (2) Three-phase three-wire system with neutral; and (3) three-phase four-wire system. The primary winding connection is not considered here. The three-phase three-wire system, shown in Figure 9-9a, can be used to supply motor loads of 240 volts or 480 volts. Its major disadvantage is that it only supplies one voltage as only three hot lines are supplied to the load. The usual insulation color code for these hot lines is black, red, or blue, as specified in the *National Electrical Code.*

The disadvantage of the three-phase three-wire system may be partially overcome by adding one center-tapped phase winding as shown in the three-phase three-wire with neutral system of Figure 9-9b. This system can be used as a supply for 120/240 volts or 240/480 volts. As-

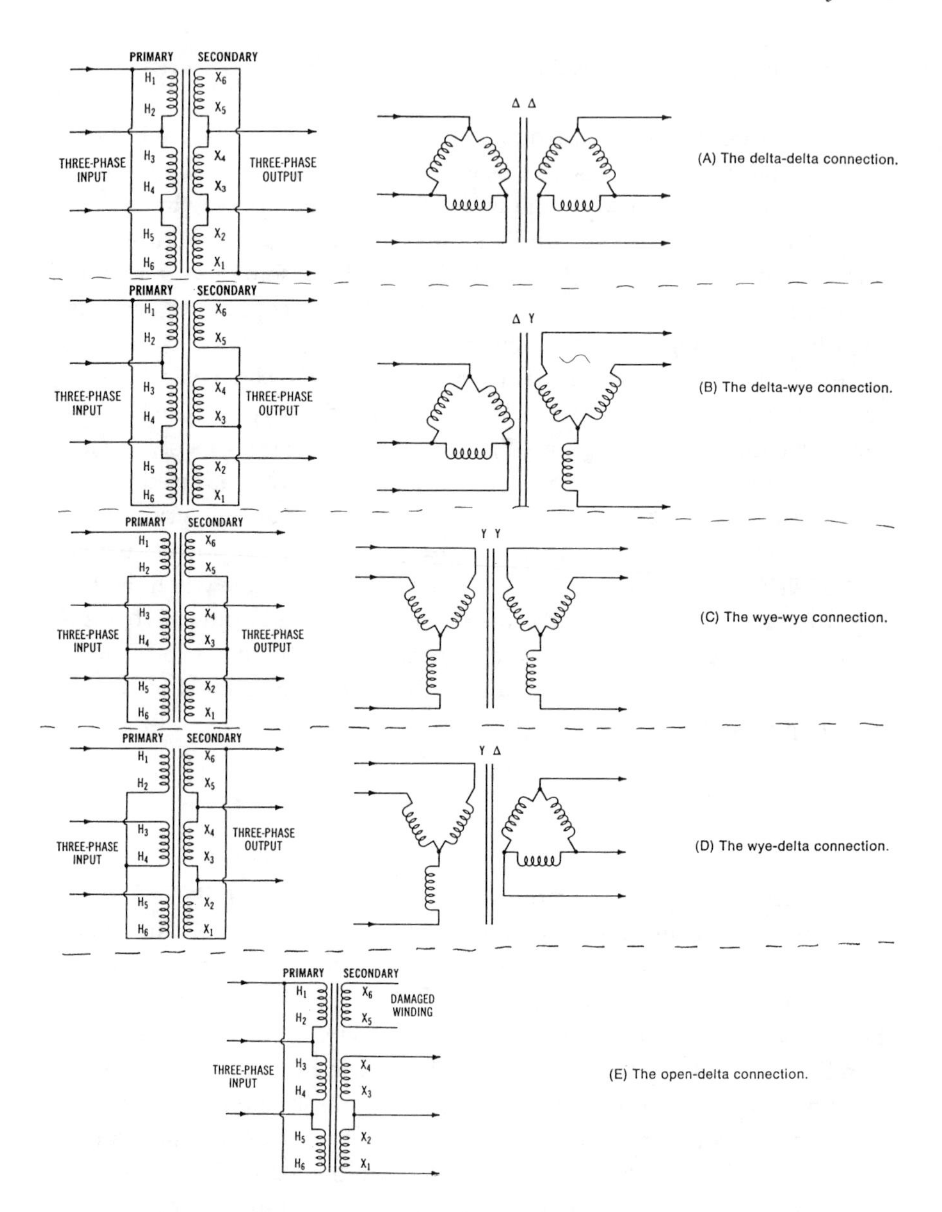

(A) The delta-delta connection.

(B) The delta-wye connection.

(C) The wye-wye connection.

(D) The wye-delta connection.

(E) The open-delta connection.

Figure 9-3. Basic three-phase transformer connection methods.

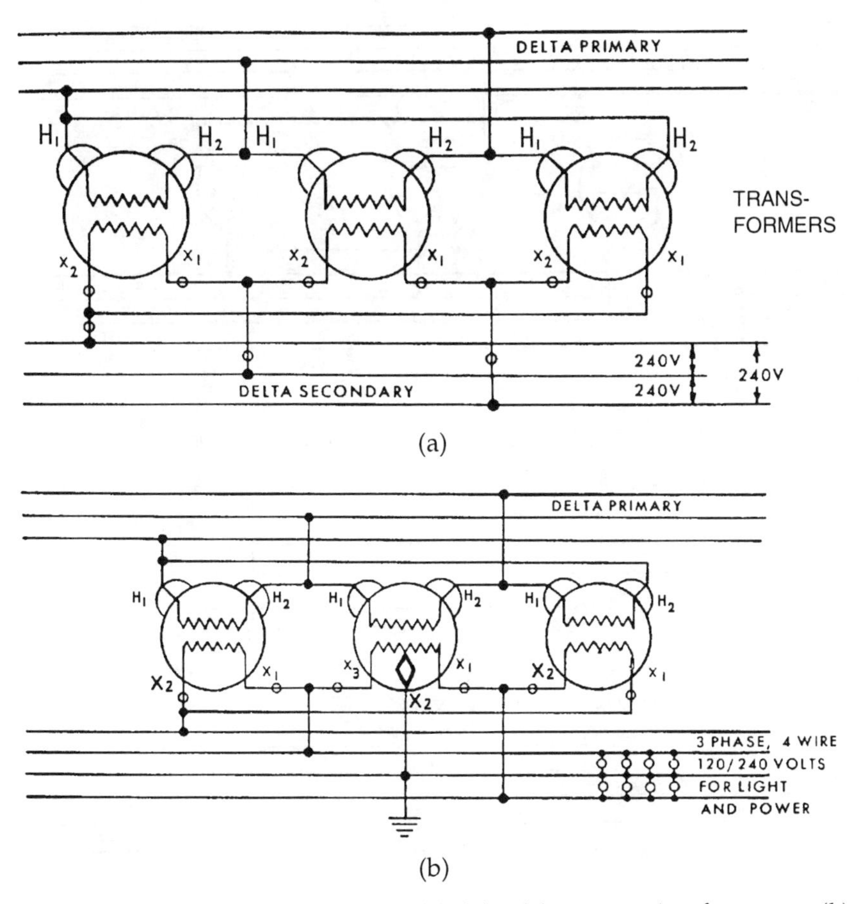

Figure 9-4. Delta-delta connections: (a) delta-delta connection for power; (b) delta-delta connection for power and light.

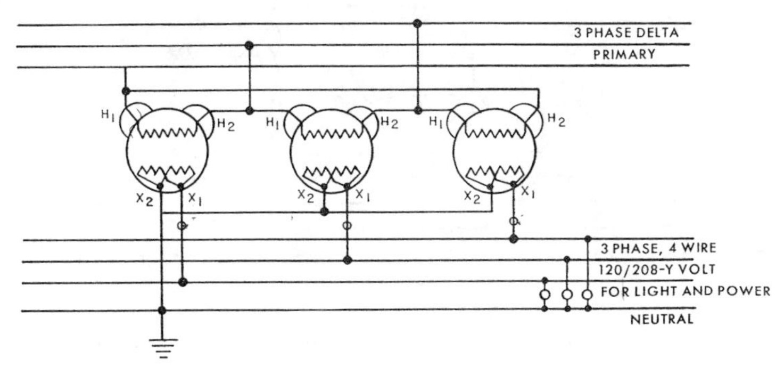

Figure 9-5. Delta-wye connection for light and power.

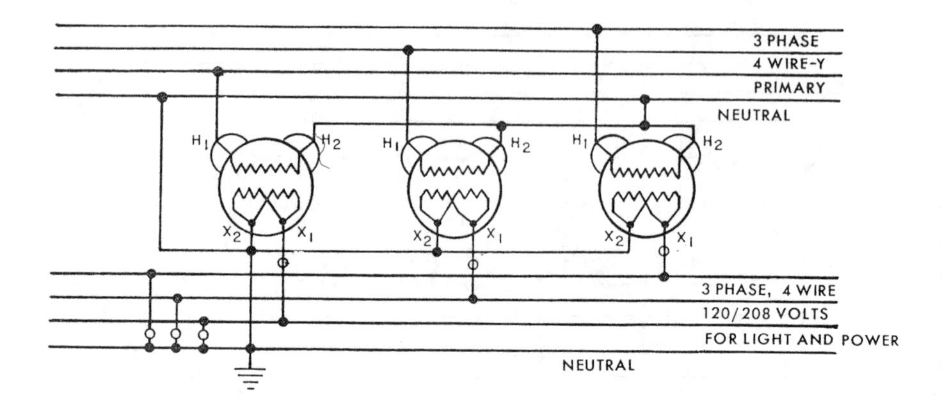

Figure 9-6. Wye-wye connection for light and power.

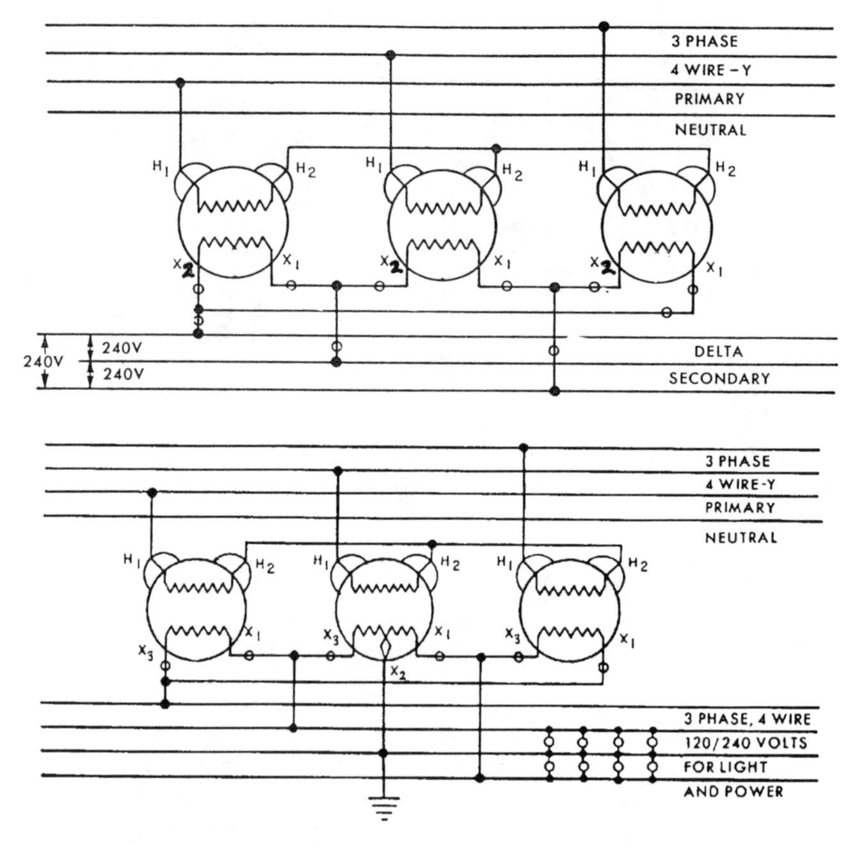

Figure 9-7. Wye-delta connections: (a-center) wye-delta connection for power; (b-bottom) wye-delta connection for light and power.

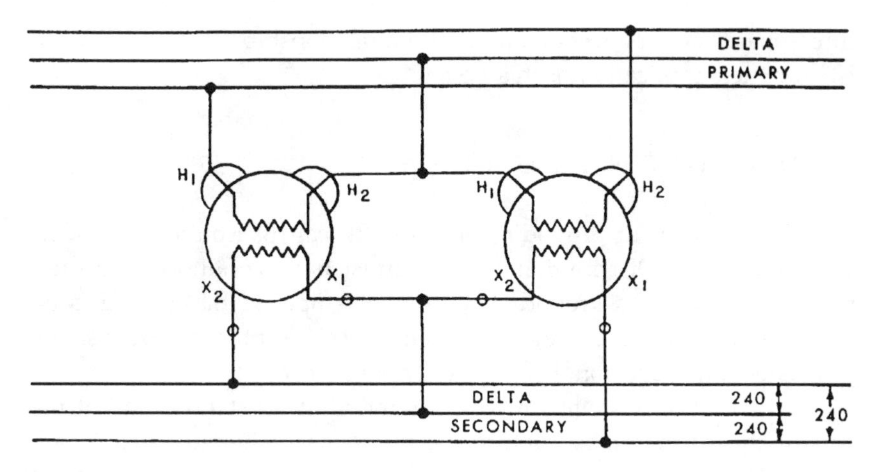

Figure 9-8. Open-delta connection for power.

sume that it is used to supply 120/240 volts. The voltage from the hot line at point 1 and the hot line at point 2 to neutral would be 120 volts, due to the center-tapped winding. However, 240 volts would still be available across any two hot lines. The neutral wire is color coded with a white or gray insulation. The disadvantage of this system is that, when making wiring changes, it would be possible to connect a 120-volt load between the neutral and point 3 (sometimes called the "wild" phase). The voltage present here would be the combination of three-phase voltages between points 1 and 4 and points 1 and 3. This would be a voltage in excess of 300 volts! Although the "wild-phase" situation exists, this system is capable of supplying both high-power loads and low-voltage loads, such as for lighting and small equipment.

The most widely used three-phase power distribution system is the three-phase four-wire system. This system, shown in Figure 9-9c, commonly supplies 120/208 volts and 277/480 volts for industrial or commercial load requirements. The 120/208-volt system is illustrated here. From neutral to any hot line, 120 volts for lighting and low-power loads may be obtained. Across any two hot lines, 208 volts is present for supplying motors or other high-power loads. The most popular system for industrial and commercial power distribution is the 277/408-volt system which is capable of supplying both three-phase and single-phase loads. A 240/416-volt system is sometimes used for industrial loads, while the 120/208-volt system is often used for underground distribution in urban areas. Note that this system is based on the voltage characteristics of the three-phase wye connection and that the relationship $V_L = V_P \times 1.73$ exists for each application of this system.

GROUNDING OF DISTRIBUTION SYSTEMS

The concept of grounding in an electrical distribution system is very important. Distribution systems must have continuous uninterrupted grounds. If a grounded conductor is opened the ground is no longer functional. An open-ground condition could present severe safety problems and cause abnormal system operation.

Distribution systems must be grounded at substations and at the end of the power lines—before the power is delivered to the load. Grounding is necessary at substations for the safety of the public and the power company maintenance personnel. Grounding also provides points for transformer neutral connections for equipment grounds. Safety and equipment grounds will be discussed in more detail later.

At substations all external metal parts must be grounded and all transformer circuit breaker and switch housings must be grounded. Also metal fences and any other metal which is part of the substation construction must be grounded. Grounding assures that any person who touches any of the metal parts will not receive a high-voltage shock. Therefore if a high-voltage line were to come in contact with any of the grounded parts the system would be opened by protective equipment. Thus the danger of high voltages at substations is substantially reduced by grounding. The actual ground connection is made by welding brazing or bolting a conductor to a metal rod or bar which is then physically placed in the earth. This rod device is called a grounding electrode. Proper grounding techniques are required for safety as well as circuit performance. There are two types of grounding: (1) system grounding and (2) equipment grounding. Another important grounding factor is ground-fault protective equipment.

SYSTEM GROUNDING

System grounding involves the actual grounding of a current-carrying conductor (usually called the neutral) of a power distribution system. Three-phase systems may be either the wye or delta type. The wye system has an obvious advantage over the delta system since one side of

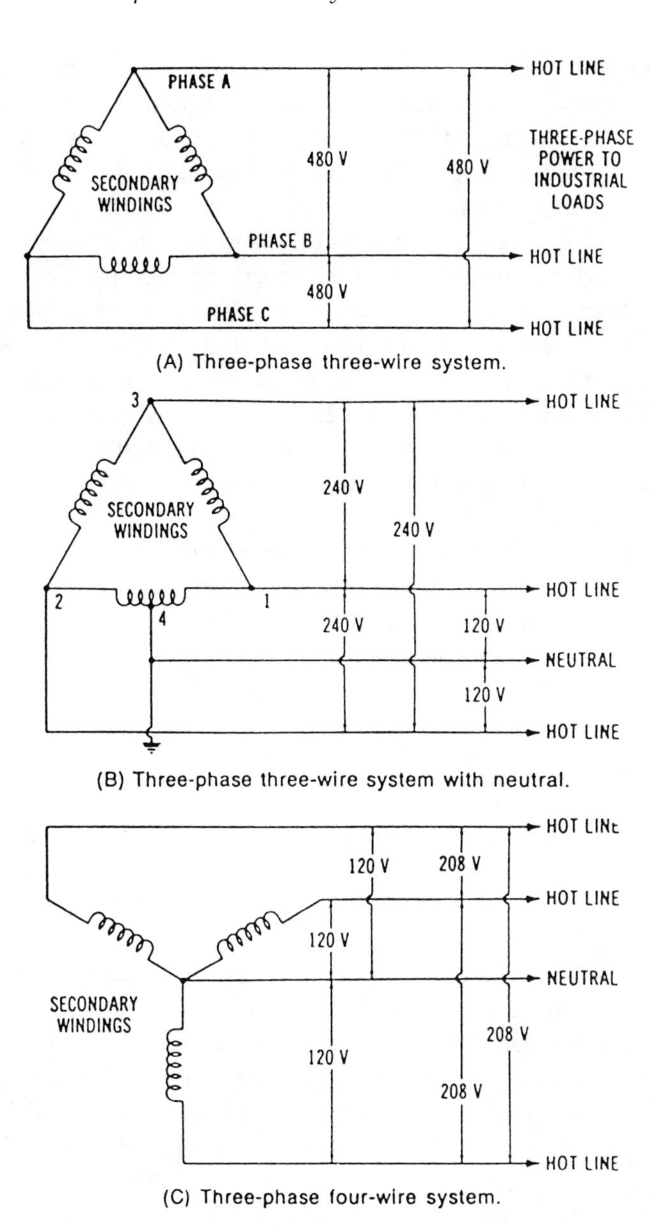

Figure 9-9. Industrial power distribution systems.

each phase winding is connected to ground. We will define a ground as a reference point of zero-volt potential, which is usually an actual connection to earth ground. The common terminals of the wye system, when connected to ground, become the neutral conductor of the three-phase four-wire system.

The delta system does not readily lend itself to grounding since it does not have a common neutral. The problem of ground faults (line-to-ground shorts) occurring in ungrounded delta systems is much greater than in wye systems. A common method of grounding a delta system is to use a wye-delta transformer connection and ground the common terminals of the wye-connected primary. However, the wye system is now used more often for industrial and commercial distribution since the secondary is easily grounded and it provides overvoltage protection from lightning or line-to-ground shorts.

Single-phase 120/240- or 240/480-volt systems are grounded in a manner similar to a three-phase ground. The neutral of the single-phase three-wire system is grounded by a metal rod (grounding electrode) driven into the earth at the transformer location. System grounding conductors are insulated with white or gray material for easy identification.

EQUIPMENT GROUNDING

The second type of ground is the equipment ground which, as the term implies, places operating equipment at ground potential. The conductor which is used for this purpose is either bare wire or a green insulated wire. The *National Electrical Code* describes conditions which require fixed-electrical equipment to be grounded. Usually all fixed electrical equipment located in industrial plants or commercial buildings should be grounded. Types of equipment that should be grounded include enclosures for switching and protective equipment for load control, transformer enclosures, electric motor frames, and fixed electronic test equipment. Industrial plants should use 120-volt single-phase duplex receptacles of the grounded type for all portable tools. The grounding of these receptacles may be checked by using a plug-in tester.

GROUND-FAULT PROTECTION

Ground-fault circuit interrupters (GFCIs), such as the one shown in Figure 9-10 are now used extensively in industrial, commercial; and resi-

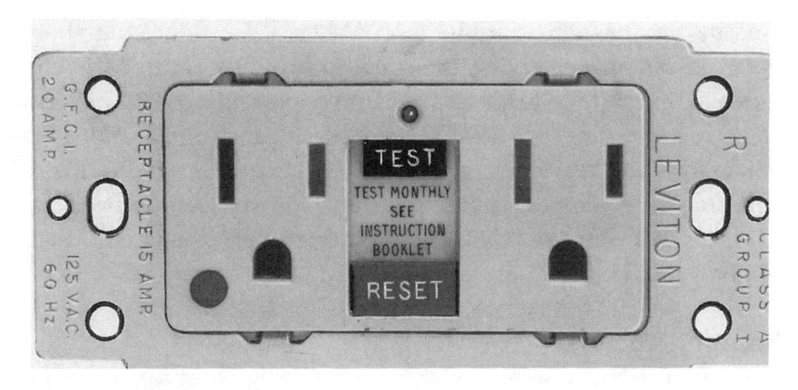

Figure 9-10. Ground fault circuit interrupter (GFCI)—mounts in a duplex power outlet to allow power cords to be plugged in (Courtesy Leviton Co.).

dential power distribution systems. It is required by the *National Electrical Code* that all 120-volt single-phase 15- or 20-ampere receptacle outlets which are installed outdoors or in bathrooms have ground-fault interrupters connected to them.

GFCI Operation

These devices are designed to eliminate electrical shock hazards resulting from individuals coming in contact with a hot AC line (line-to-ground short). The circuit interrupter is designed to sense any change in circuit conditions, such as would occur when a line-to-ground short exists. One type of GFCI has control wires which extend through a magnetic toroidal loop (see Figure 9-11). Ordinarily, the AC current flowing through the conductors inside the loop is equal in magnitude and opposite in direction. Any change in this equal and opposite condition is

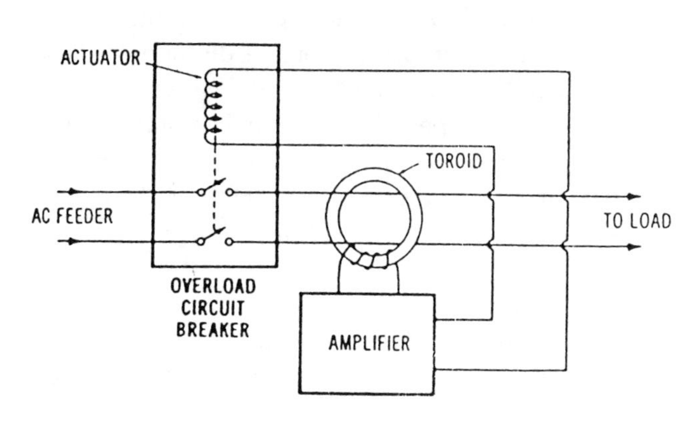

Figure 9-11. Simplified schematic of a ground-fault circuit interrupter.

sensed by the magnetic toroidal loop. When a line-to-ground short occurs, an instantaneous change in circuit conditions occurs. The change causes a magnetic field to be induced into the toroidal loop. The induced current is amplified to a level sufficient to cause the circuit-breaker mechanism to open. Thus, any line-to-ground short will cause the ground-fault circuit interrupter to open. The operating speed of the GFGI is so fast, since only a minute current opens the circuit, that the shock hazard to individuals is greatly reduced.

GFCI Applications

Construction sites, where temporary wiring is set up, are required to use GFCI for the protection of workers using electrical equipment. Ground-fault protection of individuals and commercial equipment must be provided for wye-connected systems of 150 to 600 volts for each distribution panelboard rated at over 1000 amperes. In this situation, the GFCI will open all ungrounded conductors at the panelboard when a line-to-ground short occurs. Now, GFCIs are used for all types of residential, commercial, and industrial applications.

Types of Ground-Fault Protection Systems

There are four basic types of ground-fault protection systems in use today. They are hospital applications, residential applications, motor protection applications, and specific electrical power distribution system applications. These ground-fault systems can be classified either by what they are to protect or the type of protection they are to provide. Hospital applications and residential applications are designed to protect people from excessive shock. The motor and electrical power applications are for protecting electrical equipment.

Another classification method is according to the amount of current required before an alarm system sounds or the disconnect of an electrical circuit occurs. Typical current values that will cause alarms or disconnects to activate are 0.002 amperes (2 mA) for hospital applications, 0.005 amperes (5 mA) for residential applications, 5 to 100 amperes for motor-protective circuit applications, and 200 to 1200 amperes for electrical power distribution equipment applications.

Need for Ground-fault Protection

In order to understand the need for a ground-fault circuit interrupter (for the protection of people), certain basic facts must first be understood. These facts relate to people as well as to ground faults.

One important fact is that a person's body resistance varies with the amount of moisture present on the skin, the muscular structure of the body, and the voltage to which the body is subjected. Experiments have shown that the body resistance from one hand to the other hand is somewhere between 1000 and 4000 ohms. These estimates are based upon several assumptions concerning moisture and muscular structure. We also know that resistance of the body (hand-to-hand) is lower for higher voltages. This is because higher voltages have the capability of "breaking down" the outer layers of the skin. Thus, higher voltages are more dangerous.

We can use Ohm's law to estimate that the typical current resulting from the average body resistance (from hand to hand) is about 115 mA at 240 volts AC and about 40 mA at 120 volts AC. The effects of a 60-Hz alternating current on the human body are generally accepted to be as given in Table 9-1.

Ventricular fibrillation is a contraction of the heart. Once ventricular fibrillation occurs, it will continue and death will occur within a few minutes. Resuscitation techniques, when applied immediately, could save a victim. Deaths caused by electrical shock account for a high percentage of the deaths which occur in the home and in industry. Many of these deaths are due to contact with low-voltage circuits (600 volts and under), mainly 120- and 240-volt systems.

Ground-Fault Protection for the Home

Ground-fault circuit interruptors for homes consist of three types: (1) circuit breaker, (2) receptacle, and (3) plug-in types. Ground-fault protection devices are constructed according to standards developed by the Underwriter's Laboratories. The GFCI circuit breakers combine ground-fault protection and circuit interruption in the same overcurrent and short-circuit protective,, equipment as does a standard circuit breaker. A GFCI circuit breaker fits the same space required by a standard circuit breaker. It provides the same branch-circuit wiring protection as the standard circuit breaker, as well as ground-fault protection. The GFCI sensing system continuously monitors the current balance in the ungrounded (hot) conductor and the grounded (neutral conductor. The current in the neutral wire becomes less than the current in the hot wire when a ground fault develops. This means that a portion of the circuit current is returning to ground by some means other than the neu-

Table 9-1. Body reaction to alternating current.

Amount of Current	Effect on Body
1 mA or less	No sensation (not felt).
More than 5 mA	Painful shock
More than 10 mA	Muscle contractions; could cause a "freezing" to the electrical circuit for some people.
More than 15 mA	Muscle contractions; could cause "freezing" to the electrical circuit for most people.
More than 30 mA	Breathing difficult; could cause unconsciousness.
50 to 100 mA	Ventricular fibrillation of the heart is possible.
100 to 200 mA	Ventricular fibrillation of the heart is certain.
Over 200 mA	Severe burns and muscular contractions; the heart is more apt to stop beating than to fibrillate.
1 ampere and above	Permanent damage to body tissues.

tral wire. When an imbalance in current occurs, the sensor (a differential current transformer) sends a signal to the solid-state circuitry which activates a trip mechanism. This action opens the hot line. A differential current as low as 5 mA will cause the sensor to send a fault signal and cause the circuit breaker to interrupt the circuit.

Ordinarily, GFCI receptacles provide ground-fault protection on 120-, 208-, or 240-volt AC systems. The GFCI receptacles come in 15- and 20-ampere designs. The 15-ampere unit has a receptacle configuration for use with 15-ampere plugs only. The 20-ampere device has a receptacle configuration for use with either 15- or 20-ampere plugs. These GFCI receptacles have connections for hot, neutral, and ground wires.

All GFCI receptacles have a two-pole tripping mechanism which breaks both the hot and the neutral load connections at the time a fault occurs. The GFCI shown in Figure 9-10 is this type.

The plug-in GFCI receptacles provide protection by plugging into a standard wall receptacle. Some manufacturers provide units which will fit either two-wire or three-wire receptacles. The major advantage of this type of unit is that it can be moved from one location to another.

Ground-Fault Protection for Power Distribution Equipment

Ground faults can destroy electrical equipment if allowed to continue. Phase-to-phase short circuits and some types of ground faults are usually high current. Normally, they are adequately handled by conventional overcurrent protective equipment. However, some ground faults produce an arcing effect from relatively low currents that are not large enough to trip conventional protective devices. Arcs can severely burn electrical equipment. A 480- or 600-volt system is more susceptible to arcing damage than a 120-, 208-, or 240-volt system due to the higher voltages which sustain the arcing effect. High-current faults are quickly detected by conventional overcurrent devices. Low-current values must be detested by GFCI.

Ground faults which cause an arcing in the equipment are probably the most frequent faults. They may result from damaged or deteriorated insulation, dirt, moisture, or improper connections. They usually occur between one hot conductor and the grounded equipment enclosure, conduit, or metal housing. The line-to-neutral voltage of the source will cause current to flow in the hot conductor, through the are path, and back through the ground path. The impedance of the conductor and the ground-return path (enclosure, conduit, or housing) depends on many factors. As a result, the fault-current value cannot be predicted. It can also increase or decrease as the fault condition continues.

It is apparent that many factors influence the magnitude, duration, and the effect of an arcing ground fault. Some conditions produce a large amount of fault current, while others limit the fault current to a relatively small amount. Damage to equipment that is caused by are-current magnitude and the time that the are persists can be very great. Probably the more important factor is the time period of the arcing voltage since the larger the arcing time, the more chance that the arcs will spread to different areas within the equipment.

Ground-current relaying is one method used to protect equipment from ground faults. Current- flows through a load or fault along the hot and neutral conductors and returns to the source on these conductors and, to some extent, along the ground path. The normal ground path current is very small. Therefore, essentially all the current flowing from the source is also returning on the same hot line and neutral conductors. However, if a ground fault occurs, the ground current will increase to the point where some current will escape through the fault and return via the ground path. As a result, the current returning on the hot and neutral conductors is less than the amount going out. The difference is an indication of the amount of current in the ground path. A relay, which senses this difference in currents, can act as a ground-fault protective device.

Ground-fault Protection for Electric Motors

Motor protective systems offer protection in the 5- to 100-ampere range. This type of ground-fault protective system offers a protection against ground faults in both the single-phase and the three-phase systems. Many insulation failures begin with a small leakage current which builds up with time until damage results. These ground-fault systems detect ground leakage currents while they are still small and, thus, prevent any extensive damage happening to the motors.

WIRING DESIGN CONSIDERATIONS FOR DISTRIBUTION SYSTEMS

The wiring design of electrical power distribution systems can be very complex. There are many factors which must be considered in the wiring design of a distribution system installed in a building. Wiring design standards are specified in the *National Electrical Code* (NEC) which is published by the National Fire Protection Association (NFPA). The NEC, local wiring standards, and electrical inspection policies should be considered in an electrical wiring design.

There are several distribution system wiring design considerations which are pointed out specifically in the NEC. We will deal with voltage-drop calculations, branch-circuit design, feeder-circuit design, and the design for grounding systems in this chapter.

NATIONAL ELECTRICAL CODE (NEC) USE

The *National Electrical Code* sets forth the minimum standards for electrical wiring in the United States. The standards contained in the NEC are enforced by being incorporated into the different city, and community ordinances which deal with electrical wiring in residences, industrial plants, and commercial buildings. Therefore, these local ordinances conform to the standards set forth in the NEC.

In most areas of the United States, a license must be obtained by any individual who does electrical wiring. Usually a test administered by the city, county, or state must be passed in order to obtain this license. These tests are based on local ordinances and the NEC. The rules for electrical wiring that are established by the local electrical power company are also sometimes incorporated into the license test.

ELECTRICAL INSPECTIONS

When new buildings are constructed, they must be inspected to see if the electrical wiring meets the standards of the local ordinances, the NEC, and the local power company. The organization which supplies the electrical inspectors varies from one locality to another. Ordinarily, the local power company can advise individuals about whom to contact for information about electrical inspections.

VOLTAGE DROP IN ELECTRICAL CONDUCTORS

Although the resistance of electrical conductors is very low, a long length of wire could cause a substantial voltage drop. This is illustrated in Figure 9-12. Remember that a voltage drop is current times resistance ($I \times R$). Therefore, whenever current flows through a system, a voltage drop is created. Ideally, the voltage drop caused by the resistance of a conductor will be very small.

However, a longer section of electrical conductor has a higher resistance. Therefore, it is sometimes necessary to limit the distance a conductor can extend from the power source to the load which it supplies. Many types of loads do not operate properly when a value less than the full source voltage is available.

You can also see from Figure 9-12 that as the voltage drop (V_D) increases, the voltage applied to the load (V_L) decreases. As current in the system increases, V_D increases causing V_L to decrease since the source voltage stays the same.

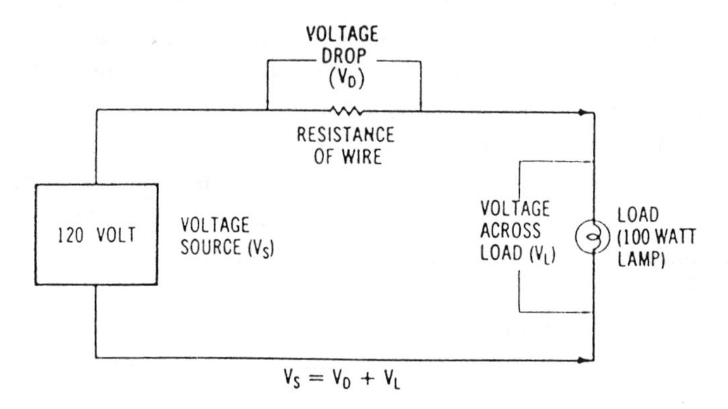

Figure 9-12. Voltage drop in an electrical circuit.

VOLTAGE-DROP CALCULATIONS USING CONDUCTOR TABLE

It is important when dealing with electrical wiring design to be able to determine the amount of voltage drop caused by conductor resistance. Table 9-2 is used to make these calculations. The *National Electrical Code* limits the amount of voltage drop that a system can have. This means that long runs of conductors must ordinarily be avoided. Remember that a conductor with a large cross-sectional area will cause a smaller voltage drop since its resistance is smaller.

To better understand how to determine the size of conductor required to limit the voltage drop in a system, we will look at a sample problem.

1. *Given*—A 200-ampere load located 400 feet from a 240-volt single-phase source. Limit the voltage drop to 2% of the source.

2. *Find*—The size of an RH copper conductor needed to limit the voltage drop of the system.

3. *Solution:*

Table 9-2. Sizes of copper and aluminum conductors.

Size (AWG or MCM)	$A = D^2$ Area (cmil)	Number of Wires	Diameter of Each Wire (in.)	DC Resistance (Ω/1000 ft) 25°C Copper	Aluminum
18	1,620	1	0.0403	6.51	10.7
16	2,580	1	0.0508	4.10	6.72
14	4,110	1	0.0641	2.57	4.22
12	6,530	1	0.0808	1.62	2.66
10	10,380	1	0.1019	1.018	1.67
8	16,510	1	0.1285	0.6404	1.05
6	26,240	7	0.0612	0.410	0.674
4	41,740	7	0.0772	0.259	0.424
3	52,620	7	0.0867	0.205	0.336
2	66,360	7	0.0974	0.162	0.266
1	83,690	19	0.0664	0.129	0.211
0	105,600	19	0.0745	0.102	0.168
00	133,100	19	0.0837	0.0811	0.133
000	167,800	19	0.0940	0.0642	0.105
0000	211,600	19	0.1055	0.0509	0.0836
250	250,000	37	0.0822	0.0431	0.0708
300	300,000	37	0.0900	0.0360	0.0590
350	350,000	37	0.0973	0.0308	0.0505
400	400,000	37	0.1040	0.0270	0.0442
500	500,000	37	0.1162	0.0216	0.0354
600	600,000	61	0.0992	0.0180	0.0295
700	700,000	61	0.1071	0.0154	0.0253
750	750,000	61	0.1109	0.0144	0.0236
800	800,000	61	0.1145	0.0135	0.0221
900	900,000	61	0.1215	0.0120	0.0197
1000	1,000,000	61	0.1280	0.0108	0.0177

a. The allowable voltage drop equals 240 volts times 0.02 (2%). This equals 4.8 volts.
b. Determine the maximum resistance for 800 feet This is the equivalent of 400 feet × 2, since there are two current-carrying conductors for a single-phase system.

$$R = \frac{V_D}{I}$$

$$= \frac{4.8\ V}{200\ A}$$

$$= 0.024 \text{ ohm, resistance for 800 feet.}$$

c. Determine the maximum resistance for 1000 feet (304.8 meters) of conductor.

$$\frac{800 \text{ feet}}{1000 \text{ feet}} = \frac{0.024 \text{ ohm}}{R}$$

$$800\,R = (1000)\ (0.024)$$

$$R = 0.030 \text{ ohm}$$

d. Use Table 9-2 to find the size of copper conductor which has the nearest DC resistance (ohms per 1000 feet) value that is equal to or less than the value calculated in 3c, above. The conductor chosen is conductor size 350 MCM, RH Copper.
e. Check this conductor with the proper ampacity table to assure that it is large enough to carry 200 amperes. Table 9-3 shows that a 350 MCM, RH copper conductor will handle 310 amperes of current; therefore, use 350 MCM conductors. (Always remember to use the largest conductor if Steps 3d and 3e produce conflicting values.)
f. If the current is larger than listed on the tables, use more than one conductor of the same size for design calculations.

Table 9-3. Ampacities of conductors in a raceway or cable (3 or less).

	Copper		Aluminum	
Wire Size	With R, T, TW Insulation	With RH, RHW, TH, THW Insulation	With R, T, TW Insulation	With RH, RHW, TH, THW Insulation
14	15	15		
12	20	20	15	15
10	30	30	25	25
8	40	45	30	40
6	55	65	40	50
4	70	85	55	65
3	80	100	65	75
2	95	115	75	90
1	110	130	85	100
0	125	150	100	120
00	145	175	115	135
000	165	200	130	155
0000	195	230	155	180
250	215	255	170	205
300	240	285	190	230
350	260	310	210	250
400	280	335	225	270
500	320	380	260	310
600	355	420	285	340
700	385	460	310	375
750	400	475	320	385
800	410	490	330	395
900	435	520	355	425
1000	455	545	375	445

ALTERNATIVE METHOD OF VOLTAGE-DROP CALCULATION

In some cases, an easier method to determine the conductor size for limiting the voltage drop is to use one of the following formulas to find the cross-sectional (cmil) area of the conductor.

$$\text{cmil} = \frac{\rho \times I \times 2d}{V_D}$$

or,

$$\text{cmil} = \frac{\rho \times I \times 1.73d}{V_D}$$

where,

ρ is the resistivity (copper - 10.4, aluminum = 17.0),
I is the load current in amperes,
V_D is the allowable voltage drop in volts,
d is the distance from source to load in feet.

The sample problem given for a single-phase system in the preceding section could be set up as follows:

$$\text{cmil} = \frac{\rho \times I \times 1.73d}{V_D}$$

$$= \frac{10.4 \times 200 \times 2 \times 400}{240 \times 0.02}$$

$$= \frac{1{,}664{,}000}{4.8}$$

$$= 346{,}666$$

$$= 347 \text{ MCM.}$$

The next largest size is a 350 MCM conductor.

BRANCH-CIRCUIT DESIGN CONSIDERATIONS

A branch circuit is defined as a circuit which extends from the last overcurrent protective device of the power system. Branch circuits, according to the *National Electrical Code,* are either 15, 20, 30, 40, or 50 amperes in capacity (see Tables 9-4 and 9-5). Loads larger than 50 amperes would not be connected to a branch circuit.

Table 9-4. Branch circuit wire sizes.

Branch Circuit Current	Wire Size (AWG)
15 Amps	No. 14
20 Amps	No. 12
30 Amps	No. 10
40 Amps	No. 8
50 Amps	No. 6

Table 9-5. Branch circuit current values.

Type of Branch Circuit	Branch Circuit Amps
General lighting	15 or 20
Small appliance	20
Clothes washer	20
Non-motor appliance	Current rating (100%)
Motor appliance	125% of current rating

There are many rules in the *National Electrical Code* (NEC) which apply to branch-circuit design. The following information is based on the NEC. First, each circuit must be designed so that accidental short circuits or grounds do not cause damage to any part of the system. Then, fuses or circuit breakers are to be used as branch-circuit overcurrent protective devices. Should a short circuit or ground condition occur, the protective device should open and intermit the flow of current in the branch circuit. One important NEC rule is that No. 16 or No. 18 (extension cord) wire may be tapped from No. 12 or No; 14 conductors but not larger than No. 12. This means that an extension cord of No. 16 wire should not be plugged into a receptacle which uses No. 10 wire. Damage to smaller wires (due to the heating effect) before the overcurrent device can open is eliminated by applying this rule. Lighting circuits are one of the most common types of branch circuits. They are usually either 15-ampere or 20-ampere circuits.

The maximum rating of an individual load (such as a portable appliance connected to a branch circuit) is 80% of the branch-circuit current rating. So, a 20-ampere circuit could not have a single load which draws more than 16 amperes. If the load is a permanently connected appliance, its current rating cannot be more than 50% of the branch-circuit capacity if portable appliances or lights are connected to the same circuit.

VOLTAGE DROP IN BRANCH CIRCUITS

Branch circuits must be designed so that sufficient voltage is supplied to all parts of the circuit. The distance that a branch circuit can extend from the voltage source or power distribution panel is, therefore, limited. A voltage drop of 3% is specified by the *National Electrical Code* as the maximum allowed for branch circuits in electrical wiring design.

The method for calculating the voltage drop in a branch circuit is a step-by-step process which is illustrated by the following problem. Refer to the circuit diagram given in Figure 9-13.

1. *Given*—A 120-volt 15-ampere branch circuit supplies a load which consists of four lamps. Each lamp draws 3 amperes of current from the source. The lamps are located at 10-foot intervals from the power distribution panel.

2. *Find*—The voltage across lamp No 4.

3. *Solution:*
   a. Find the resistance for 20 feet (6.1 meters) of conductor (same as 10-foot conductor × 2). A No. 14 copper wire is used for 15-ampere branch circuits. From Table 9-2 we find that the resistance of 1000 feet of No. 14 copper wire is 2.57 ohms. There-

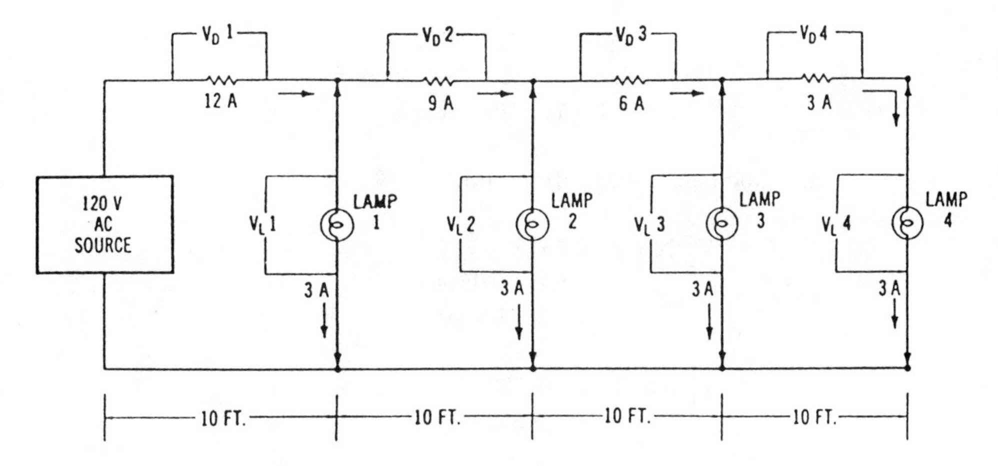

Figure 9-13. Circuit for calculating the voltage drop in a branch circuit.

fore, the resistance of 20 feet of wire is:

$$\frac{20\ \text{feet}}{1000\ \text{feet}} = \frac{R}{257\ \text{ohm}}$$

$$1000\,R = (20)\ (2.57)$$

$$R = 0.0514\ \text{ohm}$$

b. Calculate voltage drop V_D No. 1. (R equals the resistance of 20 feet of wire.)

$$\begin{aligned} V_D\ \text{No. 1} &= I \times R \\ &= 12\ A \times 0.0514\ \Omega \\ &= 0.6168\ V\ AC. \end{aligned}$$

Calculate load V_L No. 1. (The source voltage minus V_D No. 1.)

$$\begin{aligned} V_L\ \text{No. 1} &= 120V—0.6168V \\ &= 119.383\ \text{volts}. \end{aligned}$$

c. Calculate voltage drop and load No. 2.

$$\begin{aligned} V_D\ \text{No. 2} &= I \times R \\ &= 9A \times 0.0514\ \Omega \end{aligned}$$

$$= 0.4626 \text{ V AC},$$
$$V_L \text{ No. 2} = 119.383 \text{ V—}0.4626 \text{ V}$$
$$= 118.920 \text{ volts.}$$

d. Calculate voltage drop and load No. 3.

$$V_D \text{ No. 3} = I \times R$$
$$= 6\text{A} \times 0.0514 \ \Omega$$
$$= 0.3084 \text{ V AC},$$

$$V_L \text{ No. 3} = 118.920 \text{ V—}0.3084 \text{ V}$$
$$= 118.612 \text{ volts.}$$

e. Calculate voltage drop and load No. 4.

$$V_D \text{ No. 4} = I \times R$$
$$= 3 \text{ A} \times 0.0514 \ \Omega$$
$$=0.1542 \text{ V AC},$$
$$V_L \text{ No. 4} = 118.612 \text{ V—}0.1542 \text{ V}$$
$$= 118.458 \text{ volts.}$$

Notice that the voltage across lamp No. 4 is substantially reduced from the 120-volt source value due to the voltage drop of the conductors. Also, notice that the resistances used to calculate the voltage drops represented both wires (hot and neutral) of the branch circuit. Ordinarily, 120-volt branch circuits do not extend more than 100 feet (30.48 meters) from the power distribution panel. The preferred distance is 75 feet (22.86 meters). The voltage drop in branch-circuit conductors can be reduced by making the circuit shorter in length or by using larger conductors.

In residential electrical wiring design, the voltage drop in many branch circuits is difficult to calculate since the lighting and portable appliance receptacles are placed on the same branch circuits. Since portable appliances and "plug-in" lights are not used all of the time, the voltage drop will vary according to the number of lights and appliances in use. This problem is usually not encountered in an industrial or commercial wiring design for lights, since the lighting units are usually larger and are permanently installed on the branch circuits.

BRANCH-CIRCUIT WIRING

A branch circuit usually consists of a nonmetallic-sheathed cable which is connected into a power distribution panel. Each branch circuit that is wired from the power distribution panel is protected by a fuse or circuit breaker. The power panel also has a main switch which controls all of the branch circuits that are connected to it.

Single-phase Branch Circuits

A diagram of a single-phase three-wire (120/240 volt) power distributor panel is shown in Figure 9-14. Notice that eight 120-volt branch circuits and one 240-volt circuit are available from the power panel. This type of system is used in most homes where several 120-volt branch circuits and, typically, three or four 240-volt branch circuits are required. Notice in Figure 9-14 that each hot line has a circuit breaker while the neutral line connects directly to the branch circuits. Neutrals should never be opened (fused). This is a safety precaution in electrical wiring design.

Three-phase Branch Circuits

A diagram for a three-phase four-wire (120/208 volt) power distribution panel is shown in Figure 9-15. There are three single-phase 120-volt branch circuits and two 208-volt three-phase branch circuits shown. The single-phase branches are balanced (one hot line from each branch). Each hot line has an individual circuit breaker. Three-phase lines should be connected so that an overload in the branch circuit will cause all three lines to open. This is accomplished by using a three-phase circuit breaker which is arranged internally as shown in Figure 9-16.

FEEDER-CIRCUIT DESIGN CONSIDERATIONS

Feeder circuits are used to distribute electrical power to power distribution panels, as shown in Figure 9-17. Many feeder circuits extend for very long distances; therefore, voltage drop must be considered in feeder-circuit design. In higher voltage feeder circuits, the voltage drop is reduced. However, many lower voltage feeder circuits require large-diameter conductors to provide a tolerable level of voltage drop. High-current feeder circuits also present a problem in terms of the massive

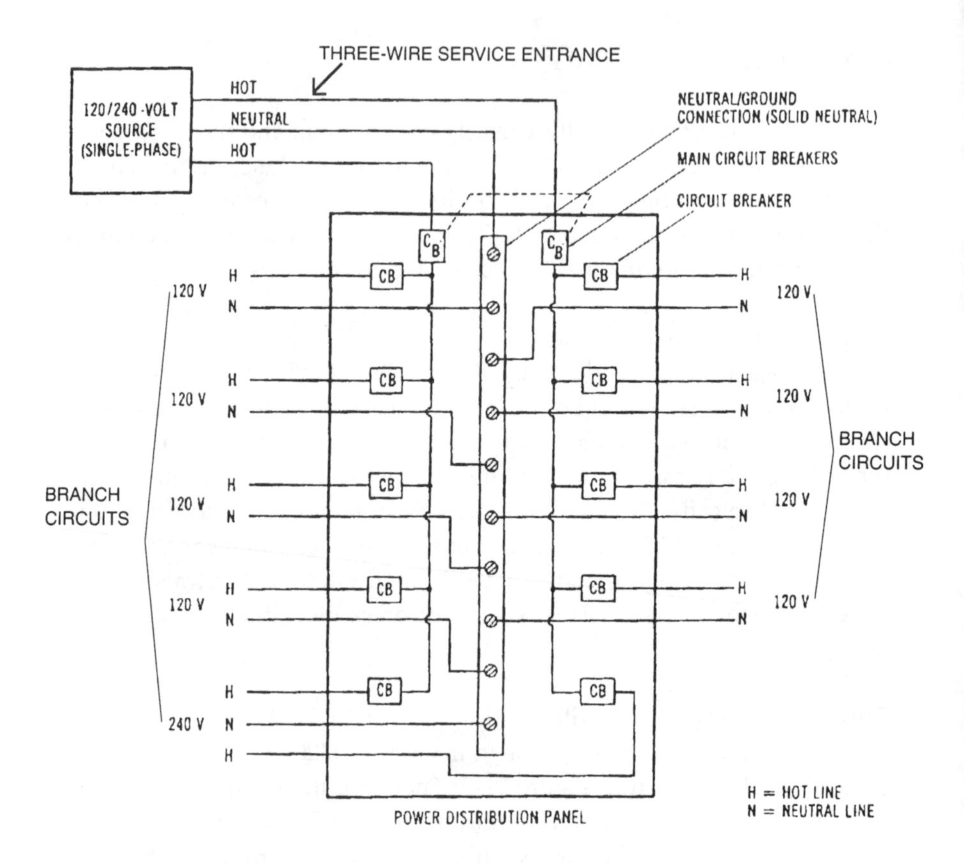

Figure 9-14. Schematic of a power distribution panel for a single-phase, three-wire branch circuit.

overload protection which is sometimes required. This protection is usually provided by system switchgear or load centers where the feeder circuits originate.

Determining the Size of Feeder Circuits

The amount of current which a feeder circuit must be designed to carry depends upon the actual load demanded by the branch-circuit power distribution panels which it supplies. Each power distribution panel will have a separate feeder circuit. Also, each feeder circuit must have its own overload protection.

The following problem is an example of size calculation for a feeder circuit.

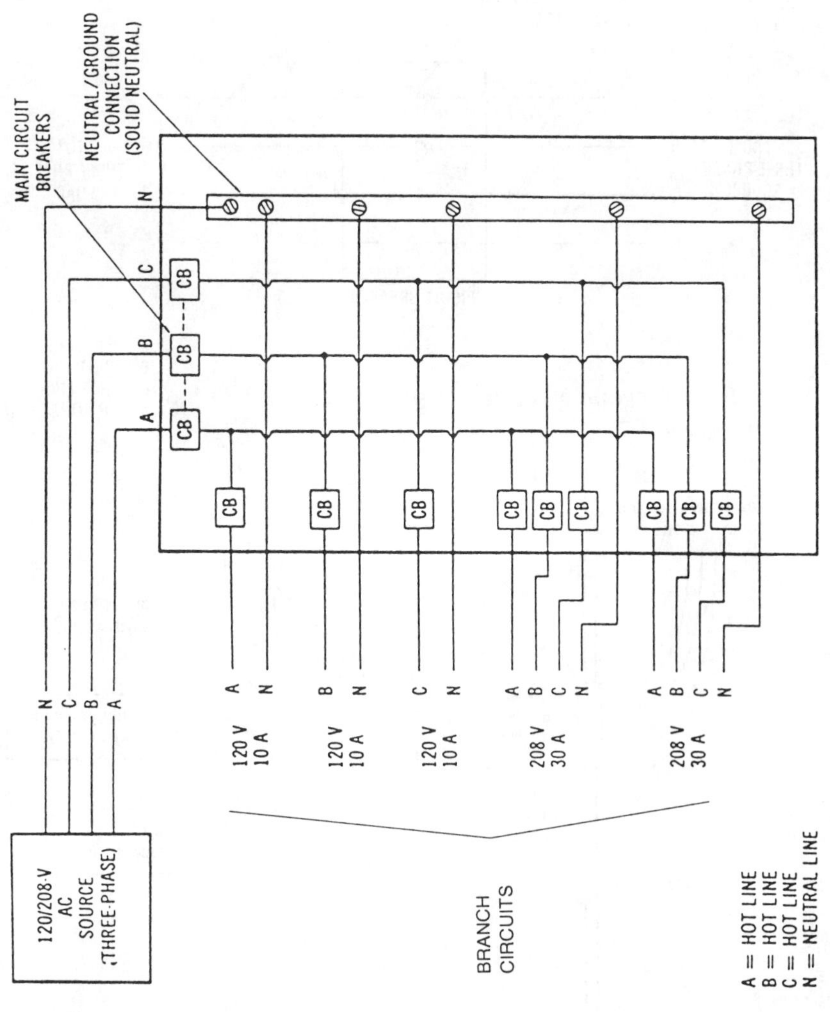

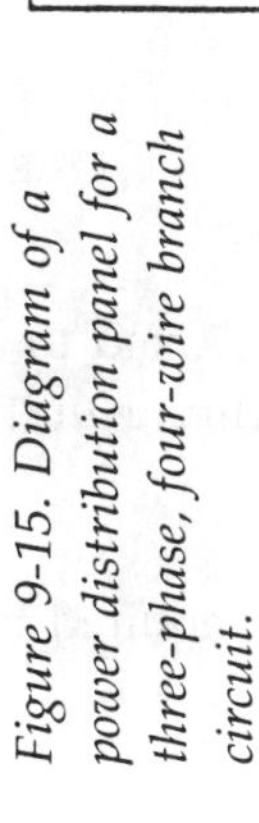

Figure 9-15. Diagram of a power distribution panel for a three-phase, four-wire branch circuit.

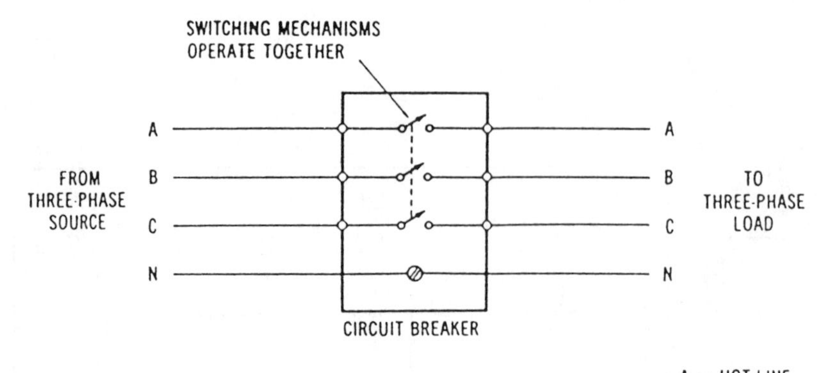

Figure 9-16. Diagram of a three-phase circuit breaker.

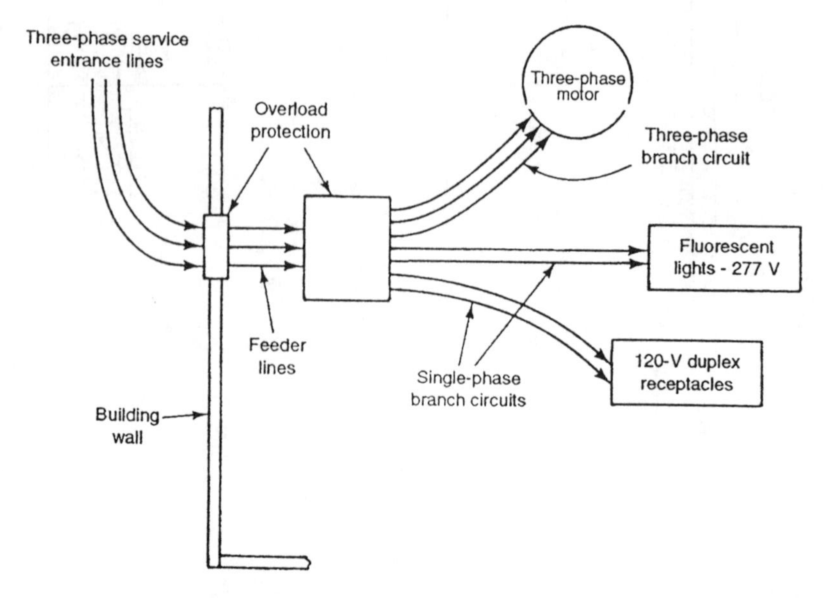

Figure 9-17. Relationship of feeders and branch circuits.

1. *Given*—Three 15-kW fluorescent lighting units are connected to a three-phase four-wire (277/480-volt) system. The lighting units have a power factor of 0.8.

2. *Find*—The size of THW aluminum feeder conductors required to supply this load.

3. *Solution:*

   a. Find the line current:

$$I_L = \frac{P_T}{1.73 \times V_L \times pf}$$

$$= \frac{45{,}000 \text{ watts}}{1.73 \times 480 \text{ volts} \times 0.8}$$

$$= 67.74 \text{ Amps}$$

   b. From Table 9-3, we find that the conductor size which will carry 67.74 amperes of current is a No. 3 AWG THW aluminum conductor.

Voltage Drop Calculation for Feeder Circuits

Feeder-circuit design must take the conductor voltage drop into consideration. The voltage drop in a feeder circuit must be kept as low as possible so that maximum power can be delivered to the loads connected to the feeder system. The NEC allows a maximum 5% voltage drop in the combination of a branch and a feeder circuit; however, a 5% voltage reduction represents a significant power loss in a circuit. We can calculate power loss due to voltage drop as V^2/R, where V^2 is the voltage drop of the circuit and R is the resistance of the conductors of the circuit.

The calculation of feeder conductor size is similar to that for a branch-circuit voltage drop. The size of the conductors must be large enough to: (1) have the required ampacity, and (2) keep the voltage drop below a specified level. If the second requirement is not met, possibly due to a long feeder circuit, the conductors chosen must be larger than the ampacity rating requires.

The following problem illustrates the calculation of feeder conductor size based upon the voltage drop in a single-phase circuit.

1. *Given*—A single-phase 240-volt load in a factory is rated at 85 kilowatts. The feeders (two hot lines) will be 260-foot lengths of RHW copper conductor. The maximum conductor voltage drop allowed is 2%.

2. *Find*—The feeder conductor size required.

3. *Solution:*

   a. Find the maximum voltage drop of the circuit.

$$V_P = \% \times \text{Load} = 0.02 \times 240 = 4.8 \text{ volts}$$

   b. Find the current drawn by the load.

$$I = \frac{\text{Power}}{\text{Voltage}} = \frac{85{,}000}{240} = 354.2 \text{ amperes}$$

   c. Find the minimum circular-mil conductor area required. Use the formula given for finding the cross-sectional area of a conductor in single-phase systems that was previously given in the "Alternative Method of Voltage-Drop Calculation" section.

$$\text{cmil} = \frac{\rho \times I \times 2d}{V_D} = \frac{10.4 \times 354.2 \times 2 \times 260}{4.8} = 399{,}065.33 \text{ cmil}$$

   d. Determine the feeder conductor size. The next larger size conductor in Table 9-2 is 400 MCM. Check Table 9-3 and you will see that a 400 MCM RHW copper conductor will carry 335 amperes. This is less than the required 354.2 amperes, so use the next larger size, which is a 500 MCM conductor.

The conductor size for a three-phase feeder circuit is determined in a similar way. In this problem, the feeder size will be determined as based upon the circuit voltage drop.

1. *Given*—A 480-volt three-phase three-wire (delta) feeder circuit supplies a 45-kilowatt balanced load to a commercial building. The load operates at a 0.75 power factor. The feeder circuit (three hot lines) will be a 300-foot length of RH copper conductor. The maximum voltage drop is 1%.

2. *Find*—The feeder size required (based on the voltage drop of the circuit).

3. *Solution:*

   a. Find the maximum voltage drop of the circuit.

$$V_D = 0.01 \times 480$$
$$= 4.8 \text{ volts}$$

   b. Find the line current drawn by the load.

$$I_L = \frac{P}{1.73 \times V \times pf}$$

$$= \frac{45{,}000}{1.73 \times 480 \times 0.75}$$

$$= 72.25 \text{ amperes}$$

   c. Find the minimum circular-mil conductor area required. Use the formula for finding cmil in three-phase systems that was given in an earlier section.

$$\text{cmil} = \frac{\rho \times I \times 1.73\,d}{V_D}$$

$$= \frac{10.4 \times 72.25 \times 1.73 \times 300}{4.8}$$

$$= 81.245 \text{ cmil}$$

d. Determine the feeder conductor size. The closest and next larger conductor size in Table 9-3 is No. 1 AWG. Check Table 9-3 and you will see that a No. 1 AWG RH copper conductor will carry 130 amperes, much more than the required 72.25 amperes. Therefore, use No. 1 AWG RH copper conductors for the feeder circuit.

DETERMINING GROUNDING CONDUCTOR SIZE

Grounding considerations in electrical wiring design were discussed previously. Another aspect in wiring design is to determine the size of the grounding conductor required in a circuit. All circuits which operate at 150 volts or less must be grounded; therefore, all residential electrical systems must be grounded. Higher voltage systems used in industrial and commercial buildings have grounding requirements that are specified by the NEC and by local codes. A ground at the service entrance of a building is usually a metal water pipe which extends uninterrupted underground or a grounding electrode that is driven into the ground near the service entrance.

The size of the grounding conductor is determined by the current rating of the system. Table 9-6 lists equipment grounding-conductor sizes for interior wiring while Table 9-7 lists the minimum grounding-conductor sizes for system grounding of service entrances. The sizes of grounding conductors listed in Table 9-6 are for equipment grounds which connect to raceways, enclosures, and metal frames for safety purposes. Note that a No. 12 or a No. 14 wiring cable, such as 12-2 WG NMC can have a No. 18 equipment ground. The ground is contained in the same cable sheathing as the hot conductors. Table 9-7 is used to find the minimum size of grounding conductors needed for service entrances, based upon the size of the hot line conductors used with the system.

PARTS OF INTERIOR ELECTRICAL WIRING SYSTEMS

Some parts of interior electrical distribution systems have been discussed previously. Such types of equipment as transformers, switchgear,

Table 9-6. Equipment grounding-conductor sizes for interior wiring.

Ampere Rating of Distribution Panel	Grounding Conductor Size (AWG)*	
	Copper	Aluminum
15	14	12
20	12	10
30	10	8
40	10	8
60	10	8
100	8	8
200	6	4
400	3	1
600	1	00
800	0	000
1000	00	0000

*No. 12 or No. 14 wiring cable can use a No. 18 equipment ground.

Table 9-7. System grounding-conductor sizes for service entrances

Conductor Size (Hot Line)		Grounding Conductor Size	
Copper	Aluminum	Copper (AWG)	Aluminum (AWG)
No. 2 AWG or smaller	No. 0 AWG or smaller	8	6
No. 1 or 0 AWG	No. 00 or 000 AWG	6	4
No. 00 or 000 AWG	No. 0000 AWG or 250 MCM	4	2
No. 000 AWG to 350 MCM	250 MCM to 500 MCM	2	0
350 MCM to 600 MCM	500 MCM to 900 MCM	0	000
600 MCM to 1100 MCM	900 MCM to 1750 MCM	00	0000

conductors, insulators, and protective equipment are parts of interior wiring systems. There are, however, certain parts of interior electrical distribution systems which are unique to the wiring system itself. These parts include the nonmetallic-sheathed cables (NMC), the metal-clad cables, the rigid conduit, and the electrical metallic tubing (EMT).

Nonmetallic-sheathed Cable (NMC)

Nonmetallic-sheathed cable is a common type of electrical cable used for interior wiring. NMC, sometimes referred to as *Romex* cable, is used in residential wiring systems almost exclusively. The most common type used is No. 12-2 WG, which is illustrated in Figure 9-18. This type of NMC comes in 250-foot rolls for interior wiring. The cable has a thin plastic outer covering with three conductors inside. The conductors have colored insulation which designates whether the conductor should be used as a hot, neutral, or equipment ground wire. For instance, the conductor connected to the hot side of the system has black or red insulation, while the neutral conductor has white or gray insulation. The equipment grounding conductor has either a green insulation or no-insulation (bare conductor). There are several different sizes of bushings and connectors used for the installation of NMC in buildings.

The designation No. 12-2 WG means that (1) the copper conductors used are No. 12 AWG as measured by an American Wire Gage (AWG), (2) there are two current-carrying conductors, and (3) the cable comes with a ground (WG) wire. A No. 14-3 WG cable, in comparison, would have three No. 14 conductors and a grounding conductor. NMC ranges in size from No. 14 to No. 1 AWG copper conductors and from No. 12 to No. 2 AWG aluminum conductors.

Metal-clad Cable

Metal-clad cable, shown in Figure 9-19, is similar to NMC except that it has a flexible spiral metal covering rather than a plastic covering. A common type of metal-clad cable is called *BX cable.* Like NMC, BX cable contains two or three conductors. There are also several sizes of connectors and bushings used in the installation of BX cable. The primary advantage of this type of metal-clad cable is that it is contained in a metal enclosure that is flexible so that it can be bent easily. Other metal enclosures are usually more difficult to bend.

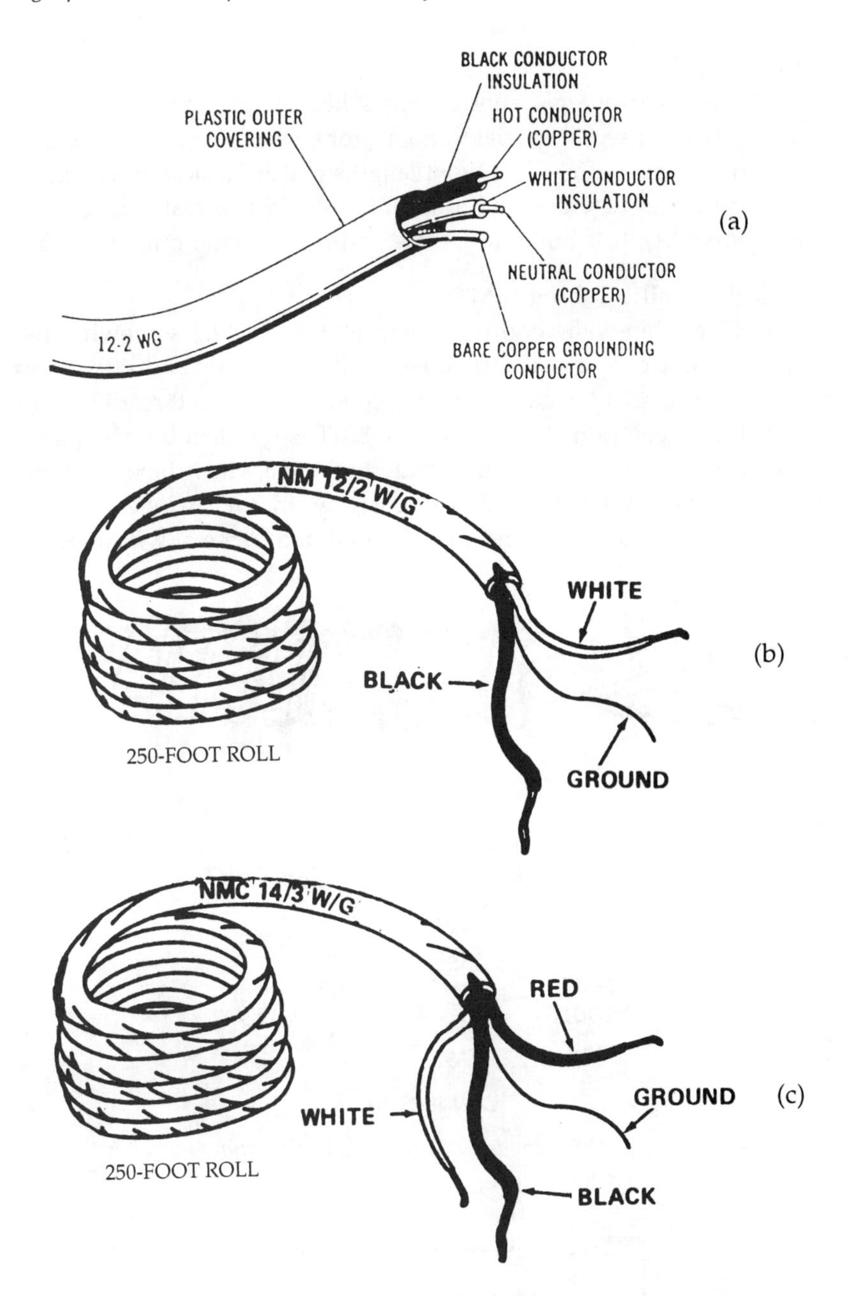

Figure 9-18. Nonmetallic-sheathed cable: (a) detail; (b) roll of NM 12-2 W/G; (c) roll of NM 14-3 W/G.

Rigid Conduit

The exterior of rigid conduit looks like water pipe, as shown in Figure 9-20. It is used in special locations for enclosing electrical conductors. Rigid conduit comes in 10-foot lengths which must be threaded for joining the pieces together. The conduit is secured to metal wiring boxes (see Figure 9-21). It is bulky to handle and takes a long time to install.

Electrical Metallic Tubing (EMT)

EMT or thin-wall conduit, shown in Figure 9-22, somewhat like rigid conduit except that it can be bent with a special conduit-bending tool. EMT is easier to install than rigid conduit since no threading is required. It also comes in 10-foot lengths. EMT is installed by using compression couplings to connect the conduit to metal wiring boxes. Interior electrical wiring systems use EMT extensively since it can be easily bent, can be connected together, and can be connected to metal wiring boxes.

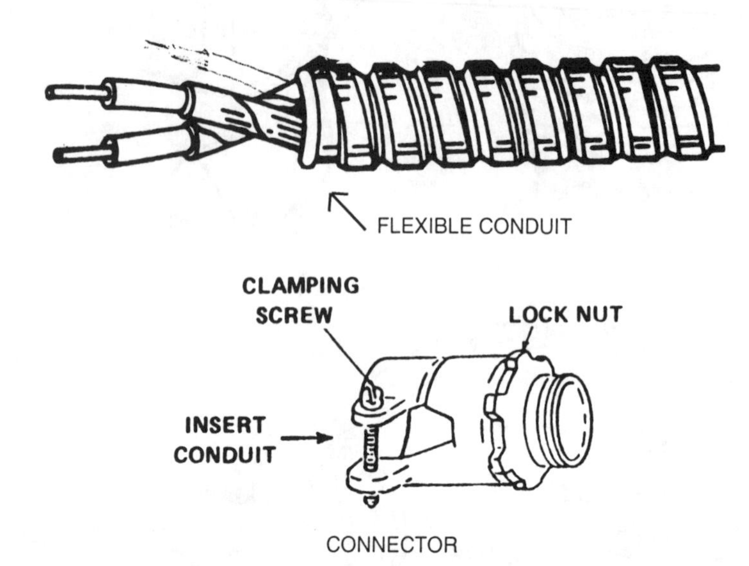

Figure 9-19. Three conductor metal-clad cable and connector.

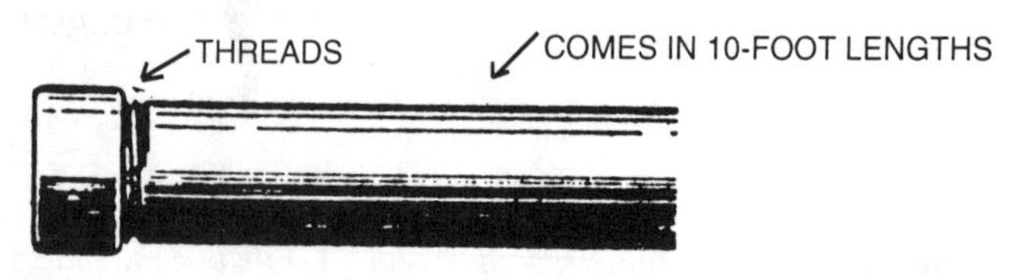

Figure 9-20. Rigid conduit.

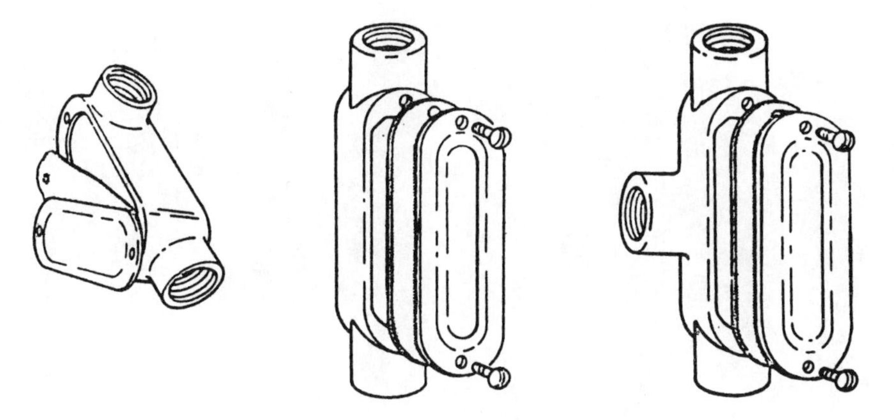

Figure 9-21. Three designs of rigid conduit boxes.

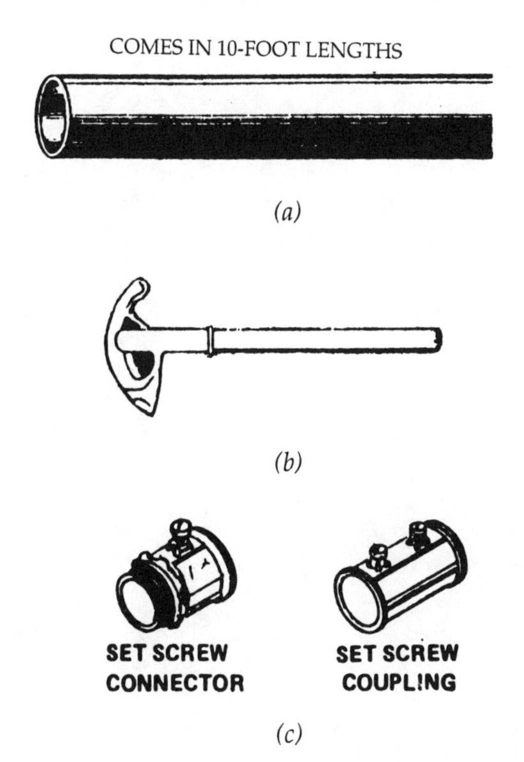

Figure 9-22. Electrical metallic tubing (EMT): (a) EMT; (b) EMT bender used to install conduit; (c) connectors used with EMT.

Chapter 10

Distribution Considerations for Electrical Loads

Power conversion systems are commonly referred to by the function they perform. Power conversion takes place in the *load* of an electrical power system. Common types of power system loads are those which convert electrical energy into heat, light, or mechanical energy. There are various types of lighting, heating, and mechanical loads used in industry, commercial buildings, and homes. Several of these power conversion systems will be discussed. Each type of electrical load has a different effect on the electrical distribution system.

BASIC HEATING LOADS

Most loads that are connected to electrical power distribution systems produce a certain amount of heat, mainly due to current flow through resistive devices. In many instances, heat represents a power loss in the circuit, since heat energy is not the type of energy that was to be produced. Lights, for instance, produce heat energy as well as light energy. The conversion of electrical energy to heat energy in a light-producing load reduces the efficiency of that load device, since not all of the available source energy is converted to light energy. There are, however, several types of power conversion systems which are mainly heating loads. Their primary function is to convert electrical energy into heat energy. Some basic systems include resistance heating, inductive heating, and dielectric (capacitive) heating.

Resistance Heating

Heat energy is produced when an electrical current flows through a resistive material. In many instances, the heat energy produced by an electrical current is undesirable; however, certain applications require controlled resistance heating. Useful heat may be transferred from a resistive element to a point of utilization by the common methods of heat transfer—convection, radiation, or conduction. A heating-element enclosure is needed to control the transfer of heat by convection and radiation. For heat transfer by conduction, the heating element is in direct contact with the material to be heated. Actual heat transfer usually involves a combination of these methods.

Figure 10-1 illustrates the principle of resistance heating. The self-contained heating element uses a coiled resistance wire which is placed inside a heat-conducting material and enclosed in a metal sheath. This principle may be used to heat water, oil, the surrounding atmosphere, or various other media. This type of heater may be employed in the open air or immersed in the media to be heated. The useful life of the resistance elements depends mainly upon the operating temperature. As the temperature increases, the heat output also increases. Basically, the heat energy produced is dependent upon the current flow and the resistance of the clement and can be calculated as current squared times resistance (I^2R).

Induction Heating

The principle of induction heating is illustrated in Figure 10-2. Heat is produced in magnetic materials when they are exposed to an alternating-current field. In the example shown, current is induced in the material heated by electromagnetic induction. This is brought about by the application of an alternating current to the heating coil. The material to be heated must be a *conductor* in order for current to be induced.

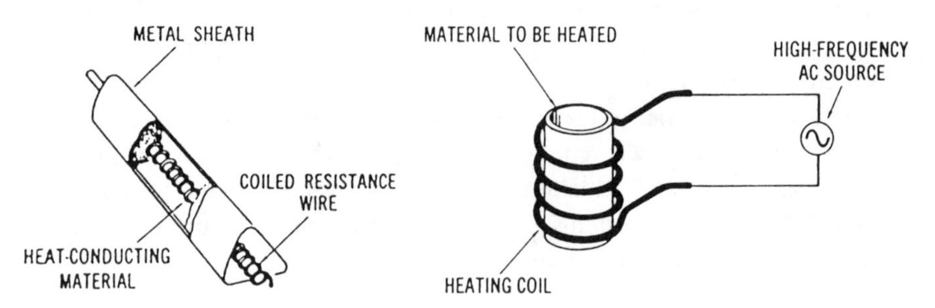

Figure 10-1. Resistance heating principle.

Figure 10-2. Induction heating principle.

Ordinarily, a high-frequency AC source in the range of 100-500 kHz is used to produce a higher heat output. This higher heat output is due to greater amounts of induced voltage.

As the magnetic field created by the high-frequency AC source moves across the material to be heated, the induced voltage causes *eddy currents* (circulating currents) to flow in the material. Heat results due to the resistance of the material to the flow of the eddy currents. The heat produced by this method is rapid and thus advantageous.

The major application of the induction-heating process is in metal-working industries for such processes as hardening, soldering, melting, and annealing of metals. The heat produced by this process is extremely rapid compared to other methods of heating. The area of the metal that is actually heated can be controlled by the size and position of the heating coils of the induction heater. This type of control is difficult to accomplish by other methods. Induction furnaces use the induction-heating principle.

By varying the frequency of the voltage applied to the induction heater windings, it is possible to vary the depth of heat penetration into the heated metal. At higher frequencies, due to the so-called "skin effect," the heat produced by the induced current from the heating coils will not penetrate as deeply. Thus, heat will penetrate more deeply at lower frequencies. When heat must be localized into the surface of a material only, such as for surface hardening of a metal, higher frequencies are used. The cost of higher frequency induction heaters is greater due to the more complex oscillator circuits required to produce these frequencies.

Dielectric (Capacitive) Heating

Induction heating can only be used with conductive materials. Therefore, some other method must be used to heat nonconductive materials. Such a method is illustrated in Figure 10-3, and is referred to as *dielectric* or *capacitive* heating. Nonconductors may be heated by placing them in an electrostatic field created between two metal electrodes that are supplied by a high-frequency AC source. The material to be heated becomes the dielectric or insulation of a capacitive device. The metal electrodes constitute the two plates.

When high-frequency AC is applied to a dielectric heating assembly, the changing nature of the applied AC causes the internal atomic structure of the dielectric material to become distorted. As the frequency

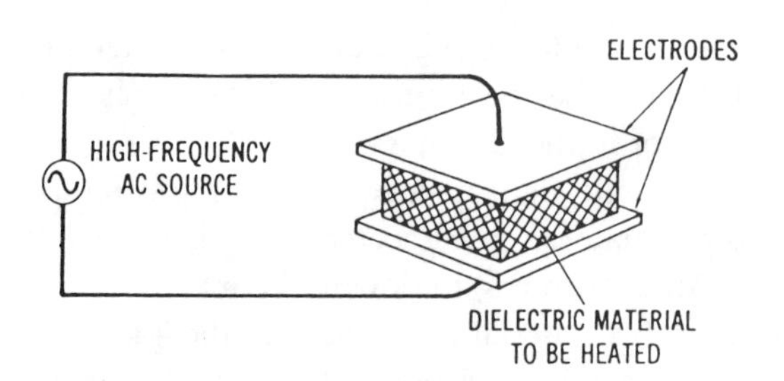

Figure 10-3. Capacitive heating principle.

of the AC increases, the amount of internal atomic distortion also increases. This internal friction produces a large amount of heat in the nonconductive material. Frequencies in the 50-MHz range may be used for dielectric heating. Dielectric heating produces a rapid heat which is spread throughout the heated material. Common applications of this heating method are the gluing of plywood and the bonding together of plastic sheets.

ELECTRICAL WELDING LOADS

Electrical welding is another common type of heat-producing power conversion system. The types of electrical welding systems include resistance welding, electric arc welding, and induction welding. One type of welder is shown in Figure 10-4.

Resistance Welding

Several familiar welding methods, such as spot welding, seam welding, and butt welding, are resistance welding processes. All of these processes rely upon the resistance heating principle. Spot welding, illustrated in Figure 10-5a, is performed on overlapping sheets of metal, which are usually less than 1/4-inch thick. The metal sheets are clamped between two electrodes and an electrical current is passed through the electrodes and metal sheets. The current causes the metals to fuse together. The instantaneous current through the electrodes is usually in excess of 5000 amperes, while the voltage between the electrodes is less than 2 volts.

Seam welding, shown in Figure 10-5b, is accomplished by passing sheets of metal through two pressure rollers while a continuously inter-

Figure 10-4. Electric welder (Courtesy Miller Electric Mfg. Co.).

rupted current is passed through the electrodes. The operational principle of scam welding is the same as spot welding. Several other similar methods, which are referred to as butt welding, edge welding, and projection welding are also commonly used.

Electric Arc Welding

While resistance welding utilizes pressure on the materials to be welded, electric arc welding produces welded metals by localized heating without pressure, as shown in Figure 10-6 An electric arc is created when the electrode of the welder is brought in contact with the metal to be welded. Carbon electrodes are used for direct-current or alternating-current arc welding of nonferrous metals and alloys. Not all metals can be welded by the arc-welding process. The welding metal is produced

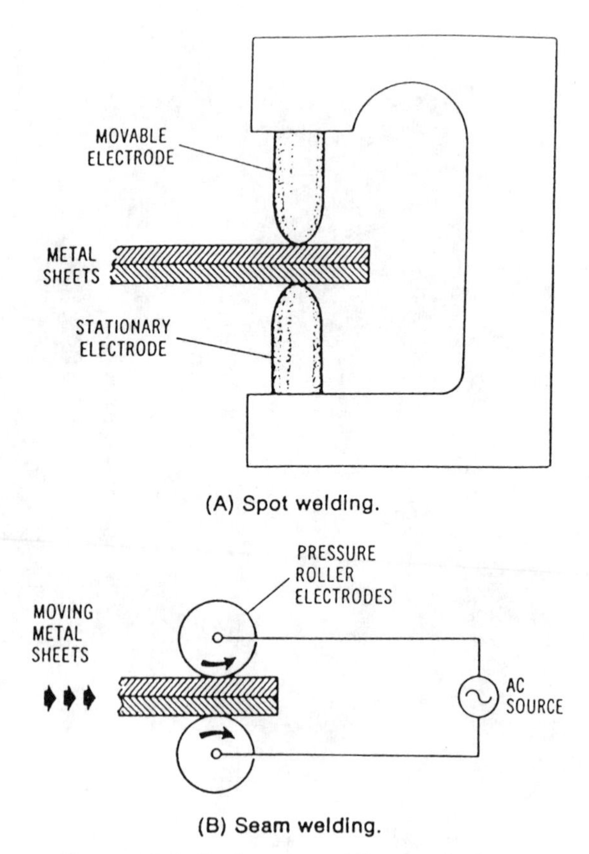

Figure 10-5. Resistance welding methods.

by inciting part of the metal that is to be welded. This method pool is added to, when necessary, by the use of a filler rod. The puddle (molten metal pool) then fills in the gap (arc crater) that was created by the arc of the electrode. Various types and various current-voltage ratings of electric arc welders are available.

A smaller amount of current is required for arc welding than for resistance welding. The currents may range from 50 to 200 amperes or higher for some applications. Voltages typically range from 10 to 50 volts. An electric arc welder may be supplied by a portable generator, a storage-battery unit, a step-down transformer, or a rectification unit.

Induction Welding

The induction welding process uses the principle of induction heating to fuse metals together. High-frequency alternating current is

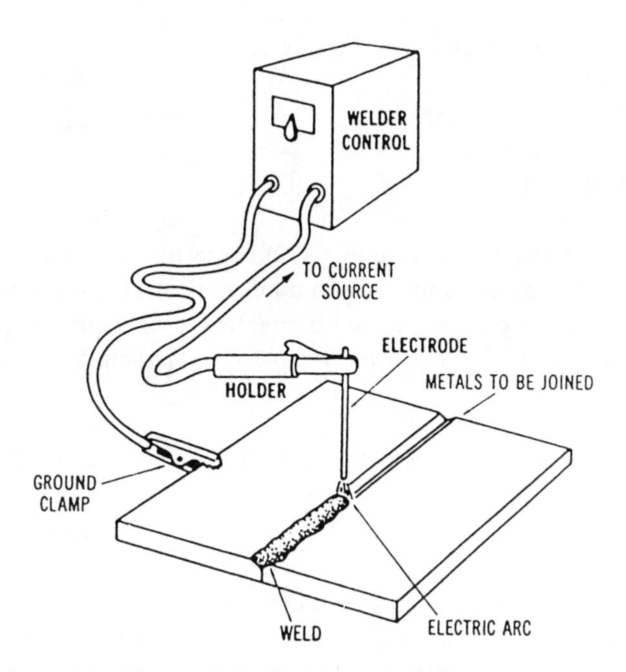

Figure 10-6. Electric arc welding.

applied to a heating coil into which the materials to be welded are placed. Tubular metal is often welded in this way.

POWER CONSIDERATIONS FOR ELECTRIC WELDERS

Electric welders are rather specialized types of equipment since they use very high amounts of current at low voltage levels. They have a peculiar effect on the power system operation. They draw large amounts of current for short periods of time. Silicon controlled rectifiers (SCRs) are commonly used to control the starting and stopping of the large currents associated with electric welders. The current rating of these devices must be very high, sometimes in the range of 1000 to 100,000 amperes, and the power distribution equipment must be able to handle these high currents.

Figure 10-7 illustrates a typical electric welding system. The AC power supplied from the branch circuit of the power system is either stepped down by a transformer to deliver AC voltage to the welder or rectified to produce DC voltage for direct-current welders. In either type of machine, an SCR contactor may be used to control the on and off time of the welder.

SCR CONTROL

Silicon-controlled rectifier (SCR) contactors are electronic control devices designed to handle large amounts of current. SCRs are triggered or turned on by pulses supplied by the timing or sequencing circuits of the welder. The SCRs are usually cooled by circulating water.

ELECTRIC HEATING AND AIR-CONDITIONING SYSTEMS

A very important type of electrical load is the heating and air-conditioning systems of homes, industries, and commercial buildings. These loads convert a high percentage of the total amount of electrical power that is distributed. Although natural gas and fuel-oil heating systems are still used electrical heating is becoming more prevalent each year. Air-conditioning systems are also more commonly used to cool buildings. The use of electric heating and air-conditioning systems has made us even more dependent upon efficient distribution of electrical power. We use electrical power to maintain a comfortable environment inside buildings.

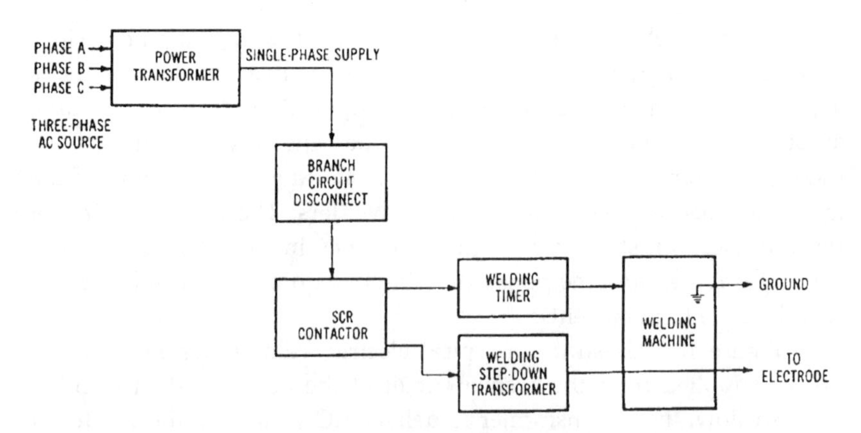

Figure 10-7. Block diagram of a typical electric welding system.

Basics of Electric Heating

There are several important factors which must be understood in order to have a knowledge of electric heating. Heat is measured in *British thermal units (Btu's).* One Btu is the amount of heat required to raise the temperature of one pound of water one degree Fahrenheit. Heat energy is the amount of 3.4 Btus per hour is equivalent to one watt of electrical energy.

Another basic factor to be considered in the study of electric heating is called *design temperature difference.* This is the difference between inside and outside temperatures in degrees Fahrenheit. The outside temperature is considered to be the *lowest* temperature which is expected to occur *several* times a year. The inside temperature is the desired temperature (thermostat setting).

A factor used in conjunction with design temperature difference is called degree *days.* The degree-day factor is used to determine the average number of degrees that the mean temperature is below 65°F. This data is averaged over seasonal periods for consideration in insulating buildings.

The insulation of a building is a very important consideration in electric heating systems. Insulation is used to oppose the escape of heat. The quality of insulation is expressed by a *thermal resistance* factor (R). The total thermal resistance of a building is found by considering the thermal resistance of the entire structure (wood concrete insulation etc.). The inverse of thermal resistance is called the *coefficient of heat trans*fer (U), which is an expression of the amount of heat flow through an are a expressed in Btu per square foot per hour per degree Fahrenheit. The following formulas are used in the conversion of either U or R to electrical units (watts):

$$\text{Thermal resistance} = \frac{1}{\text{Coefficient of heat transfer}}$$

$$\text{Watts} = \frac{\text{Coefficient of heat transfer}}{3.4}$$

or

$$\text{Watts} = 0.29 \times U$$

ELECTRIC HEATING AND COOLING SYSTEMS

Several types of electric heating systems are used today. Some common types are baseboard heaters, wall- or ceiling-mounted heaters and heat pumps. Most of these systems use forced air to circulate the heat. Some electric heaters have individual thermostats while others are connected to one central thermostat which controls the temperature in an entire building. The possibility of having temperature control in each room is an advantage of electric heating systems.

Heat Pumps

In recent years, the heat pump has become very popular as a combination heating and cooling unit for buildings. The heat pump is a heat-transfer unit. When the outside temperature is warm, the heat pump acts as an air-conditioning unit and transfers the indoor heat to the outside of the building. This operational cycle is reversed during cool outside temperatures. In the winter, the outdoor heat is transferred to the inside of the building. This process can take place during cold temperatures since there is always a certain amount of heat in the outside air, even at sub-zero temperatures. However, at the colder temperatures, there is less heat in the outside air.

Thus, heat pumps transfer heat rather than producing it. Since heat pumps do not produce heat, as resistive-heating units do, they are more economical in terms of energy conservation. Heating and cooling are reversible processes in the heat-pump unit; thus, the unit is self-contained. The reversible feature of heat pumps reduces the space requirement for separate heating and cooling units. Another advantage is that the changeover from heating to cooling can be made automatically. This feature could be desirable during the spring and autumn seasons in the many areas where temperatures are very variable. In extremely cold areas, the heat pump can be supplemented by an auxiliary resistance-heating unit. This auxiliary unit will operate when the outside temperatures are very cold, and will be useful in maintaining the inside temperatures at a comfortable level. Air is circulated past these heating elements into the heat vents of the building.

Heat pumps are used for residential as well as commercial and industrial applications and are being used more extensively each year. Figure 10-8 shows a simplified circuit arrangement of a heat pump in which a compressor takes a refrigerant from a low-temperature, low-pressure evaporator and converts it to a high temperature and a high pressure. The refrigerant is then delivered to a condenser much the same as a refrigerator functions.

Air-conditioning Systems

The increased use of air-conditioning systems provides greater comfort in homes, industrial plants, and commercial facilities. Most air-conditioning units are for the purpose of controlling the inside temperature of buildings so as to make working and living conditions more comfortable. However, many units are used to cool the insides of various types of equipment. Both air temperature and the relative humidity are changed by air-conditioning units. In the design of air-conditioning systems, all heat-producing items must be considered. Body heat, electrical appliances, and lights represent some common sources of heat. The diffusion of heat takes place through floors, walls, ceilings, and the windows of buildings. A simplified diagram of a room air-conditioning unit is shown in Figure 10-9.

Considerations for Heating Loads

Heating systems used for residential, commercial and industrial applications are usually referred to as HVAC systems. This means Heating Ventilation and Air Conditioning system. The electrical power requirement for HVAC systems is a major concern for electrical system design for a building. Electrical HVAC systems provide individual ther-

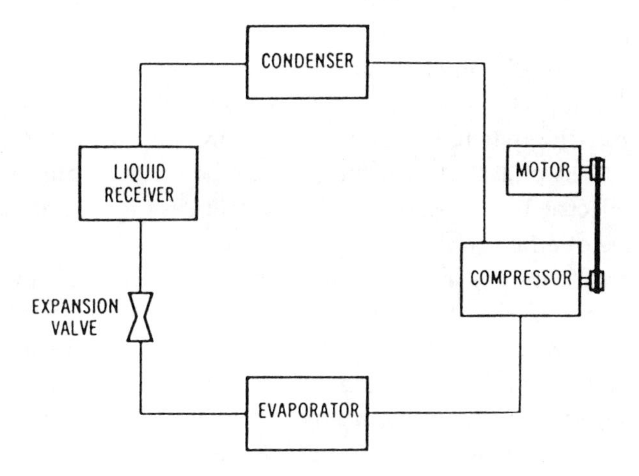

Figure 10-8. A simplified heat-pump circuit diagram.

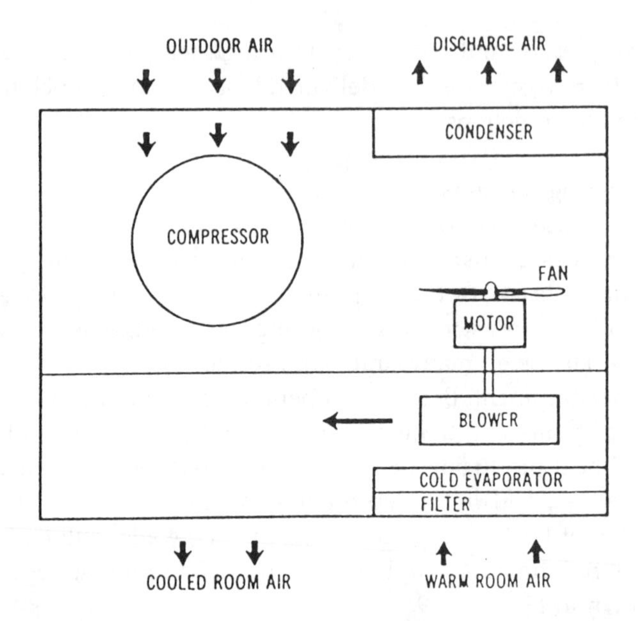

Figure 10-9. A simplified diagram of a room air-conditioning unit.

mostatic temperature control, have long equipment life, and are safe to use. A well-insulated building is necessary to reduce heat loss for economical use of HVAC systems.

Electrically energized comfort heating systems are widely used today to produce heat for commercial, industrial, and residential buildings. This type of energy is readily available at nearly any building site and has a number of advantages over fuel-burning methods of producing heat. Ecologically, any fuel needed to produce electricity is burned or consumed at a power plant that is usually located some distance away from the building where it is being used. Through this method of heating there is less pollution than there would be if fuel is burned at each building. Electric heat is also clean to use and easy to control and has a high degree of efficiency.

Electric heating is important today because of its high level of efficiency. Theoretically, when electrical energy is applied to a system, virtually all of it is transformed into heat energy. Essentially, this means that, when a specified amount of electricity is applied, it produces an equivalent Btu output. One thousand watts or 1 kW of electricity, when converted to heat, produces 3412 Btu of heat energy.

Heating can be achieved in a variety of ways through the use of electricity. Comfort heating systems contain an energy source, transmission path, control load device, and the possibility of one or more optional indicators. The primary difference in the two electrical systems is in the production of heat energy. Resistance heating is accomplished by passing an electrical current through wires or conductors to the load device. By comparison, the heat pump operates by circulating a gas or liquid through pipes that connect an inside coil to an outside coil. Electricity is needed in both cases as an energy source to make the systems operational.

Resistance Heating in Buildings

When an electric current flows through a conductive material, it encounters a type of opposition called resistance. In most circuits this opposition is unavoidable to some extent because of the material of the conductor, its length, cross-sectional area, and temperature. The conductor wires of a heating system are purposely kept low in resistance to minimize heat production between the source and the load device. Heavy-gauge insulated copper wire is used for this part of the system.

The load device of a resistance heating system is primarily responsible for the generation of heat energy. The amount of heat developed by the load is based upon the value of current that passes through the resistive element. Element resistance is purposely designed to be quite high when compared with connecting wires of the system. An alloy of nickel and chromium called Nichrome is commonly used for the heating elements.

Resistive elements may be placed under windows or at strategic locations throughout the building. In this type of installation the elements are enclosed in a housing that provides electrical safety and efficient use of the available heat. Air entering at the bottom of the unit circulates around the fins to gain heat and exits at the top. Different configurations may be selected according to the method of circulation desired, unit length, and heat-density production.

Resistive elements are also used as a heat source in forced-air central heating systems. In this application, the element is mounted directly in the main airstream of the system. The number of elements selected for a particular installation is based upon the desired heat output production. Individual elements are generally positioned in staggered configuration to provide uniform heat transfer and to eliminate hot spots. The element has spring-coil construction supported by ceramic insulators.

Units of this type provide an auxiliary source of heat when the outside temperature becomes quite cold. Air circulating around the element is warmed and forced into the duct network for distribution throughout the building.

Heat Pump Systems in Buildings

A heat pump is defined as a reversible air-conditioning system that transfers heat either into or away from an area that is being conditioned. When the outside temperature is warm, it takes indoor heat and moves it outside as an air-conditioning unit. Operation during cold weather causes it to take outdoor heat and move it indoors as a heating unit. Heating can be performed even during cold temperatures because there is always a certain amount of heat in the outside air. At 0 F (–22 C), for example, the air will have approximately 89% of the heat that it has at 100 F (38 C). Even at subzero temperatures, it is possible to develop some heat from the outside air. However, it is more difficult to develop heat when the temperature drops below 20 F (6 C). For installations that encounter temperatures colder than this, heat pumps are equipped with resistance heating coils to supplement the system.

A heat pump, like an air conditioner, consists of a compressor, an outdoor coil, an expansion device, and an indoor coil. The compressor is responsible for pumping a refrigerant between the indoor and outdoor coils. The refrigerant is alternately changed between a liquid and a gas, depending upon its location in the system. Electric fans or blowers are used to force air across the respective coils and to circulate cool or warm air throughout the building.

A majority of the heat pumps in operation today consist of indoor and outdoor units that are connected together by insulated pipes or tubes. The indoor unit houses the supplemental electric heat elements, blower and motor assembly, electronic air cleaner, humidifier, control panel, and indoor coil. The outdoor unit is covered with a heavy-gauge steel cabinet that encloses the outdoor coil, blower fan assembly, compressor, expansion device, and cycle-reversing valve. Both units are designed for maximum performance, high operational efficiency, and low electrical power consumption.

The Heating Cycle of a Heat Pump

If a unit air conditioner were turned around in a window during its operational cycle, it would be extracting heat from the outside air and pumping it into the inside. This condition, which is the operational basis of the heat pump, is often called the reverse-flow air-conditioner principle. The heat pump is essentially "turned around" from its cooling cycle by a special valve that reverses the flow of refrigerant through the system. When the heating cycle occurs, the indoor coil, outdoor coil, and fans are reversed. The outdoor coil is now responsible for extracting heat from the outside air and passes it along the indoor coil, where it is released into the duct network for distribution.

During the heating cycle, any refrigerant that is circulating in the outside coil is changed into a low-temperature gas. It is purposely made to be substantially colder than the outside air. Since heat energy always moves from hot to cold, there is a transfer of heat from the outside air to the cold refrigerant. In a sense, we can say that the heat of the cold outside air is absorbed by the much colder refrigerant gas.

The compressor of the system is responsible for squeezing together the heat-laden gas that has passed through the outside coil. This action is designed to cause an increase in the pressure of the gas that is pumped to the indoor coil. As air is blown over the indoor coil, the high-pressure gas gives up its heat to the air. Warm air is then circulated through the duct network to the respective rooms of the system.

When the refrigerant gas of the indoor coil gives up its heat, it cools and condenses into a liquid. It is then pumped back to the outside coil by compressor action. Once again it is changed into a cool gaseous state and is applied to the outside coil to repeat the cycle. If the outside temperature drops too low, the refrigerant may not be able to collect enough heat to satisfy the system. When this occurs, electric-resistance heaters are energized to supplement the heating process. The place where electric heat is supplied to the system is called the balance point.

Figure 10-10 shows an illustration of heat-pump operation during its heating cycle. At (1), the heat is absorbed from the cold outside air by the pressurized low-temperature refrigerant circulating through the outside coil. At (2), the refrigerant is applied to the compressor and compressed into a high-temperature, high-pressure gas. At (3), the heated gas is transferred to the indoor coil and released as heat. At (4), warm air is circulated through the duct network. Note that the supplemental resistance heat element is placed in this part of the system. At (5), the refrigerant is returned to the compressor and then to an expansion device, where it condenses the liquid refrigerant and returns it to the outdoor coil. The cycle repeats itself from this point.

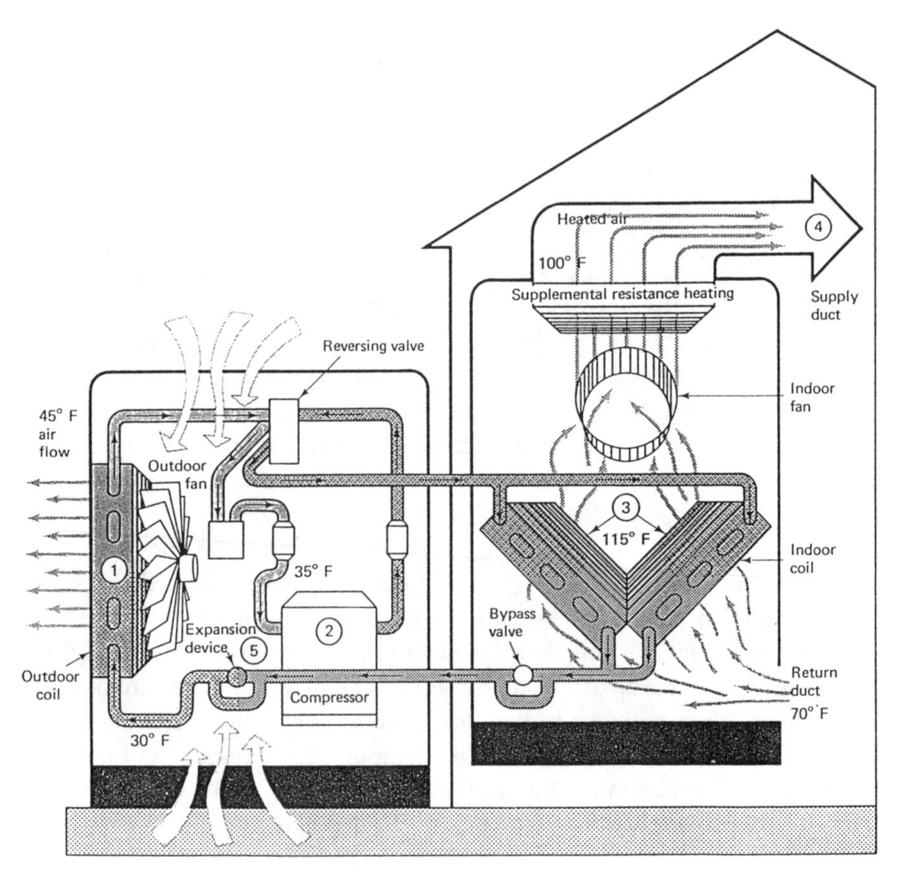

Figure 10-10. Heating-cycle operation.

The Cooling Cycle of a Heat Pump

During the summer months, a heat pump is designed to respond as an air-conditioning unit. For this to occur the reversing valve must be placed in the cooling-cycle position. In some systems this is accomplished by a manual changeover switch, whereas in others it is achieved automatically according to the thermostat setting. The operating position of the valve simply directs the flow path of the refrigerant.

When the cooling cycle is placed in operation it first causes the refrigerant to flow from the compressor into the indoor coil. During this part of the cycle the refrigerant is in a low-pressure gaseous state that is quite cool. As the circulation process continues, the indoor coil begins to heat from the inside air of the building. Air passing over the indoor coil is cooled and circulated into the duct network for distribution throughout the building.

After leaving the indoor coil, the refrigerant must pass through the reversing valve and into the compressor. The compressor is responsible for increasing refrigerant pressure and circulating it into the outdoor coil. At this point of the cycle, the refrigerant gives up its heat to the outside air, is cooled, and is changed into a liquid state. It then returns to the compressor, where it is pumped through an expansion device and returned to the indoor coil. The process then repeats itself.

Figure 10-11 shows an illustration of the heat pump during its air-conditioning cycle. At (1), heat is absorbed from the inside air and cool air is transferred into the building. At (2), the pressure of the heat-laden refrigerant is increased by the compressor and cycled into the outside coil for transfer to the air. At (3), cool dehumidified air is circulated

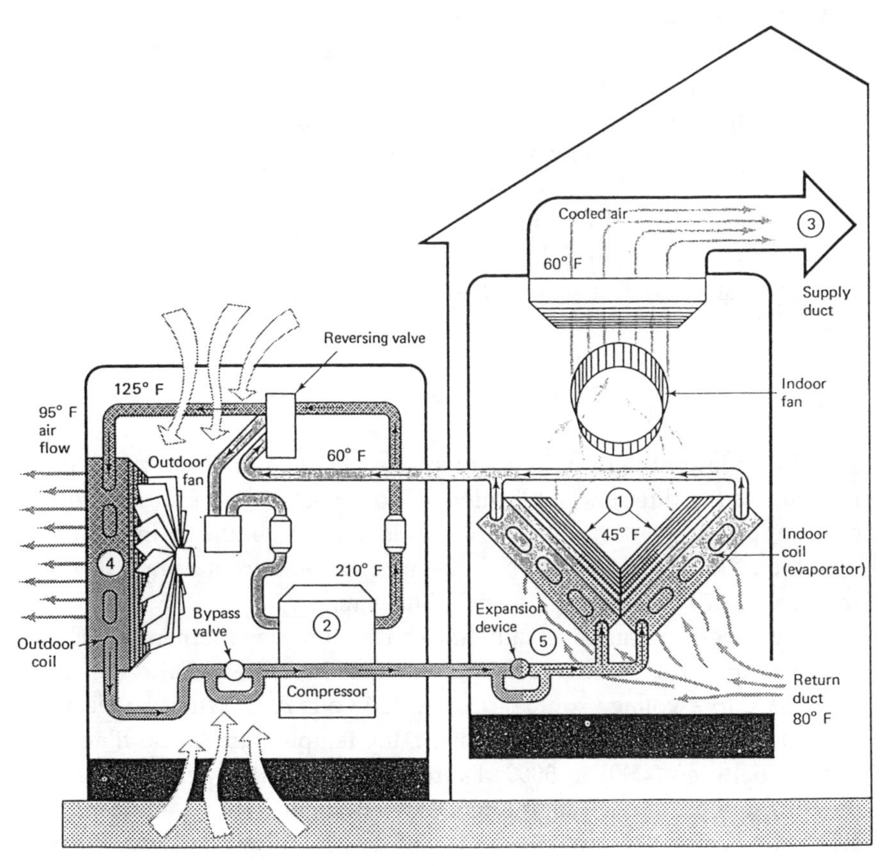

Figure 10-11. Cooling-cycle operation.

through the duct network as a result of passing through the cooled indoor coil. At (4), the refrigerant condenses back into a liquid as it circulates through the outdoor coil. At (5), the liquid refrigerant flows through the compressor and expansion device, where it is vaporized and returned to the indoor coil to complete the cycle.

LIGHTING SYSTEMS

Electrical lighting systems are designed to create a comfortable and safe home or working environment. There are several types of lighting systems in use today. Among the most popular are incandescent, fluorescent, and vapor lights, and several special-purpose types of lighting. Lighting systems are another type of electrical *load.*

The three types of lighting systems discussed in this chapter constitute most of the lighting loads placed on electrical distribution power systems. Incandescent, fluorescent, and vapor lights are used for many lighting applications. However, there are a few other specialized types of lighting that are also used. These include zirconium and xenon arc lights, glow lights, black lights, and infrared lights. The planning involved to obtain proper lighting is quite complex and may involve several types of lights and special distribution systems.

INCANDESCENT LIGHTING

Incandescent lighting is a widely used method of lighting and it is used for many different applications. The construction of an incandescent lamp is shown in Figure 10-12. This lamp is simple to install and maintain. The initial cost is low; however, incandescent lamps have a relatively low efficiency and a short life span.

Incandescent lamps usually have a thin tungsten filament which is in the shape of a coiled wire. This filament is connected through the lamp base to a voltage source (usually 120 volts AC). When an electric current is passed through the filament, the temperature of the filament rises to between 3000° to 5000° Fahrenheit. At this temperature range, the tungsten produces a high-intensity white light. When the incandescent lamp is manufactured, the air is removed from the glass envelope to prevent the filament from burning; also, an inert gas is added.

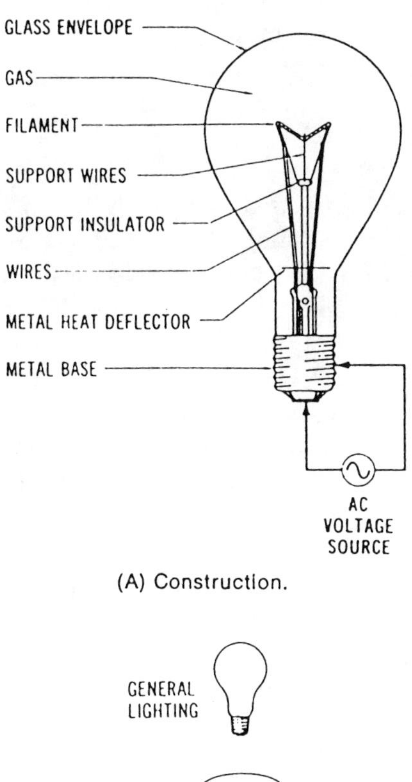

(A) Construction.

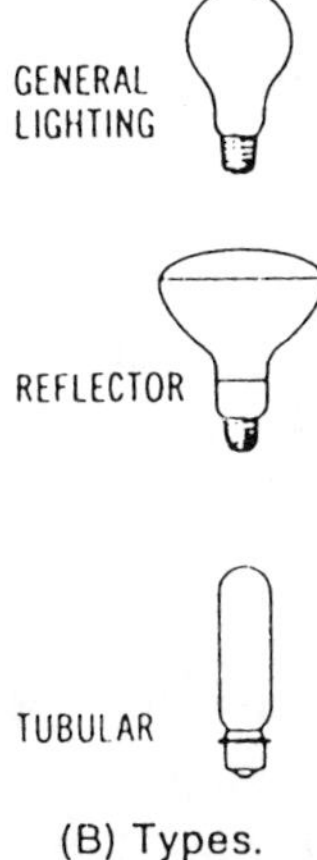

(B) Types.

Figure 10-12. The incandescent lamp.

Incandescent lamps are purely resistive devices and, thus, have a power factor of 1.0. As they deteriorate, their light output is reduced. Typically, at the time that an incandescent lamp burns out, its light output is less than 85% of its original output. A decrease in the voltage of the power system also reduces the light output. A 1% decrease in voltage will cause a 3% decrease in light output.

FLUORESCENT LIGHTING

Fluorescent lighting is used extensively today, particularly in industries and commercial buildings. Fluorescent lamps are tubular bulbs with a filament at each end. There is no electrical connection between the two filaments. The operating principle of the fluorescent lamp is shown in Figure 10-13. The tube is filled with mercury vapor; thus, when an electrical current flows through the two filaments, a continuous arc is formed between them by the mercury vapor. High-velocity electrons passing between the filaments collide with the mercury atoms, producing an ultraviolet radiation. The inside of the tube has a phosphor coating which reacts with the ultraviolet radiation to produce visible light.

One circuit for a fluorescent light is shown in Figure 10-14. Note that a thermal starter, which is basically a bimetallic strip and a heater, is connected in series with the filaments. The bimetallic strip remains closed long enough for the filaments to heat and vaporize the mercury in the tube. The bimetallic switch will, then, bend and open due to the heat produced by current flow through the heater. The filament circuit is now opened. A capacitor is connected across the bimetallic switch to reduce contact sparking. Once the contacts of the starter open, a high voltage is momentarily placed between the filaments of the lamp due to the action of the inductive ballast coil. The ballast coil has many turns of small-diameter wire and, thus, produces a high counter-electromotive force when the contacts of the starter separate. This effect is sometimes called "inductive kickback." The high voltage across the filaments causes the mercury to ionize and initiates a flow of current through the tube. There are several other methods used to start fluorescent lights; however, this method illustrates the basic operating principle.

Fluorescent lights produce more light per watt than incandescent lights; therefore, they are cheaper to operate. Since the illumination is produced by a long tube, there is also less glare. The light produced by fluorescent bulbs is very similar to natural daylight. The light is whiter and the operating temperature is much less with fluorescent lights. Various sizes and shapes of fluorescent lights are available. The bulb sizes

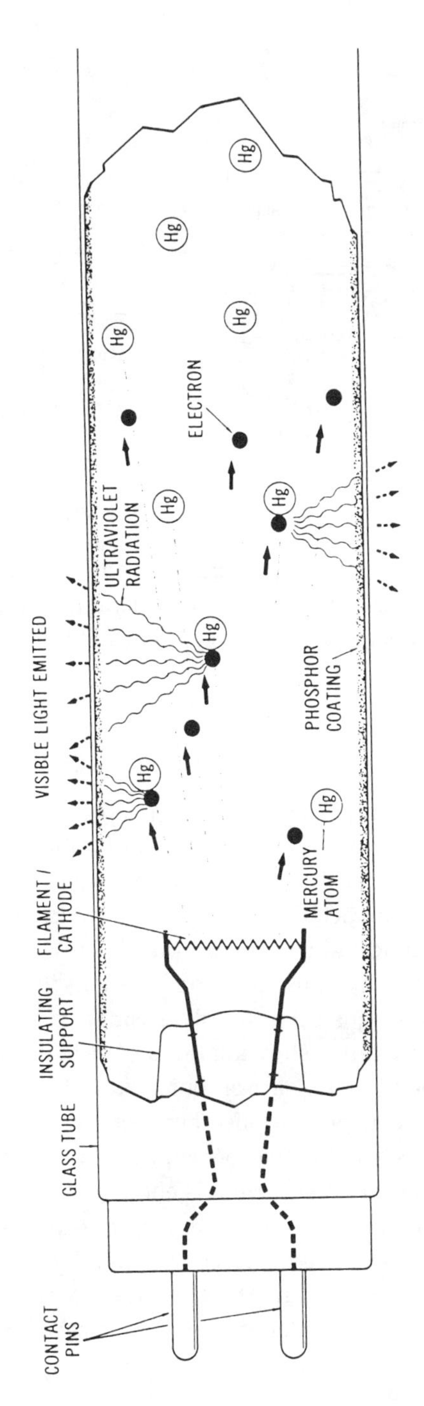

Figure 10-13. Operating principle of a fluorescent lamp.

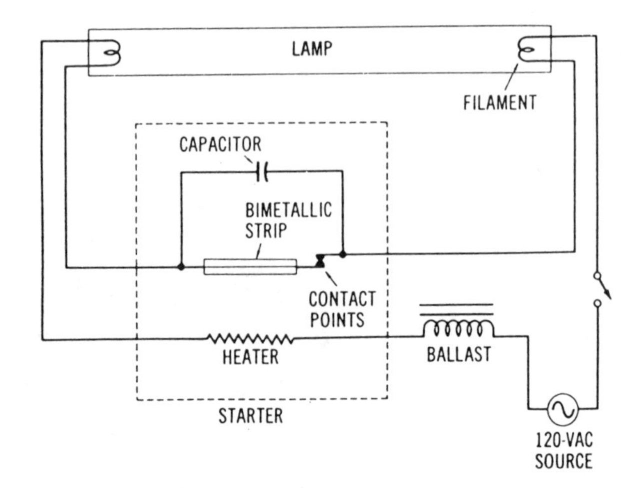

Figure 10-14. The circuit diagram of a fluorescent lamp.

are expressed in eighths of an inch with the common sizes being T-12 and T-8. (A T-12 bulb is 1-1/2 inch.) Common lengths are 24, 48, 72, and 96 inches. Figure 10-15 shows some fluorescent lighting components.

VAPOR LIGHTING

Another popular form of lighting is the vapor type. The mercury-vapor light is one of the most common types of vapor lights. Another common type is the sodium-vapor light. These lights are filled with a gas which produces a characteristic color. For instance, mercury vapor produces a greenish-blue light and argon a bluish-white light. Gases may be mixed to produce various color combinations for vapor lighting. This is often done with signs used for advertising.

A mercury-vapor lamp is shown in Figure 10-16. It consists of two tubes with an arc tube placed inside an outer bulb. The inner tube contains mercury. When a voltage is applied between the starting probe and an electrode, an arc is started between them. The arc current is limited by a series resistor; however, the current is enough to cause the mercury in the inner tube to ionize. Once the mercury has ionized, an intense greenish-blue light is produced. Mercury-vapor lights are compact, long-lasting, and easy to maintain. They are used to provide a high-intensity light output. At low voltages, mercury is slow to vaporize, so these lamps require a long starting time (sometimes 4 to 8 minutes).

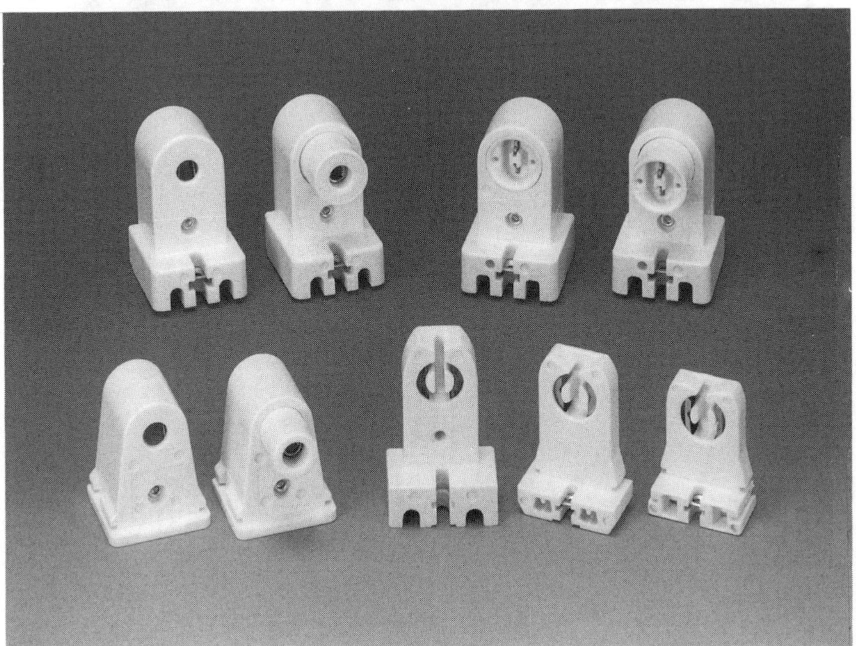

Figure 10-15. Fluorescent lighting component; (a) fluorescent lighting fixture (Courtesy Simkar Lighting); (b) fluorescent lamp holders (Courtesy Eagle Electric Mgf. Co.

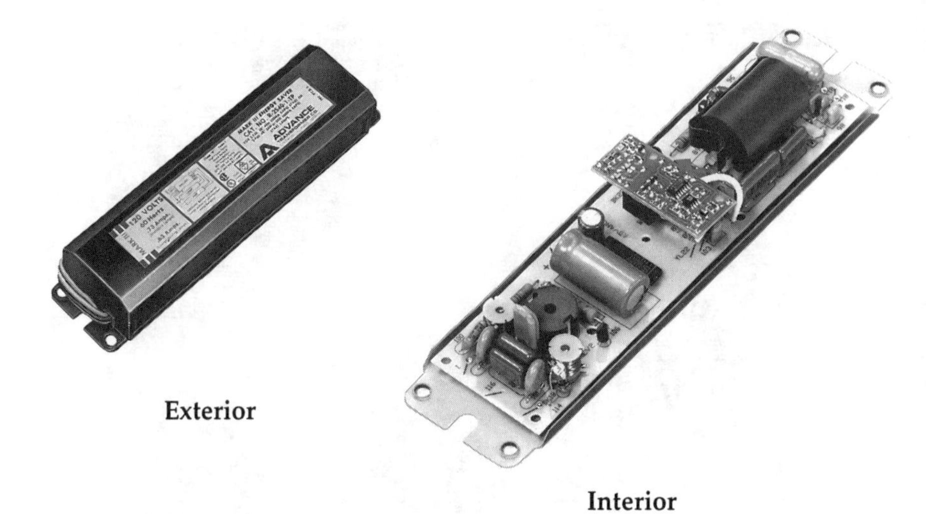

Figure 10-15. Fluorescent lighting component; (a) fluorescent lighting fixture (Courtesy Simkar Lighting); (b) fluorescent lamp holders (Courtesy Eagle Electric Mgf. Co.; (c) fluorescent lamp ballast (Courtesy Advance Transformer Co.)

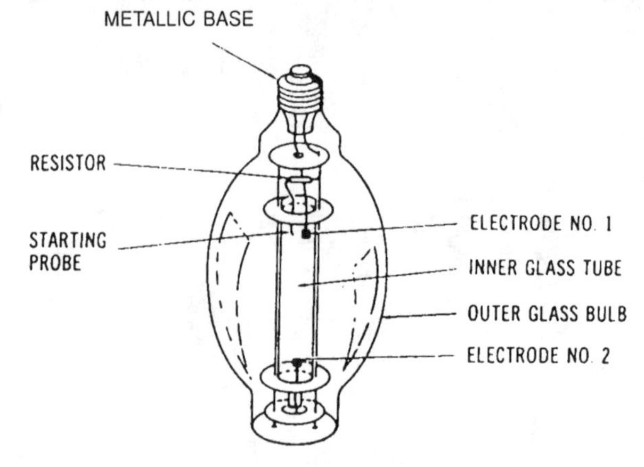

Figure 10-16. A mercury-vapor lamp.

Mercury-vapor lights can also be used for outdoor lighting.

The sodium-vapor light, shown in Figure 10-17 is popular for outdoor lighting and for highway lighting. This lamp contains some low-pressure neon gas and some sodium. When an electric current is passed through the heater, electrons are given off. The ionizing circuit causes a positive charge to be placed on the electrodes. As electrons pass from the

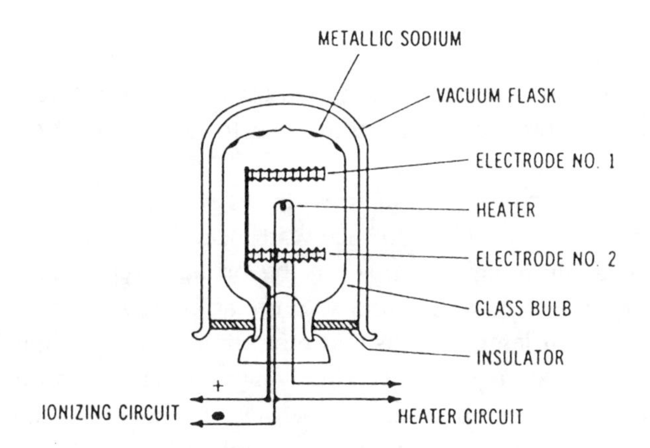

Figure 10-17. A sodium-vapor lamp.

heater to the positive electrodes, the neon gas is ionized. The ionization of the neon gas produces enough heat to cause the sodium to ionize. A yellowish light is produced by the sodium vapor. The sodium-vapor light can produce about three times the candlepower per watt as that of an incandescent light.

Another type of vapor lamp is called a metal halide type. This light source is a high intensity mercury vapor lamp which metallic substances called metal halides added to the bulb. The addition of these substances improves the efficacy of the lamps. The efficacy of a metal halide lamp is typically 75 lumens/watt compared to approximately 50 lumens/watt for mercury vapor lamps.

Another type of vapor lamp is the high pressure sodium (HPS) lamp. Sodium is the primary element used to fill the lamp's tube when it is manufactured. These lamps have very high efficacies—approximately 110 lumens/watt.

STREET LIGHTING

The lighting systems of today are highly reliable when compared to the systems that were installed many years ago. Earlier systems were either turned on and off manually or by timing devices that were regulated by the time of day rather than the natural light intensity. Some were controlled by electrical impulses that were transmitted on the power lines. Now, most systems are controlled by automatic, photoelectric circuits. The lights which are now used to illuminate streets and highways have photoelectric controls. They operate during periods of darkness and are automatically turned off when natural light is present. These street lights are usually connected to existing 120/24-volt power distribution systems.

There are many types of street lights in use. The earliest types of street lights used were 200- to 1000-watt incandescent lamps. Now mercury- and sodium-vapor lamps are the primary types used. (Mercury lamps produce a white light and sodium lamps have a yellowish color.) Several different lamp designs and mounting fixture designs are used.

Several years ago, street lights were converted from incandescent lamps to mercury-vapor lamps. Now the trend in street lighting seems to be toward the use of sodium-vapor lights, and several areas of the country have converted to the use of sodium-vapor lamps. Sodium-vapor lights produce more illumination than a similar mercury-vapor light. They also require less electrical power to produce a specific amount of illumination. Thus, the ability of sodium-vapor lights to deliver more light with less power consumption makes them more economically attractive than mercury-vapor lights.

COMPARISON OF LIGHT SOURCES

The purpose of a light source is to convert electrical energy into light energy. A measure of how well this is done is called efficacy. Efficacy is the lumens of light produced per watt of electrical power converted. A comparison of some different types of light sources is shown in Table 10-1.

Table 10-1. Comparison of light sources.

Lamp Type	Efficacy (Lumens/watt)
200 Watt incandescent lamp	20
400 Watt mercury lamp	50
40 Watt fluorescent lamp	70
400 Watt metal halide lamp	75
400 Watt high-pressure sodium lump	110

ELECTRICAL LIGHTING CIRCUITS

There are several types of electrical circuits that are used to distribute power to electrical lighting circuits. We will study some of the circuits used for incandescent and fluorescent lighting systems. Electrical lighting circuits for different areas of a building must be wired properly. Some lighting fixtures are controlled from one point by one switch while other fixtures may be controlled from two or more points by a switch at each point.

Incandescent lighting circuits are a typical type of branch circuit. The most common type of incandescent lighting circuit is a 120-volt branch that extends from a power distribution panel to a light fixture or fixtures in some area of a building. The path for the electrical distribution is controlled by one or more switches that are usually placed in small metal or plastic enclosures inside a wall. These switches are then covered by rectangular plastic plates to prevent possible shock hazard. Some components used for electrical lighting circuits are shown in Figure 10-18. Figure 10-19 shows symbols and a simple floor plan used to illustrate the layout of branch circuits for lighting.

Figure 10-18. Components of an electrical branch circuit for lighting.

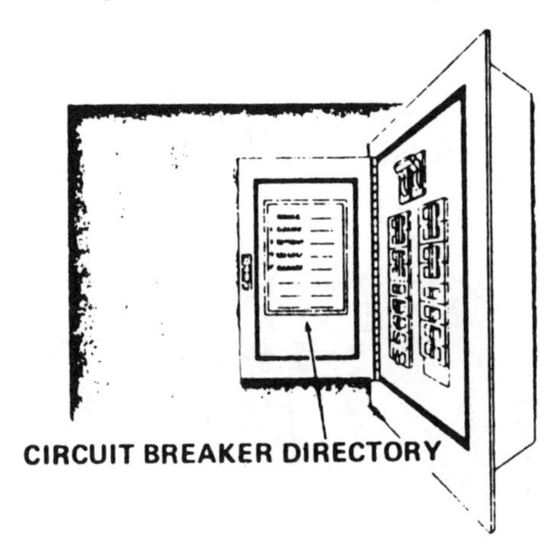

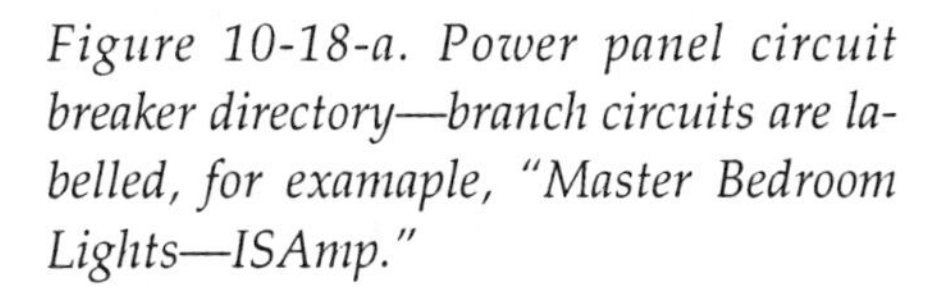

Figure 10-18-a. Power panel circuit breaker directory—branch circuits are labelled, for examaple, "Master Bedroom Lights—ISAmp."

Figure 10-18-b. Box connector and locknut—used to secure wiring cable to metal boxes.

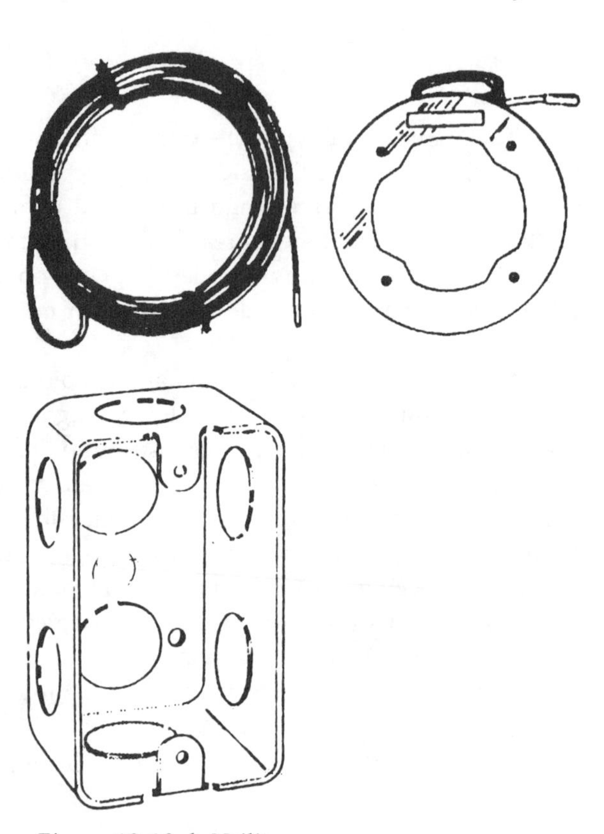

Figure 10-18-c. Fish tape—flexible steel cord used to pull wires through conduit or other enclosed spaces.

Figure 10-18-d. Utility box—used for mounting switches or receptacles.

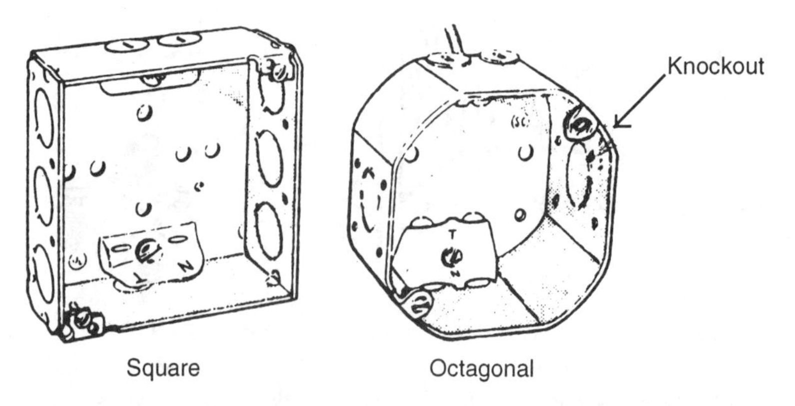

Figure 10-18-e. Ceiling boxes—used to mount lighting fixtures.

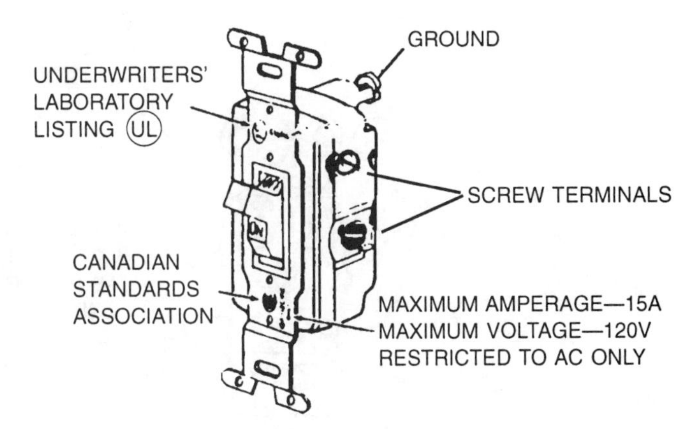

Figure 10-18-f. Single pole, single throw (SPST) switch detail—switch is mounted in a utility box; notice that an SPST has two *screw terminals.*

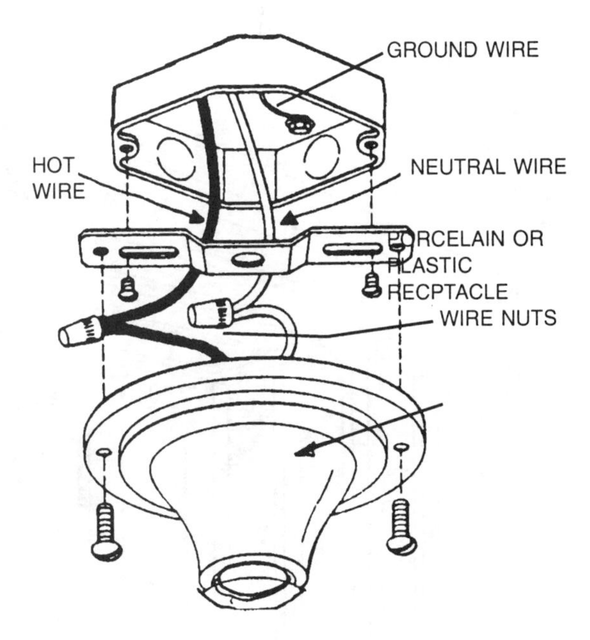

Figure 10-18-g. Attaching a light fixture to a ceiling box.

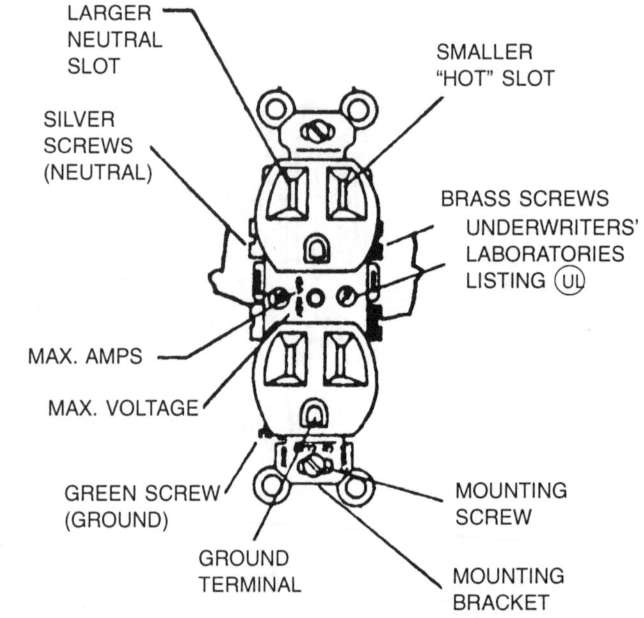

Figure 10-18-h. Duplex power receptacle—commonly used with lighting branch circuits; provides connection for 120 volts; mounted in a utility box.

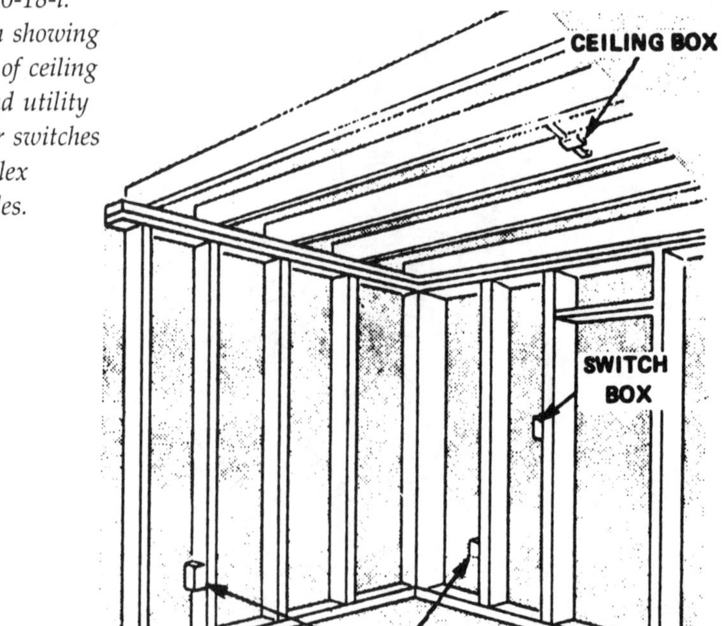

Figure 10-18-i. Diagram showing location of ceiling boxes and utility boxes for switches and duplex receptacles.

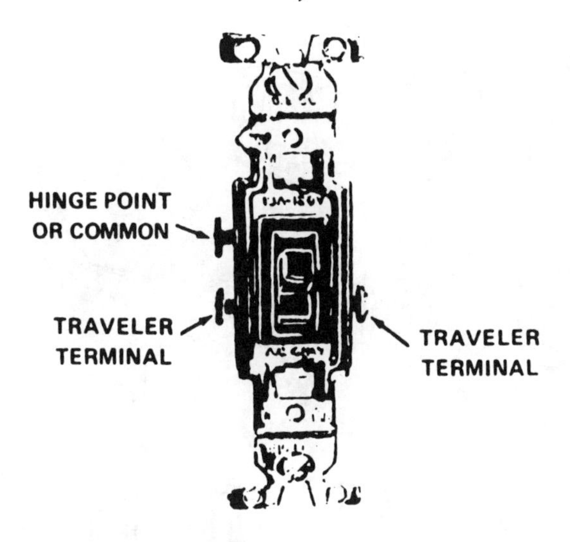

Figure 10-18-j. Three-way switch—notice that there are <u>three</u> screw terminals (one is called a "common" and the others are called "traveler" terminals.

BEND AND TIGHTEN
IN CLOCKWISE DIRECTION

Figure 10-18-k. Connecting a wire to a screw terminal properly.

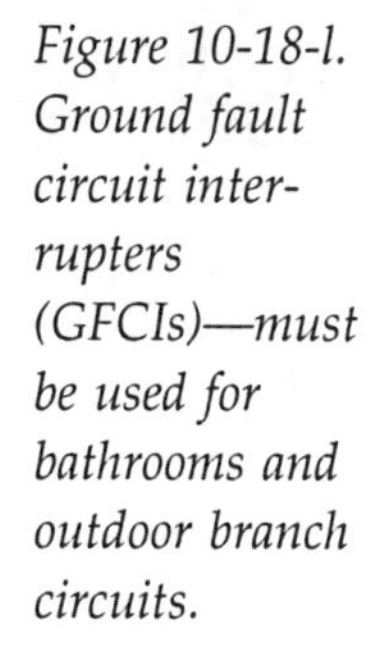
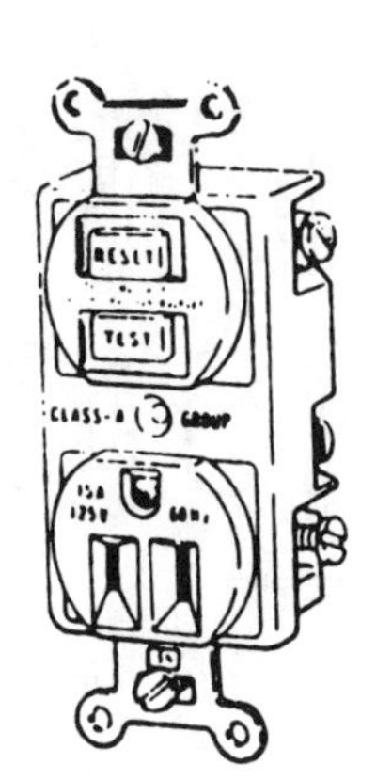
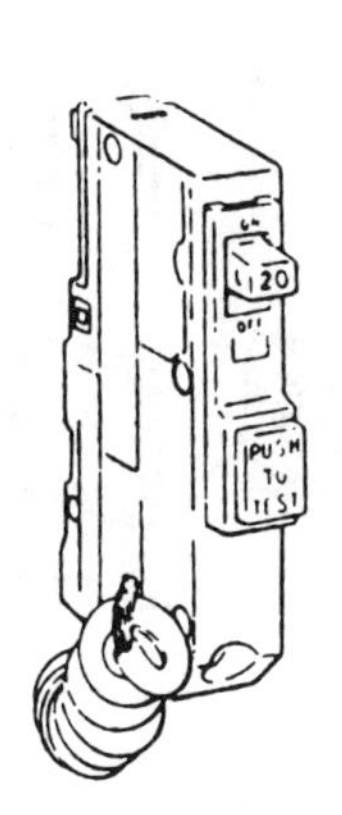
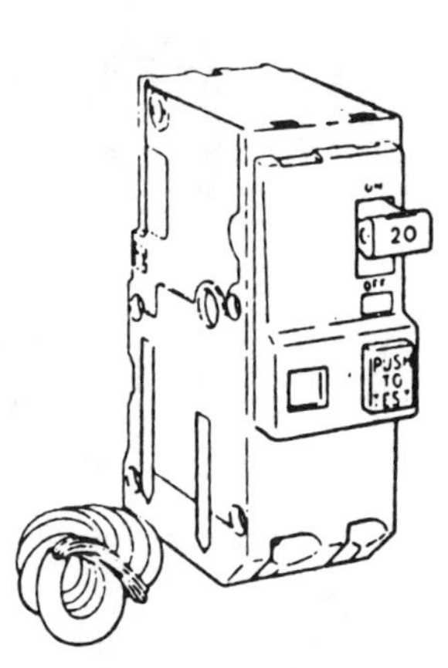

Figure 10-18-l. Ground fault circuit interrupters (GFCIs)—must be used for bathrooms and outdoor branch circuits.

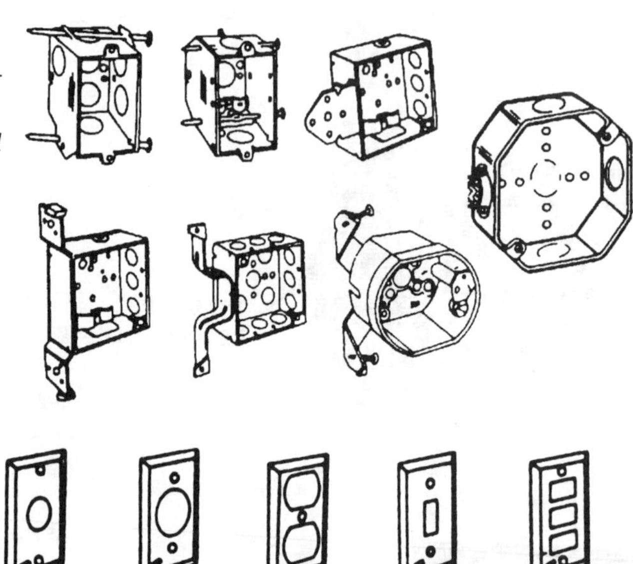

Figure 10-18-m. Electrical boxes—various shapes and sizes of metal or plastic boxes used to mount switches, receptacles or light fixtures and for splicing wires.

HANDY COVERS

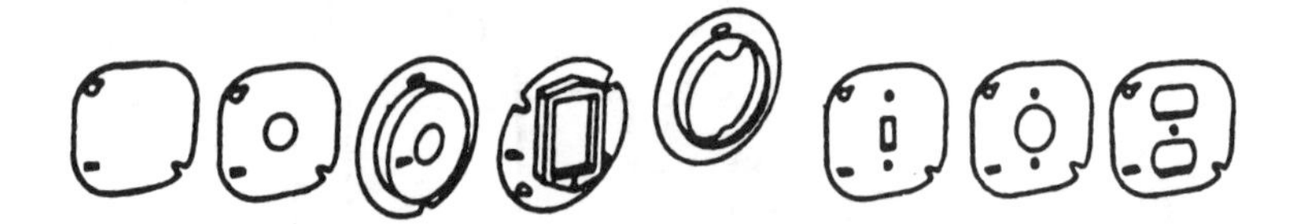

ROUND COVERS

SQUARE COVERS

Figure 10-18-n. Box covers—various designs; for safety and protection.

SYMBOL	ITEM
	Incandescent Fixture
	Fluorescent Fixture
	Junction Box
	Duplex Receptacle
	Special Outlet (as noted)
	240V. 3W. Receptacle (amps as noted)
	Wiring concealed in ceiling or wall
	Wiring concealed in floor
	Exposed branch circuit
	Branch circuit run to panel board
	Three or more wires (numbers of cross lines equals number of conductors; two conductors indicated if not otherwise noted)
	Incoming service lines: number and size of wire will be given
	Ground
S	Single pole switch
S_2	Double pole switch
S_3^3	Three-way switch
S_4	Four-way switch;
S_D	Dimmer switch
S_K	Key-operated switch
S_{MC}	Momentary contact switch

(a) Symbols used for electrical lighting branch circuits.

Figure 10-19. Symbols (a) and a simple floor plan (b) to illustrate branch circuits for lighting.

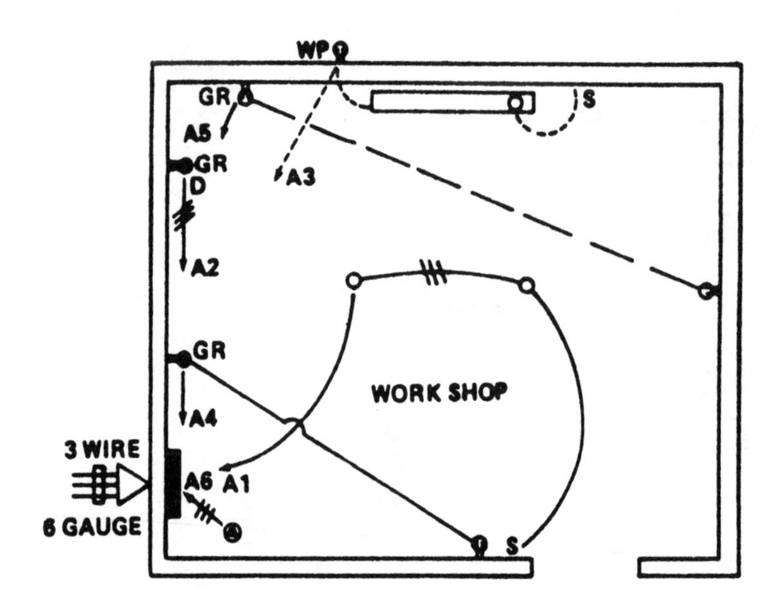

(b) Floor plan for electrical lighting branch circuits.

Specifications:

1. Fluorescent fixture is a 2-40 watt tube fixture.
2. Special purpose outlet is for an electrical saw, 240 volts, 30 amps.
3. All circuits (except A-5) will be overhead and in a straight line.
4. All receptacles are positioned 12" from the floor.
5. All switches are positioned 48" from the floor.
6. Walls are 10' in height.

Figure 10-19. Symbols (b) to illustrate branch circuits for lighting.

Switches are always placed in a wiring circuit so that they can open or close a *hot* wire. This hot wire distributes electrical power to the lighting fixtures which the switches control. These switches are referred to as "T-rated" switches. They are designed to handle the high instantaneous current drawn by lights when they are turned on or off. A switch which accomplishes control from a single location is a simple single-pole single-throw (spst) switch, such as shown in Figure 10-20. A pictorial view of the same circuit is shown in Figure 10-21.

Control from two locations, such as near the kitchen door and inside the garage door, is accomplished by two 3-way switches. This circuit is shown in Figure 10-22a, while Figure 10-23 presents a pictorial view of the same circuit. When control of a lighting fixture from more than two points is desired, two 3-way switches and one or more 4-way switches are used. For instance, control of one light from five points could be accomplished by using a combination of two 3-way switches and three 4-way switches. Figures 10-24 and 10-25 show a circuit for controlling a light from three locations. The 3-way switches are always connected to the power panel and to the light fixture, with the 4-way switch between them. Using 3-way and 4-way switch combinations, it is possible to achieve control of a light from any number of points.

Each lighting-control circuit requires an entirely different type of switching combination to adequately accomplish control of a lighting fixture. These circuits are usually wired into buildings during construction to provide the type of lighting control desired for each room. You should study the diagrams of Figures 10-20 through 10-25 to fully understand how these lighting circuits operate.

BRANCH-CIRCUIT DESIGN

The design of lighting branch circuits involves the calculation of the maximum current that can be drawn by the lights which are connected to the branch circuit. Many times, particularly in homes, a lighting branch circuit also has duplex receptacles for portable appliances. This causes the exact current calculation to be more difficult, since not all the lights or the appliances will be in use at the same time.

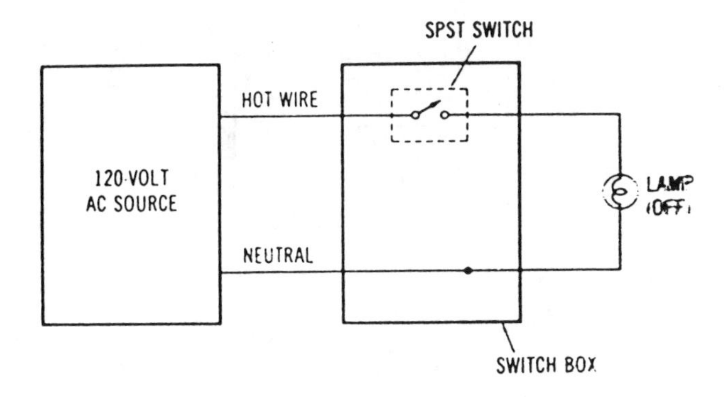

Figure 10-20. Schematic diagram for the control of a light from one location.

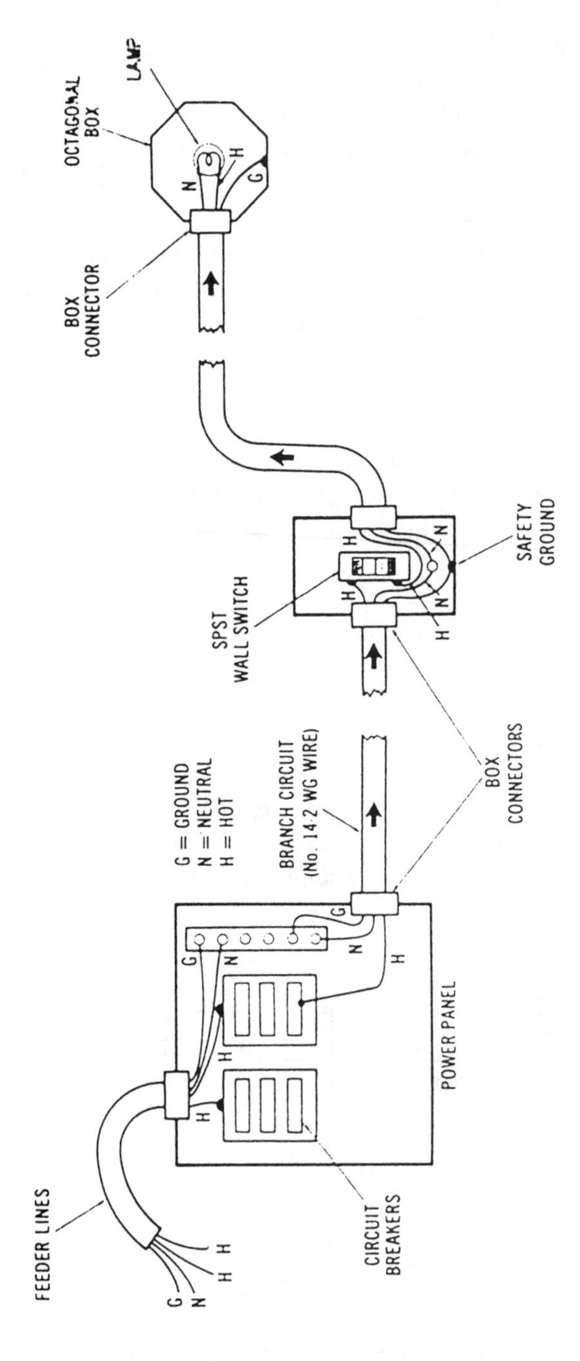

Figure 10-21. Pictorial view of diagram given in Figure 10-20.

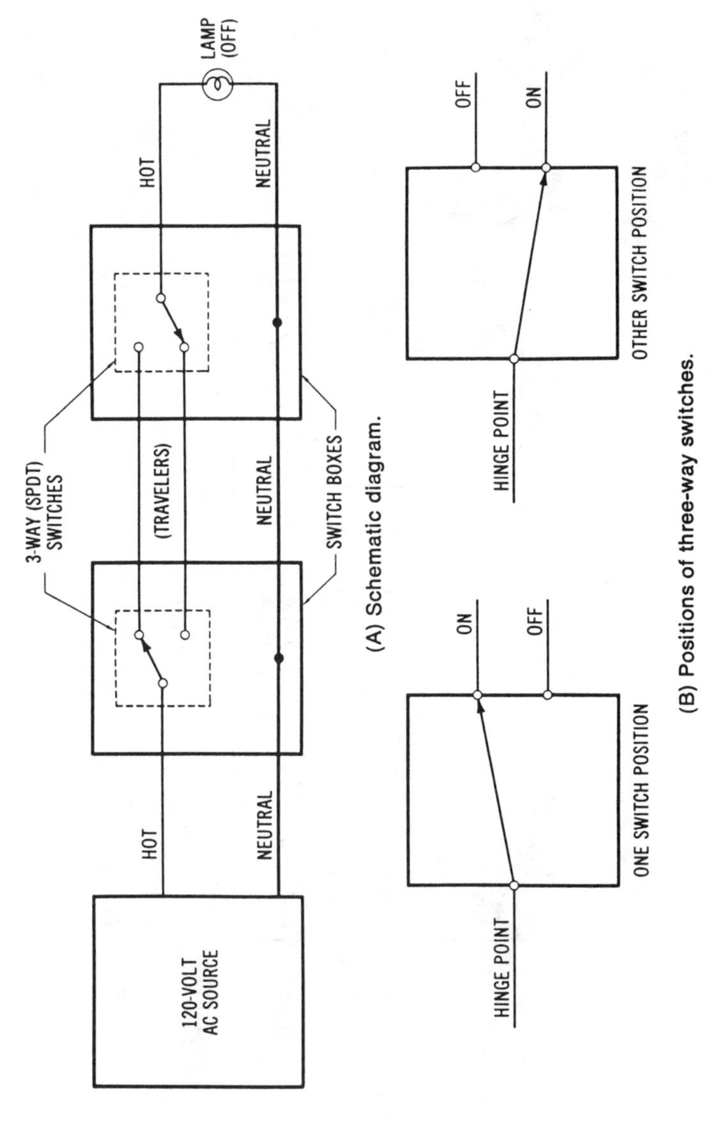

(A) Schematic diagram.

(B) Positions of three-way switches.

Figure 10-22. Circuit for controlling a light from two locations.

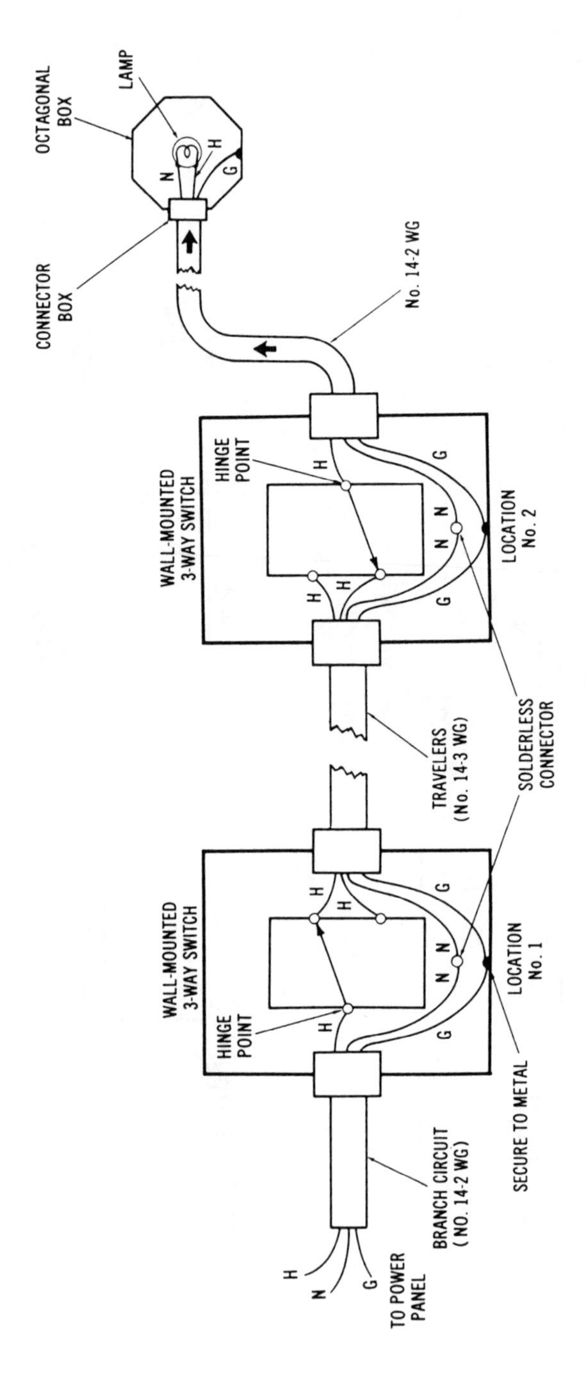

Figure 10-23. Pictorial view of circuit shown in Figure 10-22.

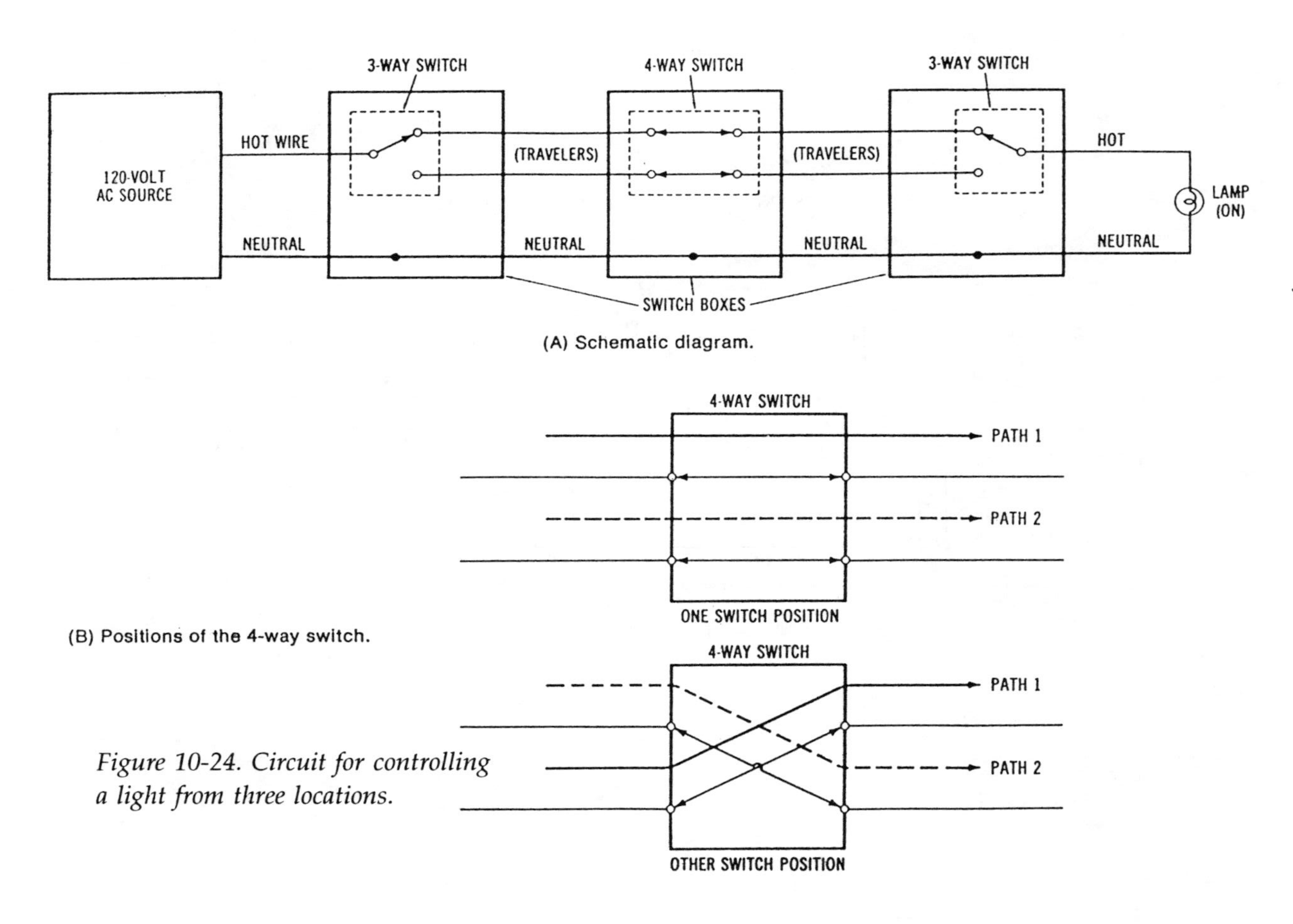

(A) Schematic diagram.

(B) Positions of the 4-way switch.

Figure 10-24. Circuit for controlling a light from three locations.

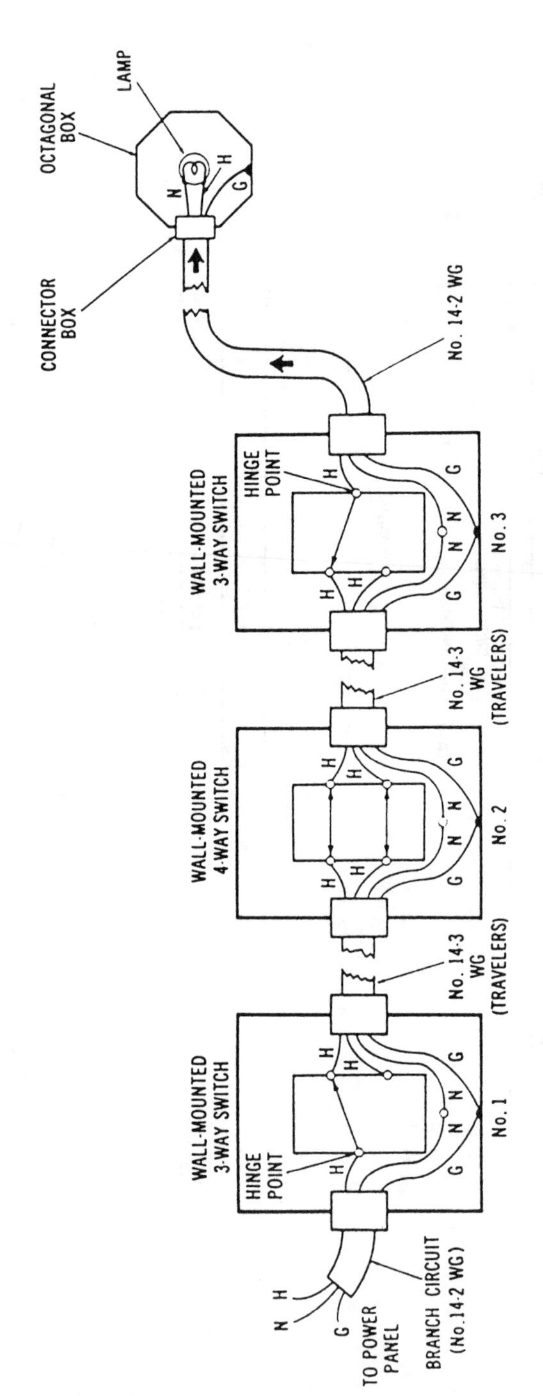

Figure 10-25. Pictorial view of circuit shown in Figure 10-24A.

The National Electrical Code (NEC) specifies that there should be at least one branch circuit for each 500 square feet of lighting area. The NEC further specifies a minimum requirement of lighting for various types of buildings. Table 10-2 lists the number of watts of light per square foot required in some buildings.

Table 10-2. Building lighting requirements in wattage per square foot.

Type of Building	Watts
Armories and auditoriums	1
Banks	2
Churches	1
Dwellings (other than hotels)	3
Garages (for commercial storage)	1/2
Hospitals	2
Hotels and motels	2
Industrial or commercial buildings	2
Office buildings	5
Restaurants	2
Schools	3
Stores (grocery)	3
Warehouses (storage)	1/4

The solution of a typical branch-circuit lighting problem follows.

1. *Given*—A room in a commercial building is 70 feet × 120 feet. It will use 125 fluorescent lighting units which draw 3 amperes each. Also, 20-ampere branch circuits will be used. The power factor of the units is 0.75. The operating voltage is 120 volts.

2. *Find*—The minimum number of branch circuits needed and the total power requirement for the lighting units.

3. *Solution* for the branch circuits:
   a. Find the number of units which should be installed on each branch circuit.

$$\text{Units} = \frac{\text{Branch current}}{\text{Load current}}$$

$$= \frac{20 \text{ amperes}}{3 \text{ amperes}}$$

$$= 6.67 \text{ units (round off at 6)}$$

-b. Find the number of branch circuits needed.

$$\text{Circuits} = \frac{\text{Lighting units}}{\text{No. units per branch circuit}}$$

$$= \frac{125}{6}$$

$$= 20.8 \text{ (round up to 21)}$$

Twenty-one 20-ampere branch circuits will be needed.

4. *Solution* for total power:
   a. Find the total current drawn by the lights.
      I = 125 units × 3 amperes
      = 375 amperes

   b. Calculate the total power.
      P = V × I × pf
      = 120 × 375 × 0.75
      = 33,750 watts (33.75 kW)

An even more common type of branch-circuit lighting problem is given in the following discussion. This problem begins with the minimum amount of lights required in a room. The number of branch circuits that are required must be determined.

1. *Given*—A large room in an industry needs to have 40,000 watts of incandescent lighting. It was decided that 120-volt 20-ampere branch circuits would be used. The lights will use 200-watt bulbs.

2. *Find*—The number of branch circuits required for the lighting.

3. *Solution:*

   a. Find the amount of current each lamp will draw.

$$I = \frac{P}{V}$$

$$= \frac{200 \text{ watts}}{120 \text{ volts}}$$

$$= 1.67 \text{ amperes}$$

   b. Find the maximum number of lamps in each branch circuit.

$$\text{Lamps} = \frac{\text{Amperage of circuit}}{\text{Amperage of lamps}}$$

$$= \frac{20 \text{ amperes}}{1.67 \text{ amperes}}$$

$$= 11.97 \text{ lamps (round off to 11)}$$

   c. Find the total number of bulbs (lamps) needed.

$$\text{Bulbs required} = \frac{\text{Wattage of room}}{\text{Wattage of bulbs}}$$

$$= \frac{40{,}0000 \text{ watts}}{200 \text{ watts}}$$

$$= 200 \text{ bulbs required}$$

   d. Find the minimum number of branch circuits needed.

$$\text{Branch circuits} = \frac{\text{Total bulbs required}}{\text{Number bulbs per circuit}}$$

$$= \frac{200}{11}$$

$$= 18.18 \text{ (round up to 19)}$$

   Nineteen branch circuits will be needed.

It should be pointed out that in these problems, we determined the *minimum* number of branch circuits. Some room for flexibility should be provided for each branch circuit. This can be done by designing the circuit to handle only 16 amperes (80%) rather than 20 amperes. Also, room should be provided on the power distribution panel to allow connecting some additional lighting branch circuits. This will allow additional loads to be connected at a later time.

LIGHTING FIXTURE DESIGN

Lighting fixtures are the devices which are used to hold lamps in place. They are commonly referred to as *luminaires.* Luminaires are used to *efficiently* transfer light from its source to a work surface. The proper designing of luminaires allows a more efficient transfer of light. It is important to keep in mind that *light intensity varies inversely as the square of the distance from the light source.* Thus, if the distance is doubled, the light intensity would be reduced four times.

Many factors must be considered in determining the amount of light which is transferred from a light bulb to a work surface. Some light is absorbed by the walls and by the light fixture itself; thus, not all light is efficiently transferred. The manufacturers of lighting systems develop charts that are used to predict the amount of light which will be transferred to a work surface. These charts consider the necessary variables for making a prediction of the quantity of light falling onto a surface.

Each luminaire has a rating that is referred to as its *coefficient of utilization.* The coefficient of utilization is a factor which expresses the percentage of light output which will be transferred from a lamp to a work area. The coefficient of a luminaire is determined by laboratory tests made by the manufacturer. These coefficiency charts also take into consideration the light absorption characteristics of walls, ceilings, and floors when determining the coefficient of utilization of a luminaire.

Another factor used for determining the coefficient of utilization is called the *room ratio.* Room ratio is very simply determined by the formula:

$$\text{Room Ratio} = \frac{W \times L}{H\ (W + L)}$$

where,

W is the room width in feet,

L is the room length in feet,

H is the distance in feet from the light source to the work surface.

Note: Work surfaces are considered to be 2.5 feet the floor, unless otherwise specified.

FACTORS IN DETERMINING LIGHT OUTPUT

There are several other factors which must be considered in determining the amount of light transferred from a light source to a work surface. We know for instance that the age of a lamp has an effect on its light output. Lamp manufacturers determine a *depreciation factor* or *maintenance factor* for luminaires. This factor expresses the percentage of light output available from a light source. A depreciation factor of 0.75 means that in the daily use of a light source, only 75% of the actual light output is available for transfer to the work surface. The depreciation factor is an average value. It considers the reduction of light output with age and the accumulation of dust and dirt on the luminaires. Some collect dust more easily than others.

The following problem shows the effect of the coefficient of utilization (CU) and the depreciation factor (DF) on the light output transferred to a work surface.

1. *Given*—A lighting system for a building has 16 luminaires. Each luminaire has two fluorescent lamps. Each lamp has a light output of 3000 lumens. The CU and DF found in the chart developed by the manufacturer are 0.45 and 0.90, respectively.

2. *Find*—The total light output from the lighting system.

3. *Solution:*

   a. Find the total light output from the lamps:
   3000 lumens per lamp × 16 luminaires ×
   2 lamps per luminaire = 96,000 lumens

b. Find the total light output:

$$\text{Light Output} = \text{Total Lumens} \times \text{CU} \times \text{DF}$$
$$= 96{,}000 \times 0.45 \times 0.90$$
$$= 38{,}880 \text{ lumens}$$

The distribution of light output onto work surfaces must also be considered. The coefficient of utilization and the depreciation factor are used to find the total light output of a lighting system. A greater light output is required for larger areas. The light output to a work surface is expressed as lumens per square foot or footcandles. A light meter is used to measure the quantity of light which reaches a surface.

The following two formulas are useful for finding the required lumens for a work area or the light output available from a particular lighting system:

$$\text{Total lumens} = \frac{\text{Desired footcandles} \times \text{Room area in ft.}}{\text{CU} \times \text{DF}}$$

and

$$\text{Footcandles available} = \frac{\text{Total lumens} \times \text{CU} \times \text{DF}}{\text{Room area in feet}}$$

Sample Problem:

Given: A lighting system with 12,000 total lumen output, luminaire CU = 0.83 and DF = 0.85 is installed in a 600-sq-ft room.

Find: The footcandles of light available on the work surface.

Solution:

$$\text{FC} = \frac{\text{Lumens} \times \text{CU} \times \text{DF}}{\text{Room Area}}$$

$$= \frac{12{,}000 \times 0.83 \times 0.85}{600}$$

$$\text{FC} = 14.11 \text{ Footcandles}$$

CONSIDERATIONS FOR ELECTRICAL LIGHTING LOADS

Incandescent lamps produce light by the passage of electrical current through a tungsten filament. The electrical current heats the filament to the point of incandescence which causes the lamp to produce light. The primary advantage of incandescent lights is their low initial cost. However, they have a very low efficacy (lumen/watt) rating. They also have a very high operating temperature and a short life expectancy. Incandescent lights are usually not good choices of light sources for commercial, industrial and outdoor lighting applications. Their major applications are for residential use.

Fluorescent light sources have a higher efficacy (lumens/watt) than incandescent lights. They have a much longer life and lower brightness and operating temperature. Fluorescent lights are used for residential, 120 volt applications and for general-purpose commercial and industrial lighting, 120 volt and 277 volt systems. Disadvantages of fluorescent lights include the necessity a ballast and a rather large luminaire. They also have a higher initial cost than incandescent lamps. It is estimated that fluorescent light sources provide approximately 70% of the lighting in the United States.

Vapor light sources also have very good efficacy ratings and long life expectancies. They have a high light output for their compact size. They are typically used for industrial, commercial and outdoor applications since they can be operated economically on higher voltage systems. The initial cost of vapor lights is high, they require a ballast, and they are a very bright light source. Their very high efficacy ratings have caused vapor light sources to gain increased use.

MECHANICAL SYSTEMS

Another broad category of electrical loads includes those devices that convert electrical energy into mechanical energy. Electric motors fall into this category of load devices. They are mechanical power-conversion systems. There are many types of motors used today. The electrical motor load is the major power-consuming load of electrical distribution systems. Motors of various sizes are used for purposes that range from large industrial machine operation to the small motors that are used to power blenders and mixers in the home.

BASIC MOTOR PRINCIPLES

The function of a motor is to convert electrical energy into mechanical energy in the form of a rotary motion. To produce a rotary motion, a motor must have an electrical power input. Generator action is brought about due to a magnetic field, a set of conductors within the magnetic field, and relative motion between the two. Motion is similarly produced in a motor due to the interaction of a magnetic field and a set of conductors.

All motors, regardless of whether they operate from an AC or a DC power line, have several basic characteristics in common. Their basic parts include (1) a stator, which is the frame and other stationary components; (2) a rotor, which is the rotating shaft and its associated parts; and (3) auxiliary equipment, such as a brush/commutator assembly for DC motors and a starting circuit for single-phase AC motors. The basic parts of a DC motor are shown in Figure 10-26.

The motor principle is illustrated in Figure 10-27. In Figure 10-27a, no current is flowing through the conductors due to the position of the brushes in relation to the commutator. During this condition, no motion is produced. When current flows through the conductor, a circular magnetic field is developed around the conductor. The direction of the current flow determines the direction of the circular magnetic fields, as shown in the cross-sectional diagram of Figure 10-27c.

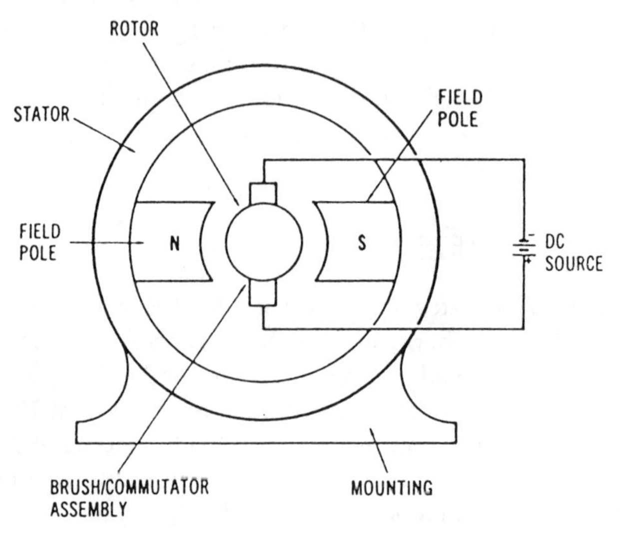

Figure 10-26. Basic parts of a DC motor.

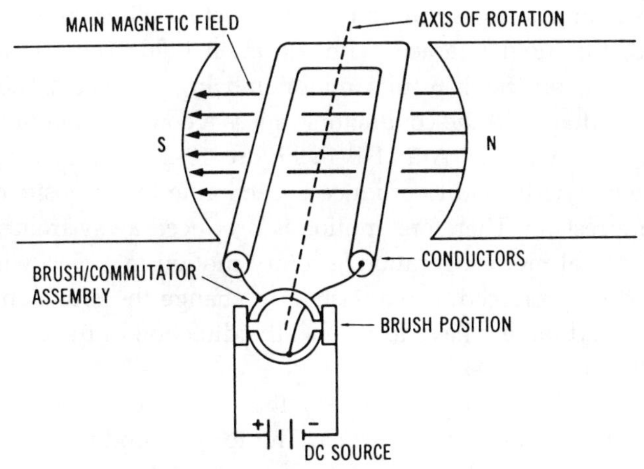

(A) Condition with no current flowing through conductors.

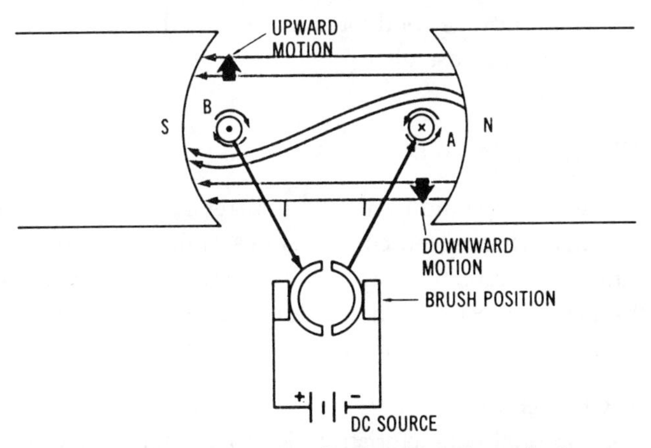

(B) Condition with current flowing through conductors.

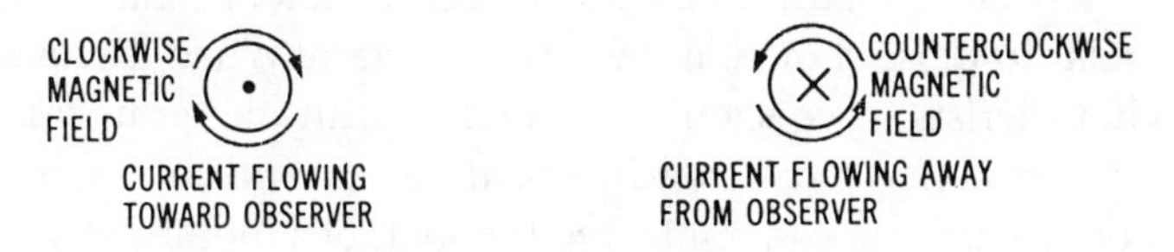

(C) Direction of current flow through conductors determines direction of magnetic field around conductors.

Figure 10-27. Illustration of the basic motor principle.

When current flows through the conductors within the main magnetic field, this field interacts with the main field. The interaction of these two magnetic fields results in motion being produced. The circular magnetic field around the conductors causes a compression of the main magnetic flux at points A and B in Figure 10-27b. This compression causes the magnetic field to produce a reaction in the opposite direction of the compression. Therefore, motion is produced away from points A and B. In actual motor operation, a rotary motion in a clockwise direction would be produced. If we wished to change the direction of rotation, we would merely have to reverse the direction of the current flow through the conductors.

The rotating effect produced by the interaction of two magnetic fields is called *torque* or *motor action*. The torque produced by a motor depends on the strength of the main magnetic field and the amount of current flowing through the conductors. As the magnetic field strength or the current through the conductors increases, the amount of torque or rotary motion will increase also.

DIRECT-CURRENT MOTORS

Motors which operate from direct-current power sources are often used in industry when speed control is desirable. They are classified as series, shunt, or compound machines depending on the method of connecting the armature and field windings. The permanent-magnet DC motor is another type of motor that is used for certain applications.

DC Motor Characteristics

The operational characteristics of all DC motors may be generalized by referring to Figure 10-28. Most electric motors exhibit characteristics similar to those shown in the block diagram. When discussing DC motor characteristics, we should be familiar with the terms *load, speed, counter-electromotive force* (cemf), *armature current,* and *torque.* The amount of mechanical load applied to the shaft of a motor determines its operational characteristics. As the mechanical load is increased, the speed of a motor tends to decrease. As speed decreases, the voltage induced into the conductors of the motor due to generator action (cemf) decreases. The generated voltage or counter-electromotive force depends upon the number of rotating conductors and the speed of rotation. Therefore, as speed of rotation decreases, so does the cemf.

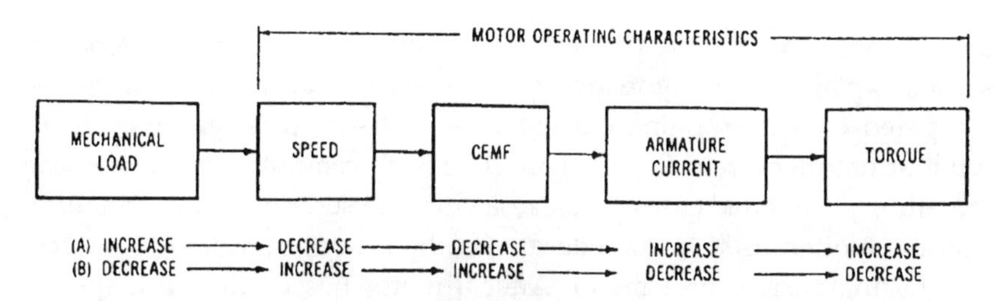

Figure 10-28. Operational characteristics of DC motors.

The counter-electromotive force generated by a DC motor is in opposition to the supply voltage. Therefore, the actual working voltage of a DC motor may be expressed as:

$$V_T = V_C + I_A R_A$$

where,

V_T is the terminal voltage of the motor in volts,
V_C is the cemf generated by the motor in volts,
$I_A R_A$ is the voltage drop across the armature of the motor in volts.

Since the cemf is in opposition to the supply voltage, the actual working voltage of a motor will increase as the cemf decreases. Due to an increase in working voltage, more current will flow through the armature conductors which are connected to the DC power supply. Since torque is directly proportional to armature current, the torque will increase as the armature current increases.

To briefly discuss the opposite situation, if the mechanical load connected to the shaft of a motor decreases, the speed of the motor would tend to increase. An increase in speed causes an increase in generated voltage. Since cemf is in opposition to the supply voltage, as cemf increases, the armature current decreases. A decrease in armature current causes a decrease in torque. We can see that torque varies with changes in load, but we need to consider each of the steps involved to understand DC motor operation. As the load on a motor is increased, its torque also increases to try to meet the increased load requirement. However; the current drawn by a motor also increases when load is increased.

The presence of a cemf to oppose armature current is very important in motor operation. The lack of any cemf when a motor is being started explains why motors draw a very large initial starting current as compared to their running current when full speed is reached. Maximum armature current flows when there is no cemf. As cemf increases, armature current decreases. Thus, resistances in series with the armature circuit are often used to compensate for the lack of cemf and to reduce the starting current of a motor. After a motor has reached full speed, these resistances may be bypassed by automatic or manual switching systems in order to allow the motor to produce maximum torque. Keep in mind that the armature current, which directly affects torque, can be expressed as:

$$I_A = \frac{V_T - V_C}{R_A}$$

where,

I_A is the armature current in amperes,
V_T is the terminal voltage of the motor in volts,
V_C is the cemf generated by the motor in volts,
R_A is the armature resistance in ohms.

In determining the functional characteristics of a motor, the torque developed can be expressed as:

$$T = K\Phi I_A$$

where,

T is the torque in foot-pounds,
K is a constant based on physical characteristics (conductor size, frame size, etc.)
Φ is the quantity of magnetic flux between poles,
I_A is the armature current in amperes.

Torque can be measured by several types of motor analysis equipment.

The horsepower rating of a motor is based on the amount of torque produced at the rated full-load values. Horsepower, which is the usual method of rating motors, can be expressed mathematically as:

$$hp = \frac{2\pi NT}{33{,}000}$$

$$= \frac{NT}{5252}$$

where,

hp is the horsepower rating,
2π is a constant,
N is the speed of the motor in revolutions per minute (rpm),
T is the torque developed by the motor in foot-pounds.

The most desirable characteristic of DC motors is their speed-control capability. By varying the applied DC voltage with a rheostat, speed can be varied from zero to the maximum rpm of the motor. Some types of DC motors have more desirable speed characteristics than others. For this reason, we can determine the comparative speed regulation for different types of motors. Speed regulation is expressed as:

$$\% R = \frac{S_{NL} - S_{FL}}{S_{FL}} \times 100$$

where,

% R is the percentage of speed regulation,
S_{NL} is the no-load speed in rpm,
S_{FL} is the rated full-load speed in rpm.

Good speed regulation (low %R) results when a motor has nearly constant speeds under varying load situations.

Types of DC Motors

The types of commercially available DC motors basically fall into four categories: (1) permanent-magnet DC motors, (2) series-wound DC motors, (3) shunt-wound DC motors, and (4) compound-wound DC motors. Each of these motors has different characteristics due to its basic circuit arrangement and physical properties.

Permanent-magnet DC Motors

The permanent-magnet DC motor, shown in Figure 10-29 is constructed in the same manner as its DC generator counterpart that was discussed in Chapter 7. The permanent-magnet motor is used for low-torque applications. When this type of motor is used, the DC power supply is connected directly to the armature conductors through the brush/commutator assembly. The magnetic field is produced by permanent magnets mounted on the stator.

This type of motor ordinarily uses either alnico or ceramic permanent magnets rather than field coils. The alnico magnets are used with high-horsepower applications. Ceramic magnets are ordinarily used for low-horsepower slow-speed motors. Ceramic magnets are highly resistant to demagnetization, yet they are relatively low in magnetic-flux level. The magnets are usually mounted in the motor frame and, then, magnetized prior to the insertion of the armature.

The permanent-magnet motor has several advantages over conventional types of DC motors. One advantage is a reduced operational cost. The speed characteristics of the permanent-magnet motor are similar to the shunt-wound DC motor. The direction of rotation of a permanent-magnet motor can be reversed by reversing the two power lines.

Series-wound DC Motors

The manner in which the armature and field circuits of a DC motor are connected determines its basic characteristics. Each of the types of DC motors are similar in construction to the type of DC generator that corresponds to it. The only difference, in most cases, is that the generator acts as a voltage source while the motor functions as a mechanical power-conversion device.

The series-wound motor, shown in Figure 10-30, has the armature and field circuits connected in a series arrangement. There is only one path for current to flow from the DC voltage source. Therefore, the field is wound of relatively few turns of large-diameter wire, giving the field a low resistance. Changes in load applied to the motor shaft causes changes in the current through the field. If the mechanical load increases, the current also increases. The increased current creates a stronger magnetic field. The speed of a series motor varies from very fast at no load to very low at heavy loads. Since large currents may flow through the low-resistance field, the series motor produces a high torque output. Series motors are used where heavy loads must be moved and

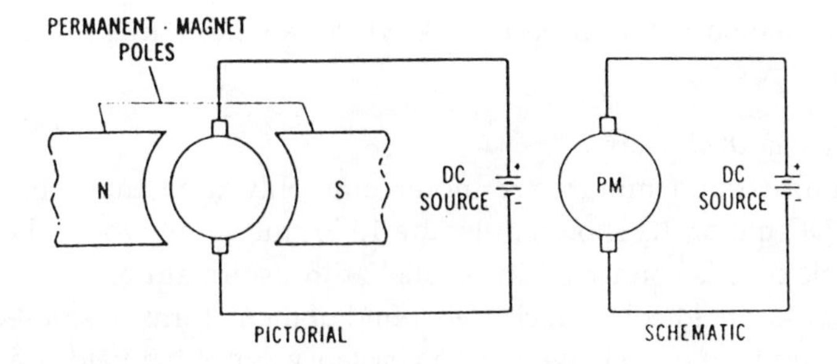

Figure 10-29. Permanent-magnet DC motor.

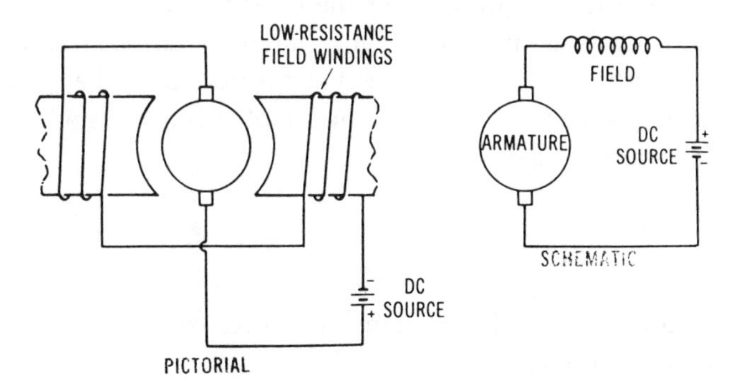

Figure 10-30. Series-wound DC motor.

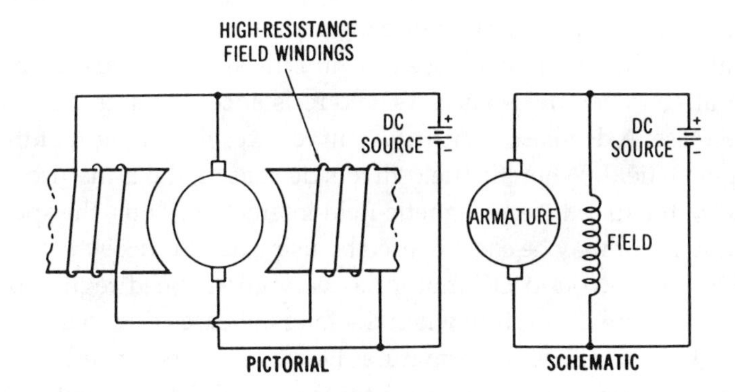

Figure 10-31. Shunt-wound DC motor.

speed regulation is not important. A typical application is for automobile starter motors.

Shunt-wound DC Motors

Shunt-wound motors are more commonly used than any other type of DC motor. As shown in Figure 10-31, the shunt-wound DC motor has field coils connected in parallel with its armature. This type of DC motor has field coils which are wound of many turns of small-diameter wire and have a relatively high resistance. Since the field is a high-resistant parallel path of the circuit of the shunt motor, a small amount of current flows through the field. A strong electromagnetic field is produced due to the many turns of wire which form the field windings.

A large majority (about 95%) of the current drawn by the shunt motor flows in the armature circuit. Since the field current has little effect on the strength of the field, motor speed is not affected appreciably by variations in load current. The relationship of the currents which flow through a DC shunt motor is as follows:

$$I_T = I_A + I_F$$

where,

I_T is the total current drawn from the power source,
I_A is the armature current,
I_F is the field current.

The field current may be varied by placing a variable resistance in series with the field windings. Since the current in the field circuit is low, a low-wattage rheostat may be used to vary the speed of the motor due to the variation in field resistance. As field resistance increases, field current will decrease. A decrease in field current reduces the strength of the electromagnetic field. When the field flux is decreased, the armature will rotate *faster,* due to reduced magnetic-field interaction. Thus, the speed of a DC shunt motor may be easily varied by using a field rheostat.

The shunt-wound DC motor has very good speed regulation. The speed does decrease slightly when the load increases due to the increase in voltage drop across the armature. Due to its good speed regulation characteristic and its ease of speed control, the DC shunt motor is commonly used for industrial applications. Many types of variable-speed machine tools are driven by DC shunt motors, such as shown in Figure 10-32.

Compound-wound DC Motors

The compound-wound DC motor, shown in Figure 10-33, has two sets of field windings, one in series with the armature and one in parallel. This motor combines the desirable characteristics of the series- and shunt-wound motors. It has high torque similar to a series-wound motor, along with good speed regulation similar to a shunt motor. Therefore, when good torque and good speed regulation are needed, the compound-wound DC motor can be used. A major disadvantage of a compound-wound motor is its expense.

Changing direction of rotation, a DC motor is designed so that its shaft will rotate in either direction. It is a very simple process to reverse the direction or rotation of any DC motor. By reversing the relationship between the connections of the armature winding and field windings, reversal of rotation is achieved. Usually, this is done by changing the terminal connections where the power source is connected to the motor. Four terminals are ordinarily used for interconnection purposes. They may be labeled A1 and A2 for the armature connections and F1 and F2 for the field connections. If either the armature connections or the field connections are reversed, the rotation of the motor will reverse. However, if both are reversed, the motor shaft will rotate in its original direc-

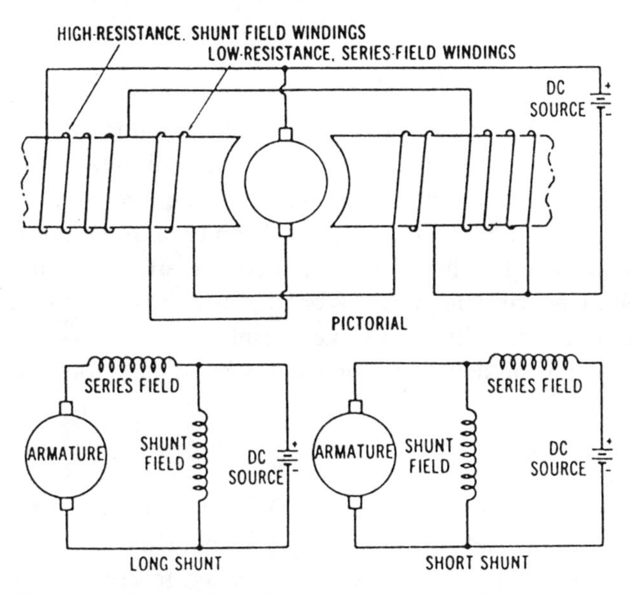

Figure 10-32. Typical direct-current motor (Courtesy of General Electric Co., DC Motor and Generator Dept.).

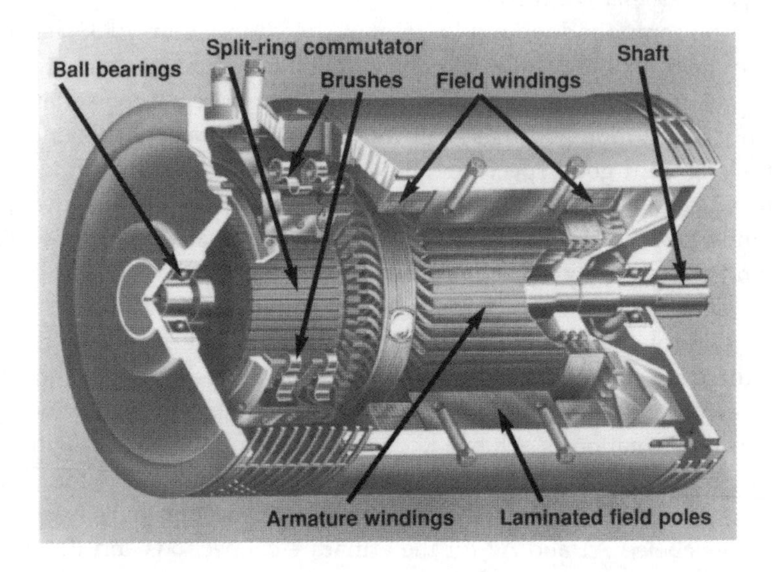

Figure 10-33. Compound-wound DC motor.

tion, since the relationship between the armature and field windings would be the same.

Brushless DC Motors

The use of solid state devices has resulted in the development of brushless DC motors which have neither brushes nor commutator assemblies. Instead they make use of solid-state switching circuits. The major problem with most DC motors is the low reliability of the commutator/brush assembly. The brushes have a limited life and cause the commutator to wear. This wearing produces brush dust which can cause other maintenance problems. Although some brushless DC motors use other methods the transistor-switched motor is the most common (see Figure 10-34).

DC Stepping Motors

Direct-current stepping motors are unique DC motors that are used to control automatic industrial processing equipment. Direct-current motors of this type are found in numerically controlled machines and robotic systems used by industry. They are very efficient and develop a

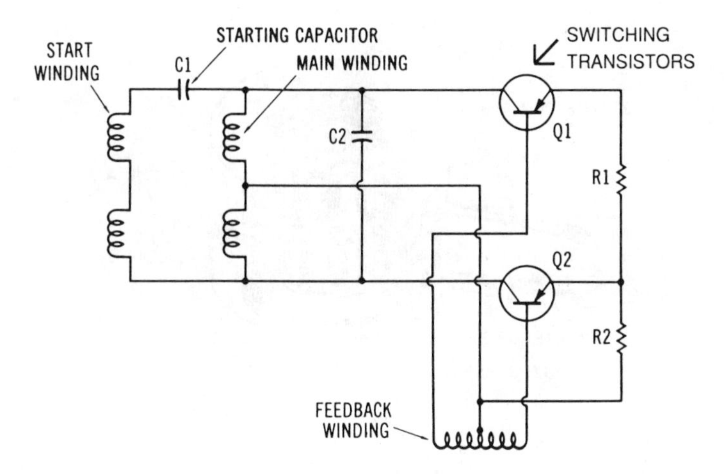

Figure 10-34. A brushless DC motor circuit.

high torque. The stepping motor is used primarily to change electrical pulses into a rotary motion that can be used to produce mechanical movements.

The shaft of a DC stepping motor rotates a specific number of mechanical degrees with each incoming pulse of electrical energy. The amount of rotary movement or angular displacement produced by each pulse can be repeated precisely with each succeeding pulse from the drive source. The resulting output of this device is used to accurately locate or position automatic process machinery.

The velocity distance and direction of movement of a specific machine can be controlled by DC stepping motors. The movement error of this device is generally less than 5% per step. This error factor is not cumulative regardless of the distance moved or the number d steps taken. Motors of this type are energized by a DC drive amplifier that is controlled by digital logic circuits. The drive-amplifier circuitry is a key factor in the overall performance of this motor. Some DC stepping motors are shown in Figure 10-35. The construction and coil layout are shown in Figure 10-35b. The rotor of a stepping motor is of a permanent-magnet type of construction. The direction of rotation can now be predicted and is determined by the polarities of the stator-coil sets. Adding more stator-coil pairs to a motor of this type Improves its rotation and makes the stepping action very accurate. Figure 10-36 shows an electrical diagram of a DC stepping motor.

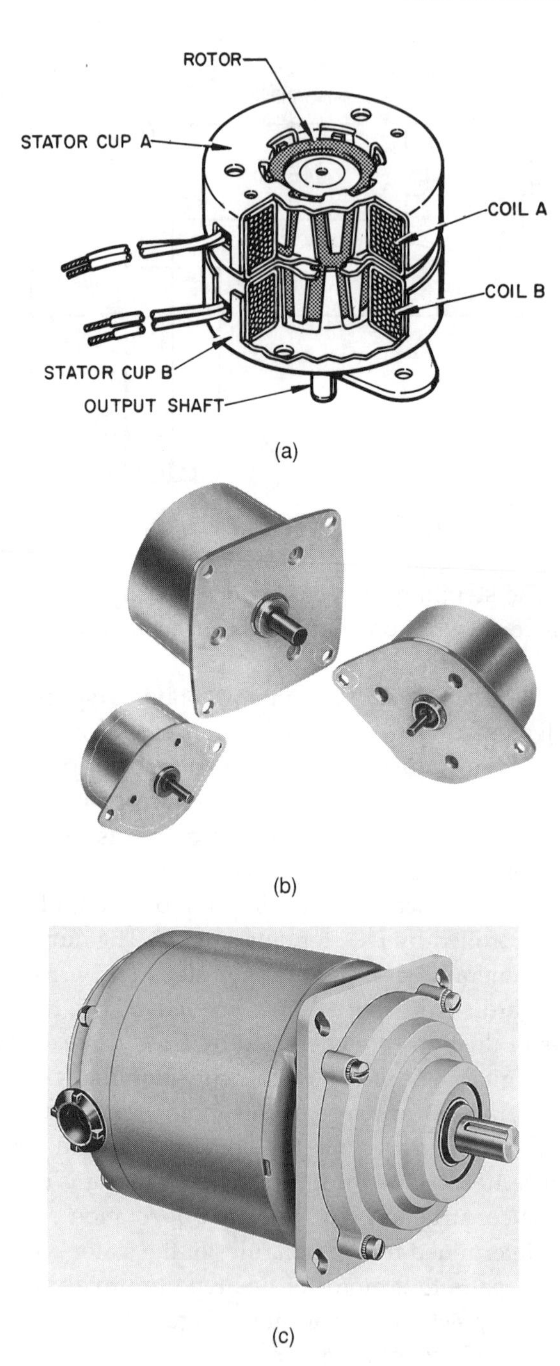

Figure 10-35. DC Stepping motors.

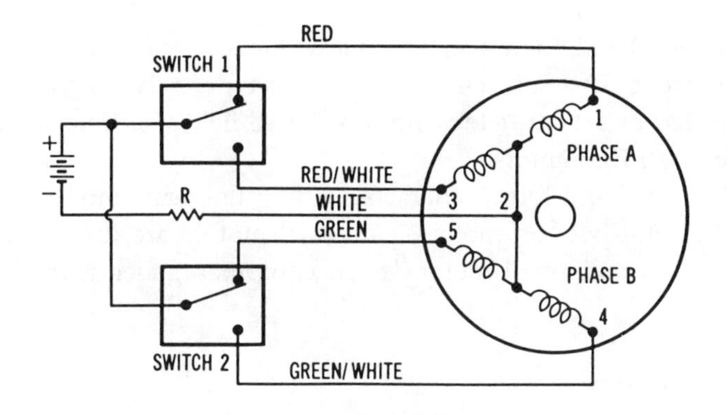

SWITCHING SEQUENCE *

STEP	SWITCH 1	SWITCH 2
1	1	5
2	1	4
3	3	4
4	3	5
1	1	5

*To reverse direction, read chart up from bottom.

Figure 10-36. Circuit diagram and switching sequence of a DC stepping motor.

SINGLE-PHASE AC MOTORS

Another classification of mechanical power-conversion equipment includes single-phase-alternating-current (AC) motors. These motors are common for industrial as well as commercial and residential usage.

They operate from a single-phase AC power source. There are three basic types of single-phase AC motors—universal motors, induction motors, and synchronous motors.

Universal Motors

Universal motors may be powered by either AC or DC power sources. The universal motor, shown in Figure 10-37, is constructed in the same way as a series-wound DC motor. However, it is designed to operate with either AC or DC applied. The series-wound motor is the only type of DC motor which will operate with AC applied. The windings of shunt-wound motors have inductance values that are too high to allow the motor to function with AC applied. However, series-wound motors have windings that have low inductances (few turns of large diameter wire) and, therefore, offer a low impedance to the flow of alternating current. The universal motor is one type of AC motor that has concentrated or salient field windings. These field windings are similar to those of all DC motors.

The speed and torque characteristics of universal motors are similar to DC series-wound motors. Universal motors are used mainly for portable tools and small motor-driven equipment, such as mixers and blenders.

Induction Motors

Another popular type of single-phase AC motor operates on the induction principle. This principle is illustrated in Figure 10-38. The coil symbols along the stator represent the field coils of an induction motor. These coils are energized by an alternating-current source; therefore, their instantaneous polarity changes 120 times per second when 60-Hz AC is applied.

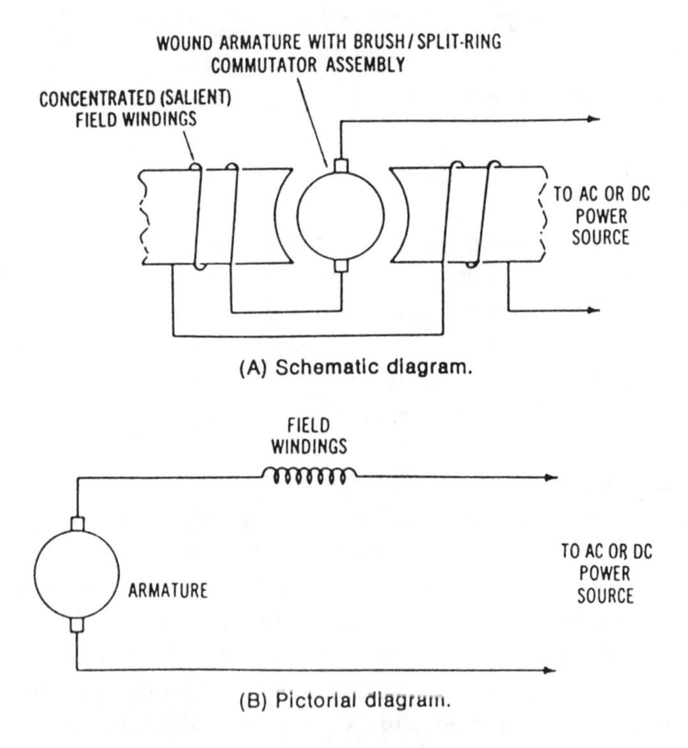

Figure 10-37. The universal motor.

Induction motors have a solid rotor which is referred to as a *squirrel-cage* motor. This type of rotor, which is illustrated in Figure 10-38b, has large-diameter copper conductors that are soldered at each end to a circular connecting plate. This plate actually short circuits the individual conductors of the rotor. When current flows in the stator windings, a current is induced in the rotor. This current is developed due to "transformer action" between the stator and rotor. The stator, which has AC voltage applied, acts as a transformer primary. The rotor is similar to a transformer secondary since current is induced into it.

It should be pointed out that the speed of an AC induction motor is based on the speed of the rotating magnetic field and the number of stator poles that the motor has. The speed of the rotor will never be as high as the speed of the rotating stator field. If the two speeds were

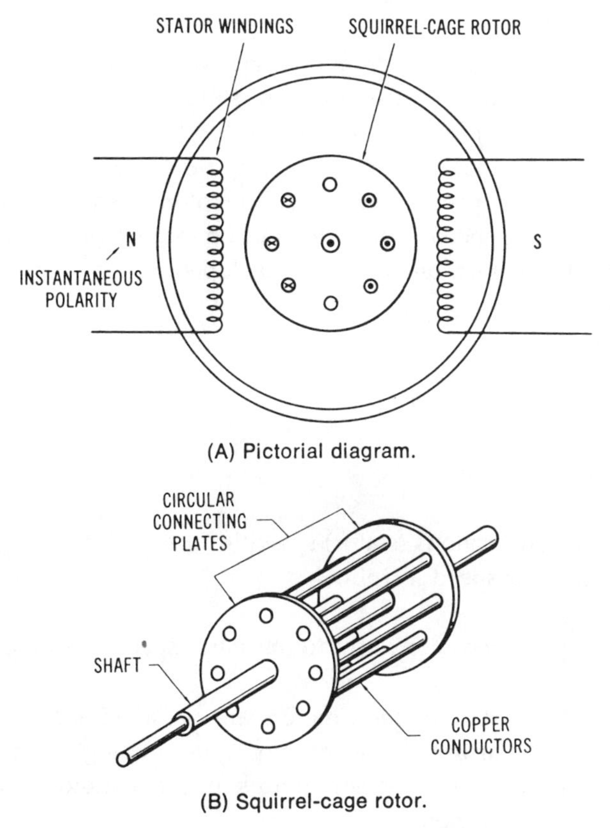

(A) Pictorial diagram.

(B) Squirrel-cage rotor.

Figure 10-38. Illustration of the induction principle.

equal, there would be no relative motion between the rotor and stator and, therefore, no induced rotor current and torque developed. The rotor speed (operating speed) of an induction motor is always somewhat less than the rotating stator field developed by the applied AC voltage.

The speed of the rotating stator field may be expressed as:

$$S = \frac{f \times 120}{n}$$

where,

S is the speed of the rotating stator field in rpm,
f is the frequency of the applied AC voltage in hertz,
n is the number of poles in the stator windings,
120 is a conversion constant.

A two-pole motor operating from a 60-Hz source would have a stator speed of 3600 revolutions per minute. The stator speed is also referred to as the synchronous speed of a motor.

The difference between the revolving stator speed of an induction motor and the rotor speed is called *slip.* The rotor speed must lag behind the revolving stator speed in order to develop torque. The more the rotor speed lags behind, the more torque is developed. Slip is expressed mathematically as:

$$\% \text{ slip} = \frac{S_S - S_r}{S_S} \times 100$$

where,

S_S is the synchronous (stator) speed in rpm,
S_r is the rotor speed in rpm.

As the rotor speed becomes closer to the stator speed, the percentage of slip becomes smaller.

Single-phase AC induction motors are classified according to the method used for starting. Some common types of single-phase AC induction motors include *split-phase* motors, *capacitor* motors, *shaded-pole* motors, and *repulsion* motors.

Split-phase Induction Motors

The split-phase AC induction motor, shown in Figure 10-39, has two sets of stator windings. One set, called the *run windings*, is connected directly across the AC line. The other set, called the *start windings*, *is* also connected across the AC line. However, the start winding is con-

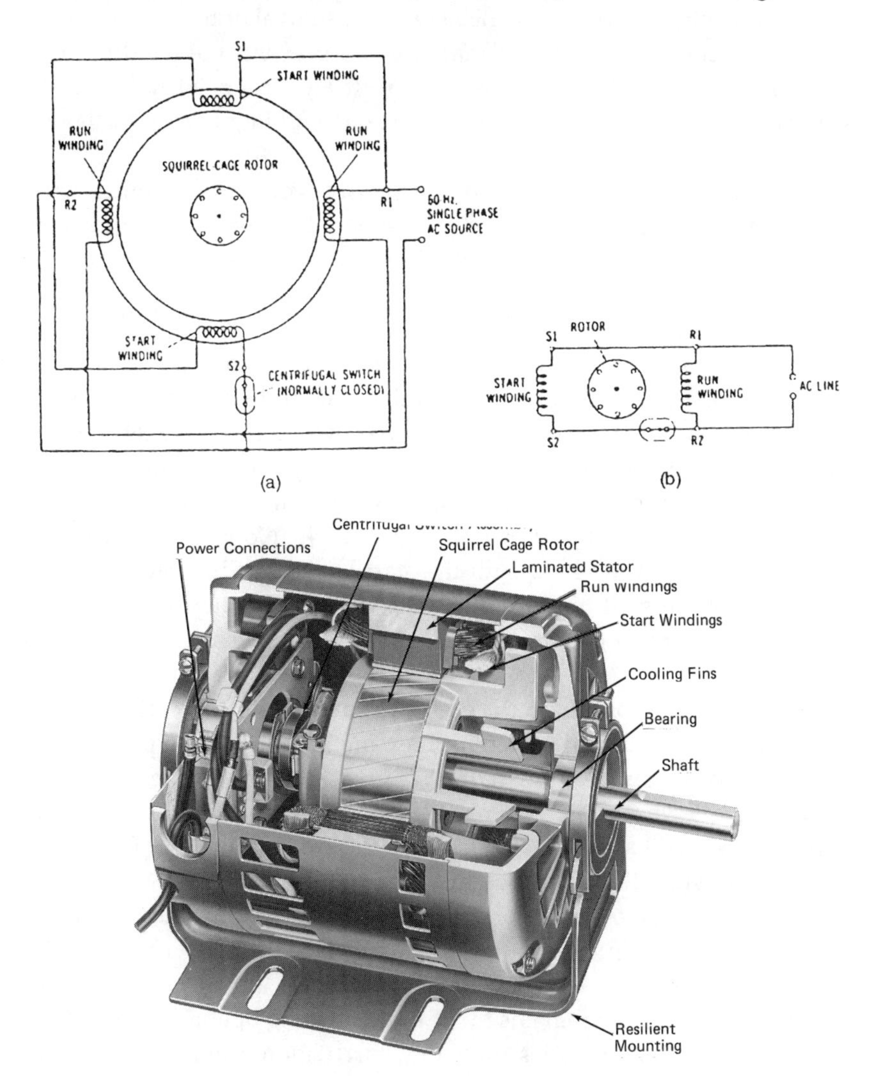

Figure 10-39. Split-phase AC induction motor: (a) pictorial diagram; (b) schematic diagram; (c) cutaway view ([c] Courtesy of Marathon Electric).

nected in series with a centrifugal switch that is mounted on the shaft of the motor. The centrifugal switch is in the closed position when the motor is not rotating.

When the split-phase AC induction motor reaches about 80% of its normal operating speed, the centrifugal switch will open the start winding circuit since it is no longer needed. The removal of the start winding minimizes energy losses in the machine and prevents the winding from overheating. When the motor is turned off and its speed reduced, the centrifugal switch closes to connect the start winding back into the circuit.

Split-phase motors are fairly inexpensive compared to other types of single-phase motors. They are used where low torque is required to drive mechanical loads such as small machinery.

Capacitor Motors

Capacitor motors are an improvement over the split-phase AC motor. Construction features are similar. All induction motors have squirrel-cage rotors that look like the one shown in Figure 10-39. You can also clearly see the centrifugal switch assembly and the start and run windings in Figure 10-39. The wiring diagram of a capacitor-start single-phase induction motor is shown in Figure 10-40. Notice that, except for a capacitor placed in series with the start winding, this diagram is the same as for the split-phase motor. The purpose of the capacitor is to cause the current in the start winding to lead (rather than lag) the applied voltage.

The starting torque produced by a capacitor-start induction motor is much greater than that of a split-phase motor. Thus, this type of motor can be used for applications requiring greater initial torque. However, they are somewhat more expensive than split-phase AC motors. Most capacitor motors, as well as split-phase motors, are used in fractional-horsepower sizes (less than one hp).

Another type of capacitor motor is called a *capacitor-start, capacitor-run,* or *two-value capacitor* motor. Its circuit is shown in Figure 10-41a. This motor employs two capacitors. One, of low value, is in series with the start winding, and remains in the circuit during operation. The other, of higher value, is in series with the start winding and a centrifugal switch. The larger capacitor is used only to increase starting torque and is removed from the circuit during normal operation by the centrifugal switch. The smaller capacitor and the entire start winding are part of the operational circuit of the motor. The smaller capacitor helps to produce a more constant-running torque, as well as quieter operation and an improved power factor.

Still another type of capacitor motor is one that is called a *permanent capacitor* motor. Its circuit is shown in Figure 10-41b. This motor has no centrifugal switch, so its capacitor is permanently connected into the circuit. These motors are only used for very low torque requirements and are made in small fractional horsepower sizes.

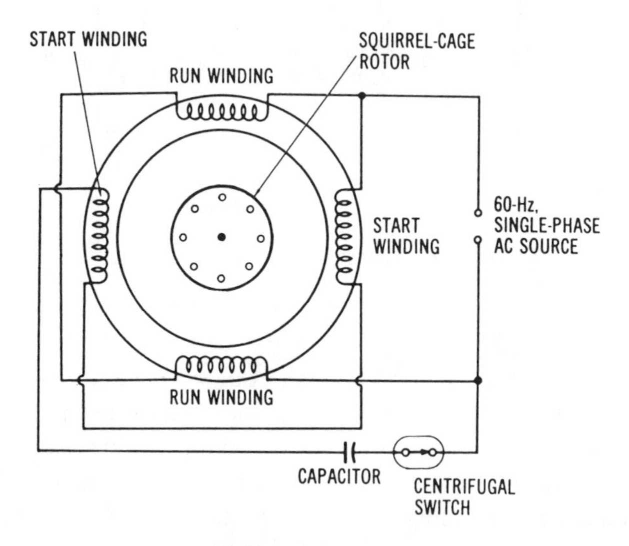

(A) Pictorial diagram.

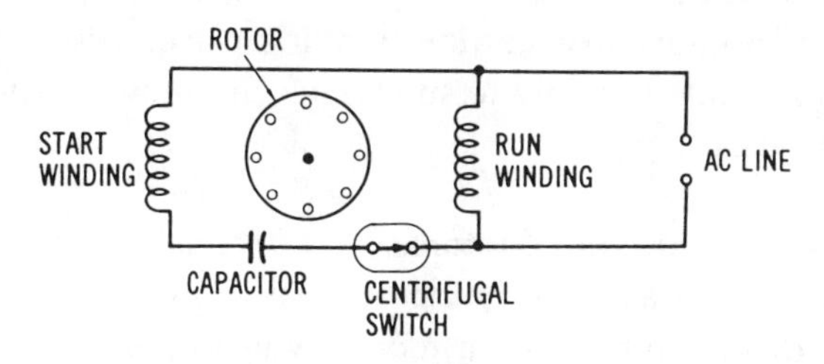

(B) Schematic diagram.

Figure 10-40. The capacitor-start single-phase induction motor.

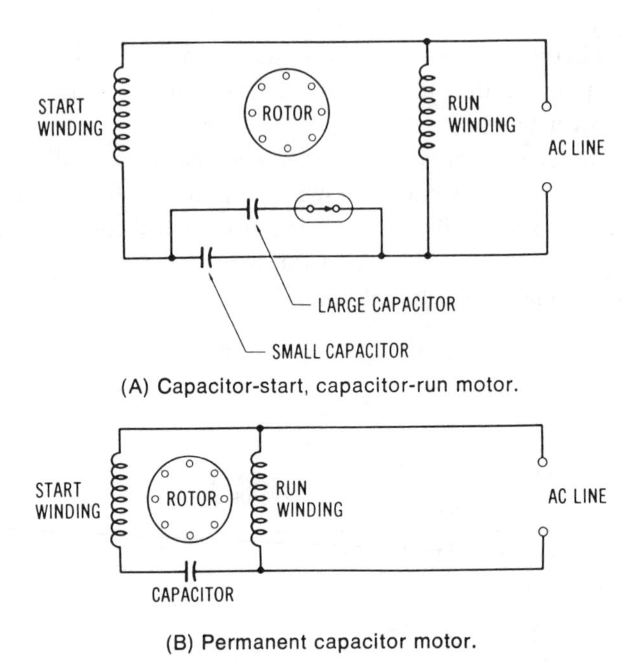

Figure 10-41. Other types of capacitor motors.

Shaded-pole Induction Motors

Another method of producing torque by a simulated two-phase method is called *pole shading.* A shaded-pole motor is shown in Figure 10-42. These motors are used for very low-torque applications such as fans and blower units. They are low-cost, rugged, and reliable motors that ordinarily come in low horsepower ratings, from 1/300 to 1/30 hp, with some exceptions.

The shaded-pole motor is inexpensive since it uses a squirrel-cage rotor and has no auxiliary starting winding or centrifugal mechanism. Application is limited mainly to small fans and blowers and other low-torque applications.

Single-phase Synchronous Motors

It is often desirable, in timing or clock applications, to use a constant-speed drive motor. Such a motor, which operates from a single-phase AC line, is called a *synchronous* motor. The single-phase synchronous motor has stator windings which are connected across the AC line. Its rotor is made of a permanent magnetic material. Once the rotor is

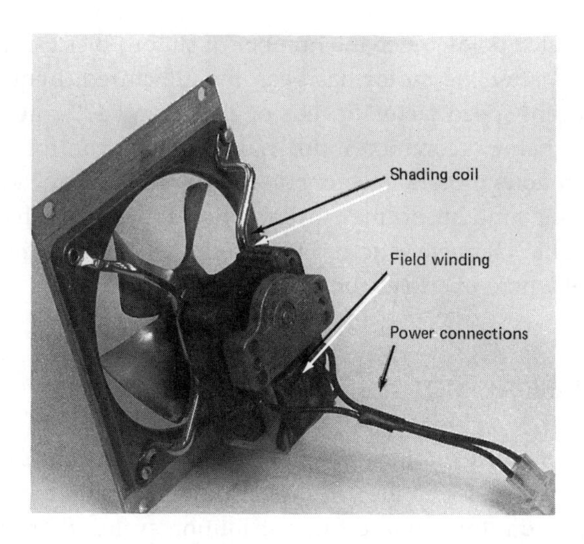

Figure 10-42. Shaded-pole induction motor.

started, it will rotate in synchronism with the revolving stator field, since it does not rely upon the induction principle. The calculation of the speeds of synchronous motors is based on the speed formula. This formula states that

$$\text{revolutions per minute} = \frac{\text{frequency} \times 60}{\text{number of pairs of poles}}$$

Therefore, for 60-Hz operation, the following synchronous speeds would be obtained:

1. Two-pole	3600 rpm.
2. Four-pole	1800 rpm.
3. Six-pole	1200 rpm.
4. Eight-pole	900 rpm.
5. Ten-pole	720 rpm.
6. Twelve-pole	600 rpm.

Small synchronous motors are used in single-phase applications for low torque applications. Such applications include clocks, phonograph drives, and timing devices which require constant speeds.

The speed; of a synchronous motor is directly proportional to the frequency of the applied AC and inversely proportional to the number of pairs of stator poles. Since the number of stator poles cannot be effectively altered after the motor has been manufactured, frequency is the most significant speed factor. Speeds of 28, 72, and 200 rpm are typical, with 72 rpm being a common industrial numerical control standard.

Synchronous motors have one very important characteristic—they draw the same amount of line current when stalled that they do when operating. This characteristic is important in automatic machine-tool applications where overloads occur frequently.

THREE-PHASE AC MOTORS

Three-phase AC motors are often called the "workhorses of industry." Most motors used in industry and several in commercial buildings are operated from three-phase AC distribution systems. There are three basic types of three-phase motors: (1) induction motors, (2) synchronous motors, and (3) wound-rotor induction motors (wrim).

Induction Motors

A pictorial diagram showing the construction of a three-phase induction motor is given in Figure 10-43. Note that the construction of this motor is very simple. It has only a distributive-wound stator which is connected in either a wye or a delta configuration, and a squirrel-cage rotor. Since three-phase voltage is applied to the stator, phase separation is already established. No external starting mechanisms are needed. Three-phase induction motors come in a variety of integral horsepower sizes and have good starting and running torque characteristics. A cutaway view of a three-phase induction motor is shown in Figure 10-44. Notice the parts of the motor which are labeled.

The direction of rotation of a three-phase motor of any type can be changed very easily. If any two power lines coming into the stator windings are reversed, the direction of rotation of the shaft will change. Three-phase induction motors are used for many applications, such as mechanical-energy sources for machine tools, pumps, elevators, hoists, conveyors, and other systems which use large amounts of power.

Synchronous Motors

Three-phase synchronous motors are unique and very specialized motors. They are considered a constant-speed motor and they can be used to correct the power factor of three-phase systems. Synchronous

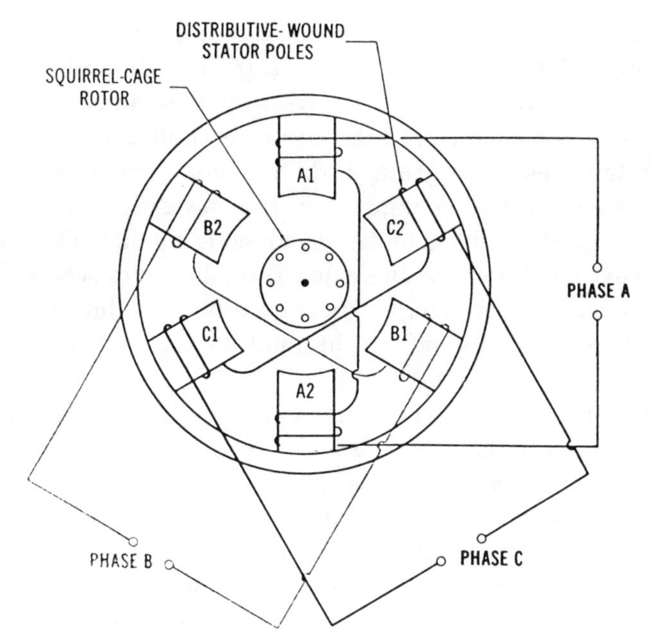

Figure 10-43. Pictorial diagram of the construction of a three-phase induction motor.

Figure 10-44. Cutaway of a three-phase AC induction motor (Courtesy of Marathon Electric).

motors are usually very large in size and horsepower rating.

Figure 10-45 shows a pictorial diagram of the construction of a three-phase synchronous motor. Physically, this motor is constructed like a three-phase alternator. Direct current is applied to the rotor to produce a rotating electromagnetic field; the stator windings are connected in either a wye or delta configuration. The only difference is that three-phase AC power is applied to the synchronous motor, while three-phase power is extracted from the alternator. Thus, the motor acts as an electrical load, while the alternator functions as a source of three-phase power. This relationship should be kept in mind during the following discussion.

The three-phase synchronous motor differs from the three-phase induction motor in that the rotor is wound and is connected through a slip ring/brush assembly to a DC power source. Three-phase synchronous motors, in their pure form, have no starting torque. Some external means must be used to initially start the motor. Synchronous motors are constructed so that they will rotate at the same speed as the revolving stator field. We can say that at synchronous speed, rotor speed equals

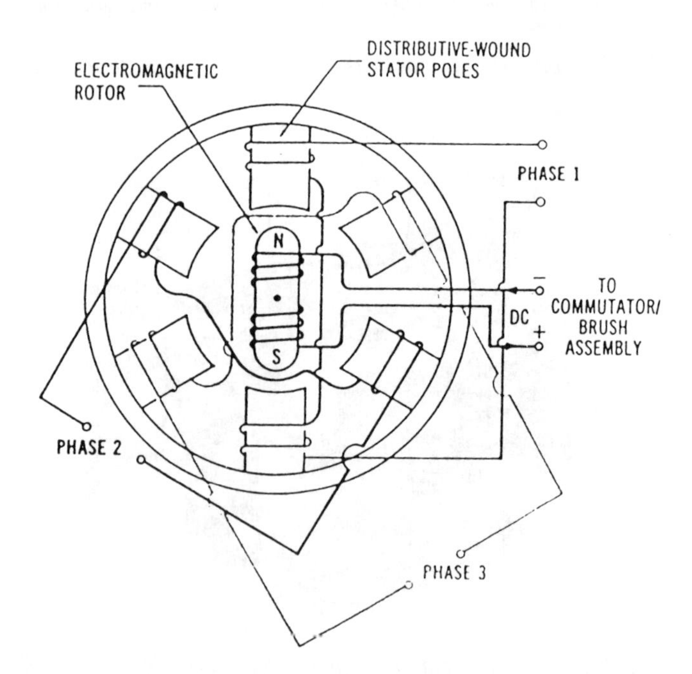

Figure 10-45. Pictorial diagram of three-phase synchronous motor construction.

stator speed and the motor has zero slip. Thus, we can determine the speed of a synchronous motor by using the following formula:

$$S = \frac{f \times 120}{n/3}$$

where,

S is the speed of a synchronous motor in r/min,
f is the frequency of the applied AC voltage in hertz,
n/3 is the number of stator poles per phase,
120 is a conversion constant.

Note that this is the same as the formula used to determine the stator speed of a single-phase motor except that the number of poles must be divided by three (the number of phases). A three-phase motor with twelve actual poles would have four poles per phase. Therefore, its stator speed would be 1800 rpm. Synchronous motors have operating speeds that are based on the number of stator poles they have.

Three-phase synchronous motors usually are employed in very large horsepower ratings. One method of starting a large synchronous motor is to use a smaller auxiliary DC machine connected to the shaft of the synchronous motor, as illustrated in Figure 10-46a. The method of starting would be as follows:

Step 1. DC power is applied to the auxiliary motor, causing it to increase in speed. Three-phase AC power is applied to the stator.

Step 2. When the speed of rotation reaches a value near the synchronous speed of the motor, the DC power circuit is opened and, at the same time, the terminals of the auxiliary machine are connected across the slip ring/brush assembly of the rotor.

Step 3. The auxiliary machine now converts to generator operation and supplies exciter current to the rotor of the synchronous motor, using the motor as its prime mover.

Step 4. Once the rotor is magnetized, it will "lock" in step or synchronize with the revolving stator, field.

Step 5. The speed of rotation will remain constant under changes in load condition.

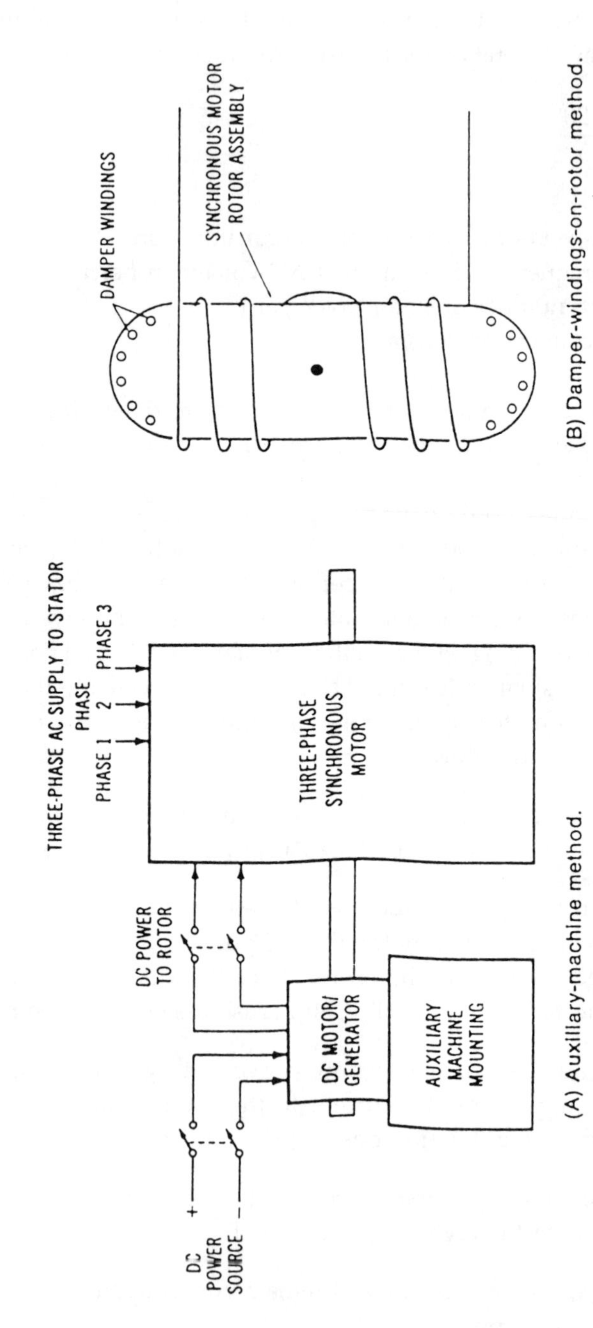

Figure 10-46. Three-phase synchronous motor starting methods.

Another starting method is shown in Figure 10-46b. This method utilizes *damper* windings, which are similar to the conductors of a squirrel-cage rotor. These windings are placed within the laminated iron of the rotor assembly. No auxiliary machine is required when damper windings are used. The starting method used is as follows:

Step 1. Three-phase AC power is applied to the stator windings.

Step 2. The motor will operate as an induction motor due to the "transformer action" of the damper windings.

Step 3. The motor speed will build up so that the rotor speed is somewhat less than the speed of the revolving stator field.

Step 4. DC power from a rotating DC machine or, more commonly, a rectification system is applied to the slip ring/brush assembly of the rotor.

Step 5. The rotor becomes magnetized and builds up speed until rotor speed is equal to stator speed.

Step 6. The speed of rotation remains constant regardless of the load placed on the shaft of the motor.

An outstanding advantage of the three-phase synchronous motor is that it can be connected to a three-phase power system to increase the overall power factor of the system. Power factor correction was discussed in Chapter 3. Three-phase synchronous motors are sometimes used only to correct the system power factor. If no load is intended to be connected to the shaft of a three-phase synchronous motor, it is called a *synchronous capacitor.* It is designed to act only as a power factor corrective machine. Of course, it might be beneficial to use this motor as a constant-speed drive connected to a load, as well as for power factor correction.

We know from previous discussions that a low power factor cannot be tolerated by an electrical power system. Thus, the expense of installing three-phase synchronous machines could be justified by industries, in order to appreciably increase the system power factor. To understand how a three-phase synchronous machine operates as a power fac-

stand how a three-phase synchronous machine operates as a power factor corrective machine, refer to the curves of Figure 10-47. We know that the synchronous motor operates at a constant speed. Variation in rotor DC-excitation current has no effect on speed. The excitation level will change the power factor at which the machine operates. Three operational conditions may exist, depending on the amount of DC excitation applied to the rotor. These conditions are:

1. Normal excitation—operates at a power factor of 1.0.

2. Under excitation—operates at a lagging power factor (inductive effect).

3. Over excitation—operates at a leading power factor (capacitive effect).

Note the variation of stator current drawn by the synchronous motor as the rotor current varies. You should also see that stator current is minimum when the power factor equals 1.0, or 100%. The situations shown on the graph in Figure 10-47a indicate stator and rotor currents under no-load, half-load, and full-load conditions. Current values when the power factors are equal to 1.0, 0.8-leading, and 0.8-lagging conditions are also shown. These curves are sometimes referred to as *V-curves* for a synchronous machine. The graph in Figure 10-47a shows the variation of power factor with changes in rotor current under three different load conditions. Thus, a three-phase synchronous motor, when over-excited, can improve the overall power factor of a three-phase system.

As the load increases, the angle between the stator pole and the corresponding rotor pole on the synchronous machine increases. The stator current will also in crease. However, the motor will remain synchronized unless the load causes "pull-out" to take place. The motor would then stop rotating due to the excessive torque required to rotate the load. Most synchronous motors are rated greater than 100 horsepower and are used for many industrial applications requiring constant-speed drives.

Wound-rotor Induction Motor

The wound-rotor induction motor (wrim), shown in Figure 10-48 is a specialized type of three-phase motor. This motor may be controlled

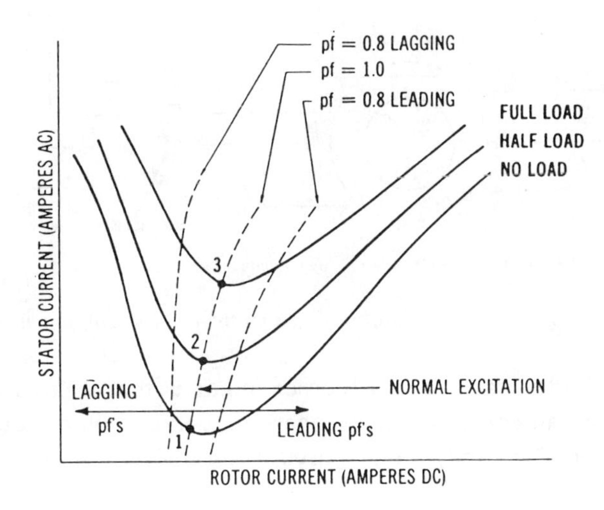

(A) Rotor current versus stator current.

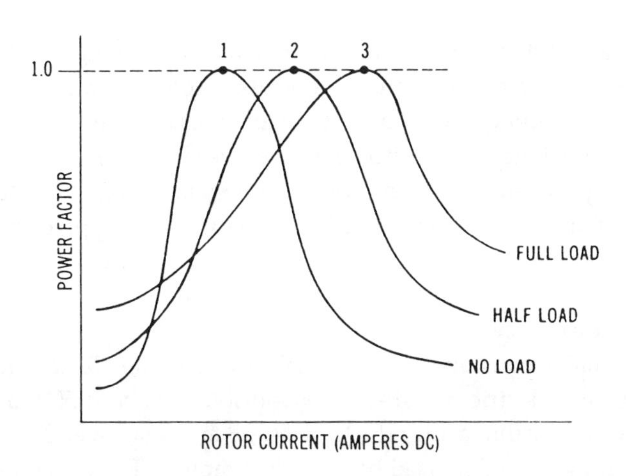

(B) Rotor current versus power factor.

Figure 10-47. Power relationships of a three-phase synchronous motor.

ing torque of a wrim motor can be varied by the value of external resistance. The advantages of this type of motor are a lower starting current, a high starting torque, smooth acceleration, and ease of control. The major disadvantage of this type of motor is that it costs a great deal

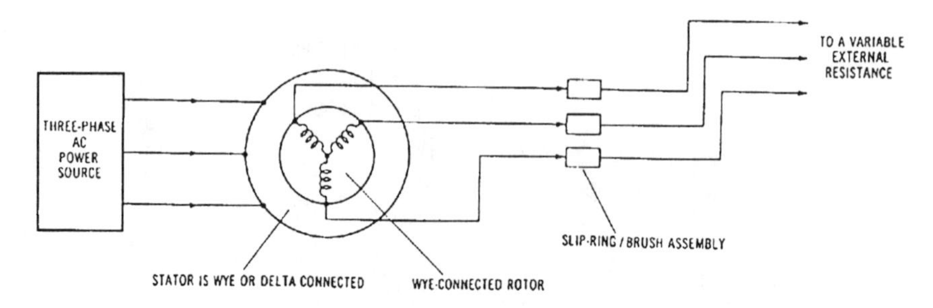

Figure 10-48. Diagram of a wound-rotor induction motor.

major disadvantage of this type of motor is that it costs a great deal more than an equivalent three-phase induction motor using a squirrel-cage rotor. Thus, they are not used as extensively as other three-phase motors.

ELECTRIC MOTOR APPLICATIONS

Certain factors must be considered when selecting an electric motor, for a specific application. Among these considerations are (1) the source voltage and power capability available, (2) the effect of the power factor and efficiency of the motor on the overall system, (3) the effect of the starting current of the motor on the system, (4) the effect of the power system on the operation of the motor, (5) the type of mechanical load, and (6) the expected maintenance it will require.

Motor Performance

The major consumer of electrical power is the electric motor. It is estimated that electric motors are responsible for about 50% of the electrical power consumed in industrial usage and that over 35% of all the electrical power used is used by electrical motors. For these reasons, we must consider the efficient operation of motors as a major part of our energy conservation efforts.

Both efficiency and the power factor must be considered in determining the effect of a motor in terms of efficient power conversion. Remember the following relationships. First,

$$\text{Efficiency (\%)} = \frac{P_{out}}{P_{in}} \times 100$$

where,

P_{in} is the power input in horsepower,

P_{out} is the power output in watts.

(To convert horsepower to watts, remember that 1 horsepower = 746 watts.) Then,

$$pf = \frac{P}{VA}$$

where,

pf is the power factor of the circuit,

P is the true power in watts,

VA is the apparent power in volt-amperes.

The maximum pf value is 1.0, or 100%, which would be obtained in a purely resistive circuit. This is referred to as *unity power factor.*

Effect of Load

Since electrical power will probably become more expensive and less abundant, the efficiency and power factor of electric motors will become increasingly more important. The efficiency of a motor shows mathematically just how well a motor converts electrical energy into mechanical energy. A mechanical load placed on a motor affects its efficiency. Thus, it is particularly important for industrial users to load motors so that their maximum efficiency is maintained.

Power factor is also affected by the mechanical load placed on a motor. A higher power factor means that a motor requires less current to produce a given amount of torque or mechanical energy. Lower current levels mean that less energy is being wasted (converted to heat) in the equipment and circuits connected to the motor. Penalties are assessed on industrial users by the electrical utility companies for having low system power factors (usually less than 0.8 or 0.85 values). By operating at higher power factors, industrial users can save money on penalties and help, on a larger scale, with mom efficient utilization of electrical power. Motor load affects the power factor to a much greater extent than it does the efficiency. Therefore, motor applications should be carefully studied to assure that motors (particularly very large ones) are not overloaded or underloaded, so that the available electrical power will be used more effectively.

Effect of Voltage Variations

Voltage variation also has an effect on the power factor and efficiency. Even slight changes in voltage produce a distinct effect on the power factor. However, a less distinct effect results when the voltage causes a variation in the efficiency. As proper power utilization is becoming more and more important, motor users should make sure that their motors do not operate at undervoltage or overvoltage conditions.

Considerations for Mechanical (Motor) Loads

Their are three basic types of mechanical (motor) loads connected to electrical power systems. These are direct current (DC), single phase alternating current (AC) and three phase AC systems. Direct current motors are ordinarily used for special applications since they are more expensive than other types and require a DC power source. Typically, they are used for small, portable applications and powered by batteries or for industrial t commercial applications with alternating current converted to direct current by rectification systems. A major advantage of DC motors is their ease of speed control. The shunt-wound DC motor can be used accurate speed control and good speed regulation. A disadvantage is the increased maintenance caused by the brushes and commutator of the machine. DC shunt-wound motors are used for variable speed drives on printing presses, rolling mills, elevators, hoists and automated industrial machine tools.

Series-wound DC motors have very high starting torque. Their speed regulation is not as good a the shunt-wound DC motor. The series-wound motor also requires periodic maintenance due to the brush/commutator assembly. Typical applications of series-wound DC motors are automobile starters, traction motors for trains and electric buses, and mobile equipment operated by batteries. Compound-wound DC motors have very few applications today.

Single phase AC motors are relatively inexpensive. Most types have good starting torque and are easily provided 120-volt and 240-volt electrical power. Disadvantages include maintenance problems due to centrifugal switches, pulsating torque and rather noisy operation. They are used in fractional horsepower sizes (less than one horsepower) for residential, commercial and industrial applications. Some integral horsepower sizes are available in capacitor start types. Uses include machine tools, pumps, washing machines, refrigerators, exhaust fans, forced-air heating/cooling system blowers, and clothes dryer motors.

Specialized applications for single phase motors include:

(1) Shaded-pole motors used for portable fans, record players, dishwasher pumps, and electric typewriters. They are low-cost and small, but inefficient.

(2) Single phase synchronous motors used for clocks, appliance timers, and recording instruments (compact disk players). They operate at a constant speed.

(3) Universal motors (AC/DC) motors are used for many types of portable tools and appliances such as electric drills, saws, office machines, mixers, blenders, sewing machines, and vacuum cleaners. They operate at speeds up to 20,000 r/min and have easy speed control. Remember that AC induction motors do not have speed control capability without expensive auxiliary equipment.

Three phase AC motors of the induction type are very simple in construction, rugged and reliable in operation. They are less expensive (per horsepower) than other motors. Applications of three phase induction motors include industrial and commercial equipment and machine tools. Three phase AC synchronous motors run at constant speeds and may be used for power factor correction of electrical power systems. However, they are more expensive, require maintenance of brushes/slip rings, and need a separate DC power supply.

Chapter 11

Control Equipment Used With Distribution Systems

The control of electrical power is a very important part of the electrical power system operation. Control is the most complex part of the electrical power system. Some of the common types of electromechanical equipment and devices which are used for electrical power control are discussed in this chapter. Control equipment and devices are used in conjunction with many types of electrical loads. Since electrical motors (mechanical loads) use about 50% of all electrical power, we will concentrate mainly on electrical motor control equipment. In most cases, similar equipment is used to control electrical lighting and heating loads.

POWER CONTROL STANDARDS, SYMBOLS, AND DEFINITIONS

There is a great amount of fundamental knowledge that is needed in order to fully understand electrical power control. Individuals who are concerned with electrical power systems should have a good knowledge of the standards, symbols, and definitions which govern electrical power control.

NEMA Standards

The National Electrical Manufacturers' Association (NEMA) has developed standards for electrical control systems. NEMA standards are used extensively to obtain information about the construction and performance of various electrical power control equipment. These standards provide information concerning the voltage, frequency, power, and current ratings for various equipment.

ANSI, IEEE, and Other Standards

There are several well-known organizations which publish standards that are of importance. Two of these organizations are the American National Standards Institute (ANSI) and the Institute of Electrical and Electronic Engineers (IEEE). The standards published by these organizations are for the use of the manufacturing industries as well as consumers and, in some eases, the general public. These standards are subject to review periodically; therefore, industrial users should keep up-to-date by obtaining the revisions.

These standards were developed through input from manufacturers, consumers, government agencies, and scientific, technical, and professional organizations. Often these published standards are used by industry as well as governmental agencies. It should be noted that the National Electrical Code discussed previously is an American National Standards Institute standard. A list of professional organizations of interest in electrical power system operation is included in Appendix D.

Definitions

There are several basic definitions which are used when dealing with electrical power control. These definitions are particularly important when interpreting control diagrams and standards. A listing of several of the important power control definitions follows. You should study these definitions.

Across-the-line control—A method of motor starting in which a motor is connected directly across the power lines when it is started.

Automatic starter—A self-acting starter which is completely controlled by control switches or some other sensing mechanism.

Auxiliary Contact—A contact that is part of a switching system. It is in addition to the main contacts and is operated by the main contacts.

Braking—A control method that is used to rapidly stop and hold a motor.

Circuit breaker—An automatic device that opens under abnormally high-current conditions and can be manually or automatically reset.

Contact—A current-conducting part of a control device which is used to open or close a circuit.

Contactor—A control device that is used to repeatedly open or close an electric power circuit.

Controller—A device or group of devices that systematically control the delivery of electric power to the load or loads connected to it.

Disconnect—A control device or group of devices that will open so that electrical current in a circuit will be interrupted.

Drum controller—Sets of electrical contacts that are mounted on the surface of a rotating cylinder; usually for controlling the on-off forward-reverse condition of a load.

Electronic control—Usually solid-state devices which perform part of the control function of a system.

Full voltage control—A control system that connects equipment directly across the power lines when it is started.

Fuse—A circuit-protection device that disconnects a circuit when an overcurrent condition occurs. However, it is self-destructive and must be replaced.

Horsepower—The power output or mechanical work rating of an electrical motor.

Jogging—Momentary operation which causes a small movement of the load that is being controlled.

Magnetic contactor—A contactor which is operated electromagnetically and usually controlled by push buttons activated by an operator.

Manual controller—An electric control device that functions when operated by mechanical means, usually by an operator.

Master switch—A switch that controls the power delivered to other parts of a system.

Motor—A device (mechanical load) for converting electrical power to mechanical power in the form of rotary motion.

Multispeed starter—An electrical power-control device which provides for varying the speed of a motor.

Overload relay—Overcurrent protection for a load. While it is in operation, it maintains the interruption of the load from the power supply until it is reset or replaced.

Overload relay reset—A push button that is used to reset a thermal overload relay after the relay has been overloaded.

Pilot device—A control device which directs the operation of another device or devices.

Plugging—A braking method which causes a motor to develop a retarding force in the reverse direction.

Push button—A button-type control switch which is manually operated for actuating or disconnecting some load device.

Push-button station—A housing for the push buttons that are used to control equipment.

Reduced-voltage starter—A control device that applies a reduced voltage to a motor when it is started.

Relay—A control device that is operated by one electrical circuit to control a load which is part of another electrical circuit.

Remote control—A system in which the control of an electrical load takes place from some distant location.

Safety switch—An enclosed manually operated disconnecting switch used to turn a load off when necessary.

Solenoid—An electromagnetically actuated control device that is used to produce linear motion for performing various control functions.

Starter—An electric controller that is used to start, stop, and protect the motor which is connected to it.

Timer—A control device that provides variable time periods so that a control function may be performed.

Symbols

One should have an understanding of the electrical symbols which are commonly used with power control systems. A few of these are somewhat different from basic electrical symbols. Some typical power control symbols are shown in Figure 11-1. You should especially observe the symbols that are used for the various types of switches and push buttons.

POWER CONTROL USING SWITCHES

An important, but often overlooked, part of electrical power control is the various types of switches used. This section will examine the many types of switches that are used to control electrical power. The primary function of a switch is to turn a circuit on or off; however, many more complex switching functions can be performed using switches. The emphasis in this section will be on switches that are used for motor control. Keep in mind that other load devices can also be controlled in a similar manner by switches.

Toggle Switches

Among the simplest types of switches are toggle switches. The symbols for several kinds of toggle switches are shown in Figure 11-2. You should become familiar with the symbols that are used for various types of toggle switches and with the control functions that they can accomplish.

Rocker Switches

Another type of switch is called a rocker switch. Rocker switches are also used for on-off control. They may be either *momentary-contact* to accomplish temporary control or *sustained contact* to cause a load to remain in an on or off condition until the switch position is manually changed.

Push-button Switches

Push-button switches are commonly used. Many motor-control

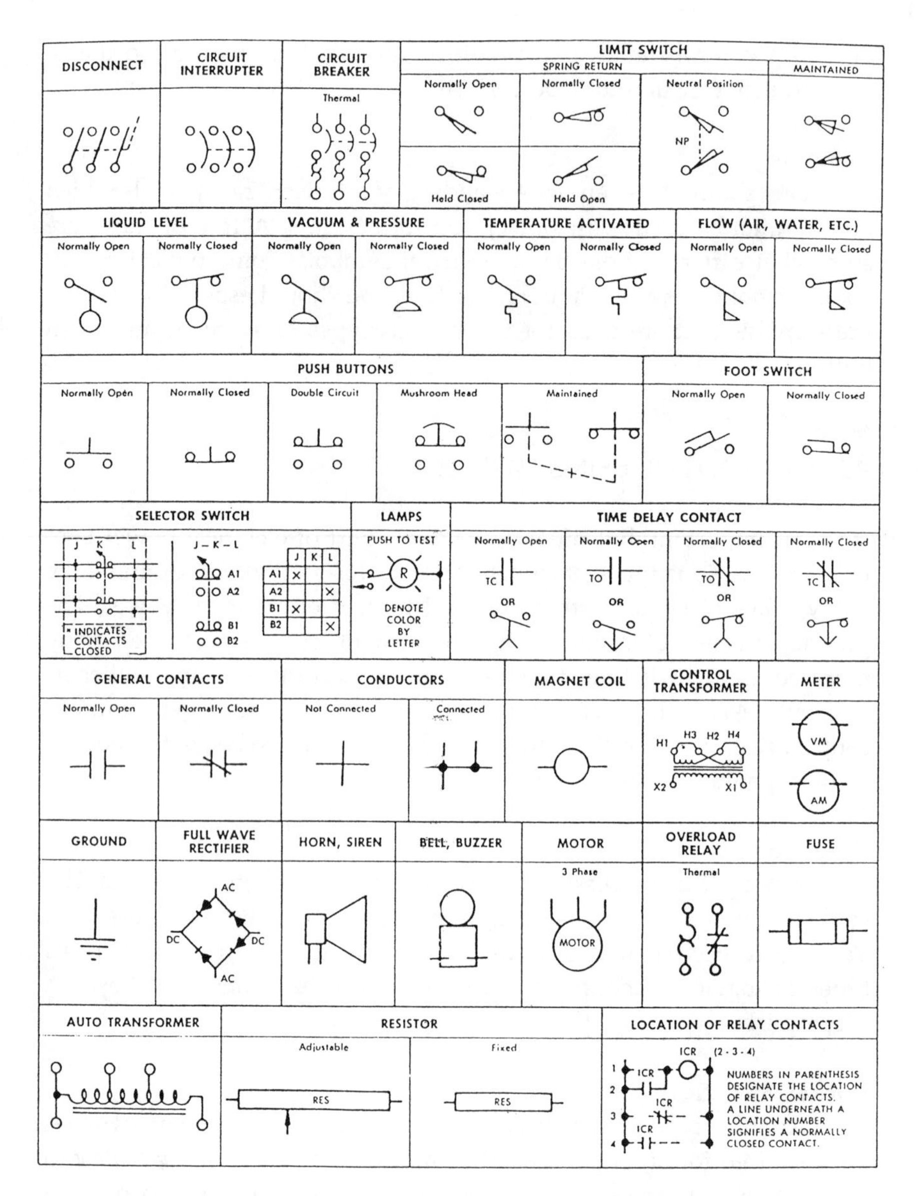

Figure 11-1. Common power control symbols (Courtesy Furnas Electric Co.).

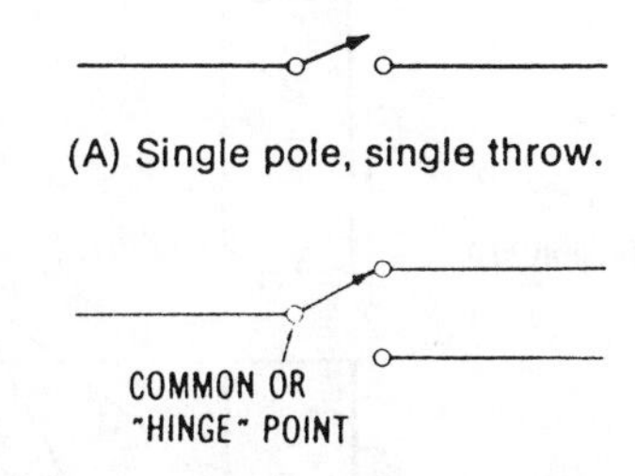

(A) Single pole, single throw.

(B) Three-way, or single pole, double throw.

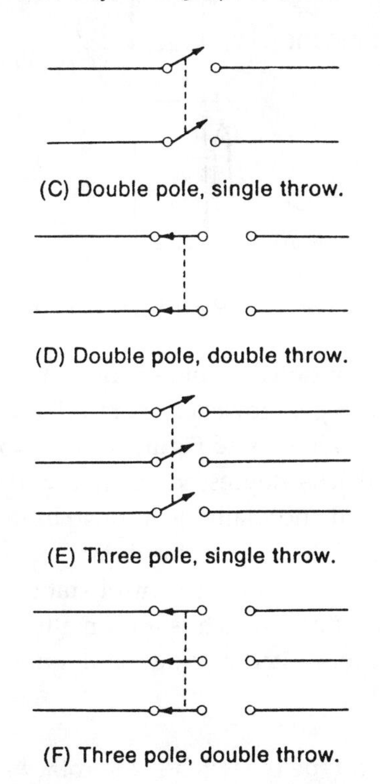

(C) Double pole, single throw.

(D) Double pole, double throw.

(E) Three pole, single throw.

(F) Three pole, double throw.

Figure 11-2. Types of toggle switches.

applications use push buttons as a means of starting, stopping, or reversing a motor; the push buttons are manually operated to close or open the control circuit of the motor. There are several types of push buttons used in the control of motors. Figure 11-3 diagrams some push-button styles. Push buttons are usually mounted in metal or plastic enclosures.

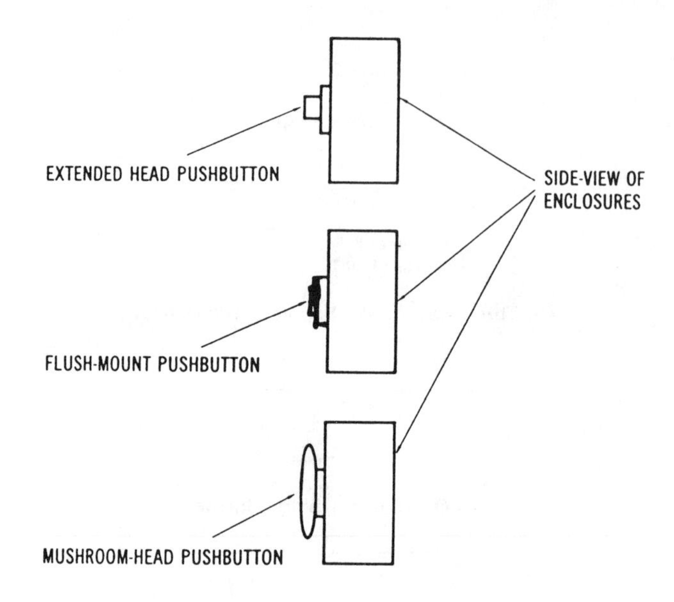

Figure 11-3. Push-button styles.

Ordinarily, push buttons are either the normally closed (N.C.) or normally open (N.O.) type. However, there are a few modifications of these as shown by the symbols of Figure 11-1. A normally closed push button is closed until it is depressed manually. It will open a circuit when it is depressed. The normally open push button is open until it is manually depressed and, then, once it is depressed, it will close a circuit. The "start" push button of a motor control station is a normally open (N.O.) type while the "stop" switch is a normally closed (N.C.) type. A "start-stop" push-button switch is shown in Figure 11-4.

Rotary Switches

Another common type of switch is the rotary switch. Many different switching combinations can be wired by using a rotary switch. The shaft of a rotary switch is attached to sets of moving contacts. These moving contacts touch different sets of stationary contacts which the mounted on ceramic segments when the rotary shaft is turned to different positions. The shaft can lock into place in any of several positions. A common type of rotary switch is shown in Figure 11-5. Rotary switches the usually controlled by manually turning the rotary shaft clockwise or counterclockwise. A knob is normally fastened to the end of the rotary shaft to permit easier turning of the shaft.

Figure 11-4. A "start-stop" push-button switch (Courtesy Eaton Corp.—Cutler-Hammer Products).

Limit Switches

Some typical limit switches are shown in Figure 11-6. They the made in a variety of sizes. Limit switches the merely on/off switches which use a mechanical movement to cause a change in the operation of the electrical control circuit of a motor or other load device. The electrical current developed as a result of the mechanical movement is used to *limit* movement of the machine or to make some change in its operational sequence Limit switches are often used in sequencing, routing, sorting, or counting operations in industry. Ordinarily, they are used in conjunction with hydraulic or pneumatic controls, electrical relays, or other motor-operated machinery such as drill presses, lathes, or conveyor systems.

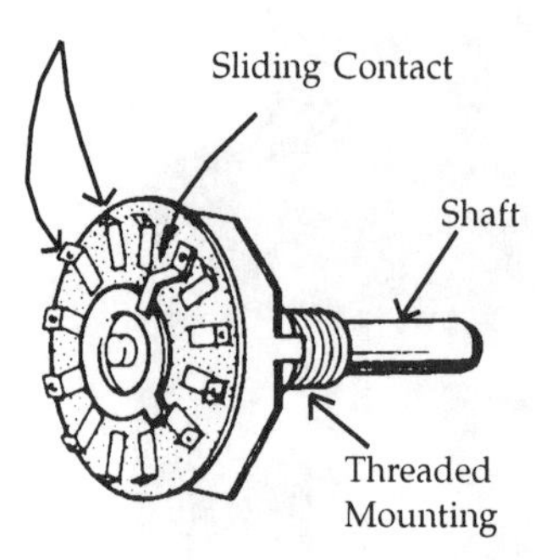

Figure 11-5. Rotary switch

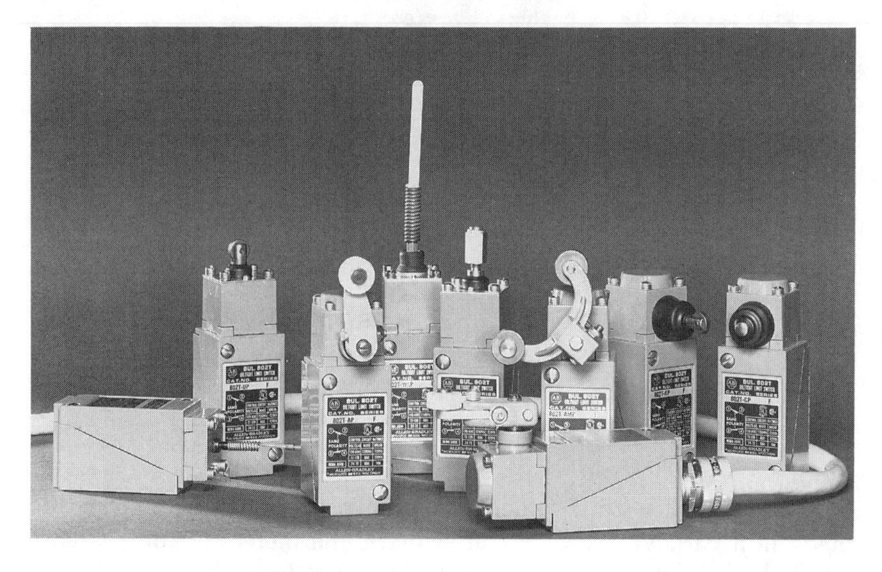

Figure 11-6. Limit switches (Courtesy Allen-Bradley Co.).

Temperature Switches

Temperature switches are among the most common types of control devices used in industry. The control element of a temperature switch contains a specific amount of liquid. The liquid increases in volume when the temperature increases. Thus, changes in temperature can be used to change the position of a set of contacts within the temperature-switch enclosure. Temperature switches may be adjusted throughout a range of temperature settings.

Float Switches

Float switches are used when it is necessary to control the level of a liquid. The float switch, shown in Figure 11-7, has its operating lever connected to a rod and float assembly. The float assembly is placed into a tank of liquid where the motion of the liquid controls the movement of the operating lever of the float switch. The float switch usually has a set of normally open and normally closed contacts which are controlled by the position of the operating lever. The contacts are connected to a pump-motor circuit. In operation, the normally open contacts would be connected in series with a pump-motor control circuit. When the liquid level is reduced, the float switch would be lowered to a point where the operating lever would be moved far enough that the contacts would be caused to change to a closed state. The closing of the contacts would cause the pump motor to turn on. More liquid would, then, be pumped into the tank until the liquid level had risen high enough to cause the float switch to turn the pump motor off.

Pressure Switches

Another type of electrical control device is called a pressure switch. A pressure switch, shown in Figure 11-8, has a set of electrical contacts

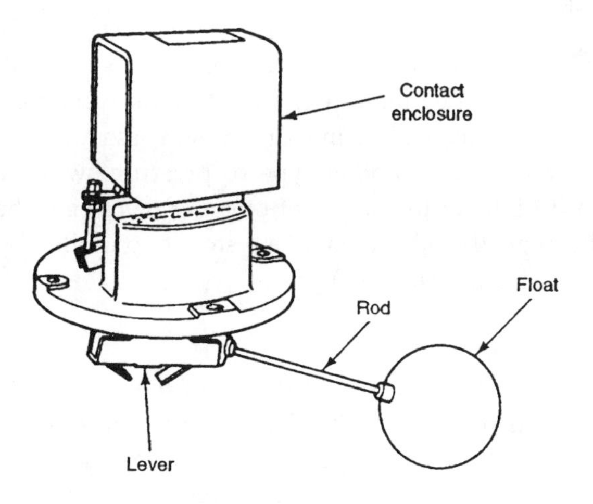

Figure 11-7. Float switch.

Figure 11-8. Pressure switches (Courtesy Furnas Electric Co.).

which change states due to a variation in the pressure of air, hydraulic fluid, water, or some other medium. Some pressure switches are diaphragm operated. They rely upon the intake or expelling of a medium, such as air, which takes place in a diaphragm assembly within the pressure-switch enclosure. Another type of pressure switch uses a piston mechanism to initiate the action of opening or closing the switch contacts. In this type, the movement of a piston is controlled by the pressure of the medium (air, water, etc.).

Foot Switches

A foot switch is a switch that is controlled by a foot pedal. This type of switch is used for applications where a machine operator has to use both hands during the operation of the machine. The foot switch provides an additional control position for the operation of a machine for such time as when the hands cannot be used.

Drum-Controller Switches

Drum controllers are special-purpose switches which are ordinarily used to control large motors. They may be used with either single-phase or three-phase motors. The usual functions of a drum-controller switch are for start/stop control or for the forward/reverse/stop control of electrical motors. A drum-controller switch is illustrated in Figure 11-9. Contacts are moved as the handle of the controller is turned to provide machine control.

Pilot Lights for Switches

Pilot lights usually operate in conjunction with switches. Motor-control devices often require that some visual indication of the operating condition of the motor be provided. Pilot lights of various types are used to provide such a visual indication. They usually indicate either an *on* or *off* condition. For instance, a pilot light could be wired in parallel with a motor to indicate when it is on. Some types of pilot lights are:

1. Full-voltage across-the-line lights—These are the less expensive lights but they do not last as long.

Figure 11-9. Drum controller.

2. Transformer-operated lights—These use a lower voltage to activate the lamp, but they require the expense of an additional transformer which ordinarily reduces the operating voltage to 6 volts.

3. Resistor-type lights—This light uses a series resistor to reduce the voltage across the lamp.

4. Illuminated push button—The functions of the push button and the pilot-light device are combined into one item of equipment which reduces the mounting-space requirement. Figure 11-10 shows illuminated push button switches.

CONTROL EQUIPMENT FOR ELECTRIC MOTORS

There are several types of electromechanical equipment used for the control of power distributed to electric motors. The selection of power-control equipment will affect the efficiency of the power-system operation and the performance of the machinery. It is very important to use the proper type of equipment for each power-control application. This section will concentrate on the types of equipment used for motor control. A control console for electrical machine control is shown in Figure 11-11.

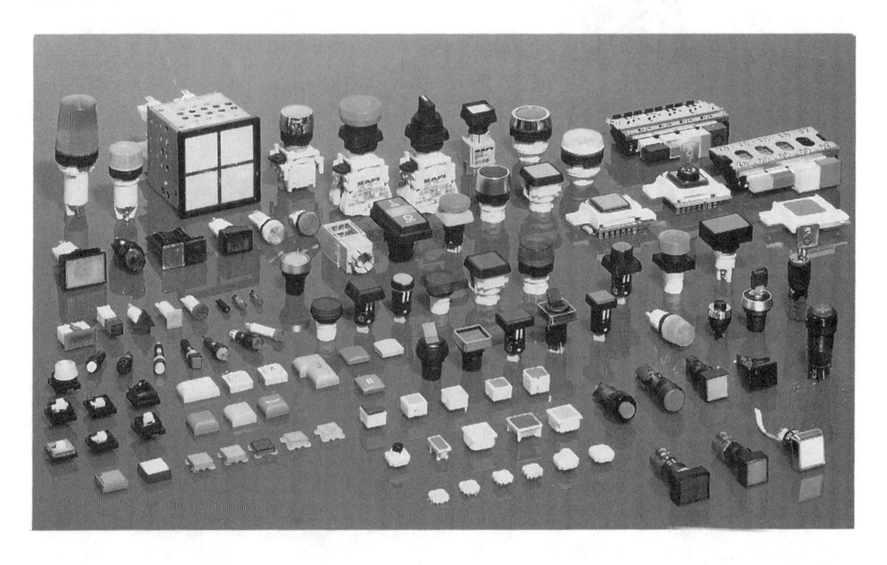

Figure 11-10. Illuminated push-button switches (Courtesy Rafi Gmbtl & Co.).

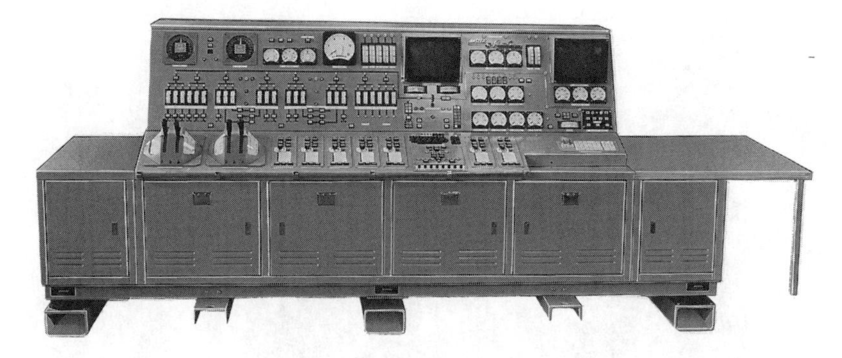

Figure 11-11. Power control center (Courtesy Basler Electric).

Motor-Starting Control

A motor-starting device is a type of power control when used to accelerate a motor from a "stopped" condition to its normal operating speed. There are many variations in motor-starter design, the simplest being a manually operated on/off switch connected in series with one or more power lines. This type of starter, shown in Figure 11-12, is used only for smaller motors which do not draw an excessive amount of current.

Another type of motor starter is the magnetic starter which relies upon an electromagnetic effect to open or close the power-source circuit of the motor. This type of starter is shown in Figure 11-13. Often, motor starters are grouped together for the control of adjacent equipment in an industrial plant. Such groupings of motor starters and associated control equipment are called control centers. Control centers provide an easier access to the power distribution system since they are more compact and the control equipment is not scattered throughout a large area.

Functions of Motor Starters

Various types of motor starters are used for control of motor power. The functions of a starter vary in complexity; however, motor starters usually perform one or more of the following functions:

1. On and off control.

2. Acceleration.

3. Overload protection.

4. Reversing direction of rotation.

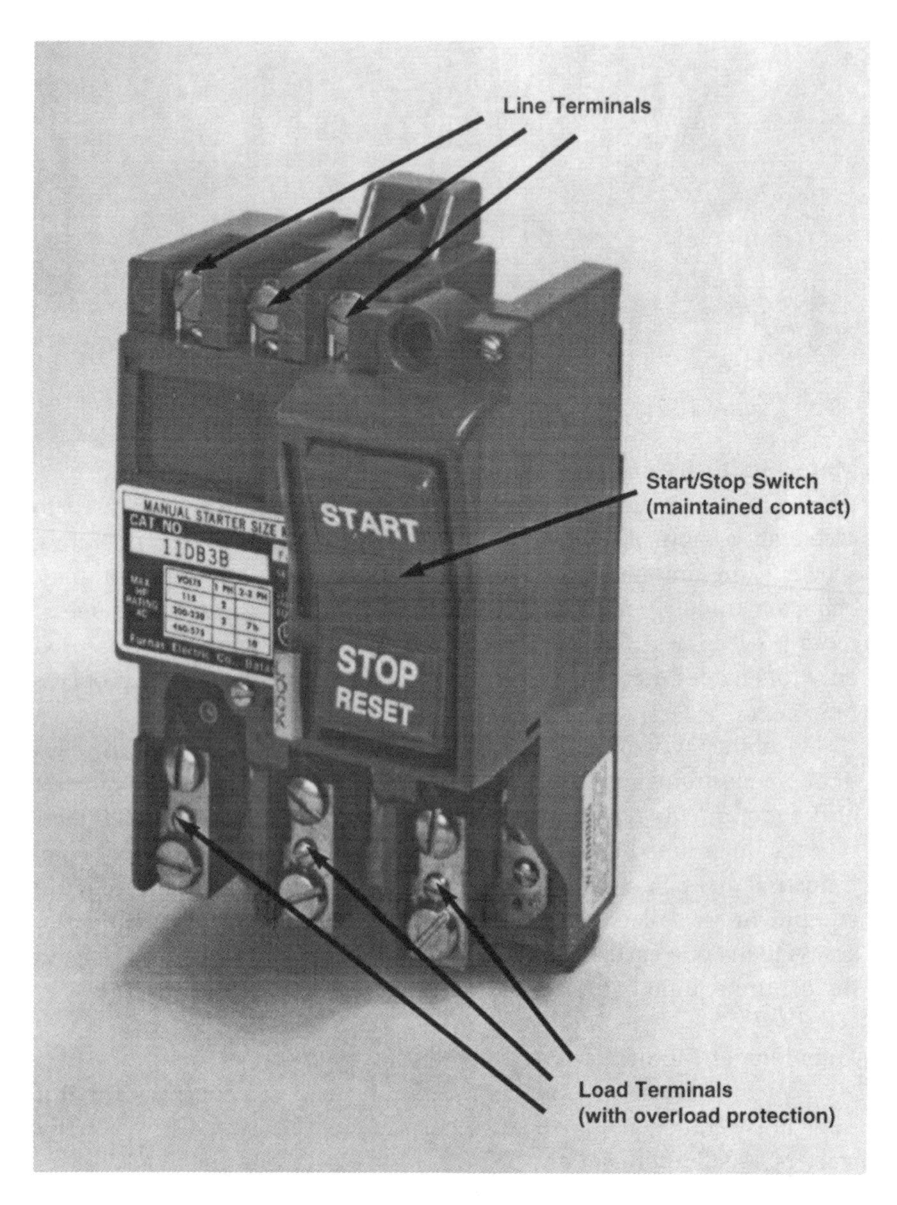

Figure 11-12. Manual motor starters ([a] Courtesy Furnas Electric Co.)

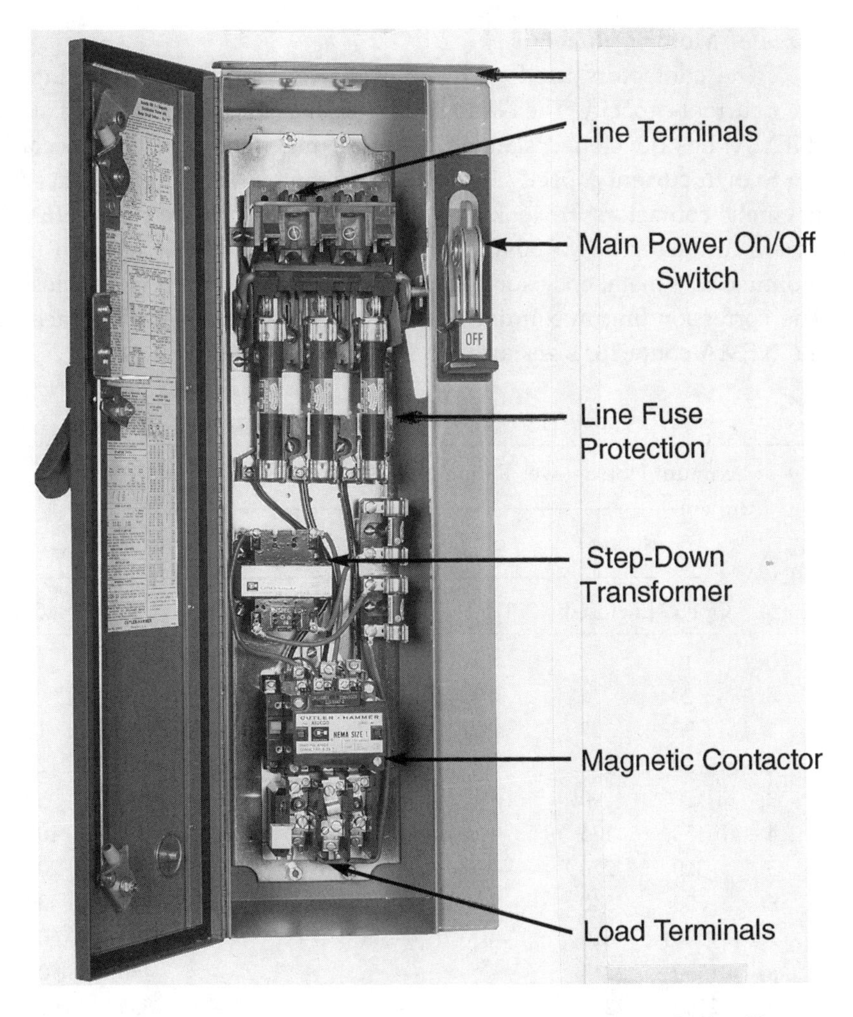

Figure 11-13. Combination starter (Courtesy Eaton Corp.—Cutler-Hammer Products.

Some starters control a motor by being connected directly across the power input lines. Other starters reduce the level of input voltage that is applied to the motor when it is started so as to reduce the value of the starting current. Ordinarily, motor overload protection is contained in the same enclosure as the magnetic contactor.

Sizes of Motor Contactors

The contactors used with motor starters are rated according to their current capacity. The National Electrical Manufacturers Association (NEMA) has developed standard sizes for magnetic contactors according to their current capacity. Table 11-1 lists the NEMA standard sizes for magnetic contactors. By looking at this chart, you can see that a NEMA size 1 contactor has a 30-ampere current capacity if it is open (not mounted in a metal enclosure) and a 27-ampere capacity if it is enclosed. The corresponding maximum horsepower ratings of loads for each of the NEMA contactor sizes are also shown in Table 11-1.

Table 11-1. Sizes of magnetic contactors.

NEMA Size	Ampere Rating		Maximum Horsepower Rating of Load					
			Single-phase		Three-phase			
	Open	Enclosed	115 V	230 V	115 V	200 V	230 V	460 V
00	10	9	0.33	1	0.75	1.5	1.5	2
0	20	18	1	2	2	3	3	5
1	30	27	2	3	3	7.5	7.5	10
2	50	45	3	7.5	7.5	10	15	25
3	100	90	7.5	15	15	25	30	50
4	150	135	—	—	25	40	50	100
5	300	270	—	—	—	75	100	200
6	600	540	—	—	—	150	200	400
7	900	810	—	—	—	—	300	600
8	1350	1215	-	—	—	—	450	900
9	2500	2250	—	—	—	—	800	1600

Manual Starters

Some motors use manual starters to control their operation. This type of starter provides starting, stopping, and overload protection similar to that of a magnetic contactor. However, manual starters must be mounted near the motor that is being controlled. Remote-control operation is not possible as it would be with a magnetic contactor. This is due to the small control current that is required by the magnetic contactor. Magnetic contactors also provide a low-voltage protection by the dropout of the contacts when a low-voltage level occurs. Manual starters remain closed until they are manually turned off. They are usually limited to low current ratings (30 Amps or less).

Motor Overload Protection

Both manual and magnetic starters can have overload protection contained in their enclosures. It is common practice to place either bimetalic or melting-alloy overload relays in series with the motor branch-circuit power lines. These devices are commonly called heaters. An overload protective relay is selected according to the current rating of the motor circuit to which it is connected. An identification number is used on overload protective devices. This number is used to determine the current that will cause the overload device to "trip" or open the branch circuit. Some typical overload protection or heater tables the given in Tables 11-2 and 11-3. Table 11-3 is used for melting-alloy-type devices. The Table Number (26142) is selected according to the type and size of controller, the size of motor, and the type of power distribution (single-phase or three-phase). For example, a heater with an H37 code number is used with a motor that has a full-load current of 18.6 to 21.1 amperes.

Table 11-3 is used with bimetalic overload relays. They are selected in the same manner as melting-alloy relays. However, this table shows the trip amperes of the heater. A heater with a K53 code number has a 13.9 trip-ampere rating. Thus, a current in excess of 13.9 amperes would cause the heater element to trip. This would open the motor branch circuit by removing the power from the motor.

Table 11-2. Melting-alloy devices.

Table 26142

Full Load Motor Amps.	Heater Code	Full Load Motor Amps.	Heater Code	Full Load Motor Amps.	Heater Code
12.6-14.5	H33	21.2-22.2	H38	34.6-38.9	H43
14.6-16.0	H34	22.3-23.6	H39	39.0-44.7	H44
16.1-16.9	H35	23.7-26.2	H40	44.8-48.8	H45
17.0-18.5	H36	26.3-30.1	H41	48.9-54.2	H46
18.6-21.1	H37	30.2-34.5	H42	54.3-60.0	H48

Courtesy Furnas Electric Co.

Table 11-3. Bimetalic devices.

Table 62K					
Trip Amperes	Heater Code	Trip Amperes	Heater Code	Trip Amperes	Heater Code
1.95	K21	5.59	K36	17.0	K56
2.12	K22	6.39	K37	18.3	K57
2.31	K23	6.88	K39	19.8	K58
2.49	K24	7.78	K41	21.0	K60
2.75	K26	8.42	K42	22.5	K61
3.05	K27	9.54	K43	24.1	K62
3.29	K28	10.1	K49	25.7	K63
3.69	K29	11.5	K50	28.3	K64
4.01	K31	12.6	K52	31.1	K67
4.32	K32	13.9	K53	34.6	K68
4.77	K33	15.1	K54		
5.14	K34	16.0	K55		

Courtesy Furnas Electric Co.

Classes of Motor Starters

The types of motor starters which the commercially available are divided into five classes. These classes, which were established by the National Electrical Manufacturers Association (NEMA), are:

1. *Class A*—Alternating-current, manual or magnetic, air-break or oil-immersed starters which operate on 600 volts or less.

2. *Class B*—Direct-current, manual or magnetic, air-break starters which operate on 600 volts or less.

3. *Class C*—Alternating-current, intermediate voltage starters.

4. *Class D*—Direct-current, intermediate voltage starters.

5. *Class E*—Alternating-current, magnetic starters which operate on 2200 volts to 4600 volts. Class E1 uses contacts and Class E2 uses fuses.

Sizes of Manual Starters

A uniform method has also been established for sizing manual motor starters. Some examples of the sizes of full-voltage manual starters are:

1. Size M-0—For single-phase 115-volt motors up to 1 horsepower.
2. Size M-1—For single-phase 115-volt motors up to 2 horsepower.
3. Size M-1P—For single-phase 115-volt motors up to 3 horsepower.

Three sizes are summarized in Table 11-4.

Table 11-4. Sizes of manual starters.

Starter Size	Maximum Horsepower Rating of Load Single-Phase 115 V	Single-Phase 230 V	Three-Phase 200-230 V	Three-Phase 460-575 V	Ampere Rating
M-0	1	2	3	5	20
M-1	2	3	7.5	10	30
M-1P	3	5	—	—	30

Combination Starters

A popular type of motor starter used in control applications is the combination starter. These starters are made with protective devices such as fused-disconnect switches, air-type circuit breakers, or a system of fuses and circuit breakers mounted in a common enclosure. They are used on systems of 600 volts or less. A combination magnetic starter is shown in Figure 11-13.

Criteria for Selecting Motor Controllers

There are several important criteria which should be considered when selecting electric motor controllers. Among these are:

1. The type of motor—AC or DC, induction or wound rotor.
2. The motor ratings—Voltage, current, duty cycle, and service factor.

3. Motor operating conditions—Ambient temperature and type of atmosphere.

4. Utility company regulations—Power factor, demand factor, load requirements, and the local codes.

5. Type of mechanical load connected to motor—Torque requirement.

In order to become more familiar with the criteria listed above, you must be able to interpret the data on a motor nameplate. The information contained on a typical nameplate is summarized as follows:

Manufacturing Co.—The company that built the motor.

Motor Type—A specific type of motor; i.e., split-phase AC, universal, three-phase induction, etc.

Identification Number—Number assigned by the manufacturer.

Model Number—Number assigned by the manufacturer.

Frame Type—Frame size defined by NEMA.

Number of Phases (AC) single-phase or three-phase.

Horsepower—The amount produced at rated speed.

Cycles (AC)—Frequency the motor should be used with (usually 60 Hz).

Speed (r/min)—The amount at rated hp, voltage, and frequency.

Voltage Rating—operating voltage of motor.

Current Rating (Amperes)—Current drawn at rated load, voltage, and frequency.

Thermal Protection—Indicates the type of overload protection used.

Temperature Rating (°C)—Amount of temperature that the motor will rise over ambient temperature, when operated.

Time Rating—Time the motor can be operated without overheating (usually continuous).

Amps—Current drawn at rated load, voltage, and frequency.

Motor-Controller Enclosures

The purpose of a motor-controller enclosure is obvious. The operator is protected against accidental contact with high voltages which could cause death or shock. In some cases, however, the enclosures are used to protect the control equipment from its operating environment, which may contain water, heavy dust, or combustible materials. The categories of motor-controller enclosures were standardized by the National Electrical Manufacturers Association (NEMA). The following list, courtesy of Furnas Electric Company, summarizes various classifications.

NEMA 1 General-purpose enclosures protect personnel from accidental electrical contact with the enclosed apparatus. These enclosures satisfy indoor applications in normal atmospheres which are free of excessive moisture, dust, and explosive materials.

NEMA 3 (3R) Weatherproof enclosures protect the control from weather hazards. These enclosures are suitable for applications on ship docks, canal locks, and construction work, and for application in subways and tunnels. The door seals with a rubber gasket They are furnished with conduit hub and pole mounting bracket.

NEMA 4 Weathertight enclosures are suitable for application outdoors on ship docks and in dairies, breweries, etc. This type meets standard hose test requirements. They are sealed with a rubber gasket in the door.

NEMA 4X Corrosion Resistant fiberglass enclosures are virtually maintenance free. These U.L. listed units are adaptable to any conduit system by using readily available metal, fiberglass or PVC conduit hubs and are suitable for NEMA 3, 3R, 3S, 4 and 12 applications because they are dust-tight, rain-tight, watertight, and oil-tight.

NEMA 12 Industrial-Use enclosure also satisfies dust-tight applications. These enclosures exclude dust, lint, fibers, and oil or coolant seepage. The hinged cover seals with a rubber gasket.

NEMA 13 Oil-tight push-button enclosures protect against dust, seepage, external condensation, oil, water, and coolant spray.

NEMA 7. For atmospheres containing hazardous gas, NEMA 7 enclosures satisfy Class 1, Group C or D applications as outlined in Section

500 of the "National Electric Code Standard of the National Board *of* Fire Underwriters for Electrical Wiring and Apparatus." These cast aluminum enclosures are for use in atmospheres containing ethylether vapors, ethylene, cyclopropane, gasoline, hexane, naptha, benzine, butane, propane, alcohol, acetone, benzol, lacquer solvent vapors or natural gas. Machined surfaces between the cover and the base provide the seal.

NEMA 9. For atmospheres containing explosive dust, NEMA 9 enclosures satisfy Class II, Group E, F, and G applications as outlined by the National Electrical Code for metal dust, carbon black, coal, coke, flour, starch, or grain dusts. Enclosure is east aluminum with machined surfaces between cover and base to provide seal.

NEMA TYPES 1 B1, 1 B2, 1 B3 Flush Types provide behind the panel mounting into machine bases, columns, or plaster walls to conserve space and to provide a more pleasant appearance. NEMA 1B1 mounts into an enclosed machine cavity. NEMA 1B2 includes its own enclosure behind the panel to exclude shavings and chips which might fall from above. NEMA 1B3 for plaster walls includes an adjustment to compensate for wall irregularities.

OTHER ELECTROMECHANICAL POWER CONTROL EQUIPMENT

There are so many types of electromechanical power control equipment used today that it is almost impossible to discuss each type. However, some of the very important types will be discussed in the following paragraphs.

Relays

Relays represent one of the most widely used control devices available today. The electromagnet of a relay contains a stationary core. Mounted close to one end of the core is a movable piece of magnetic material called the armature. When the coil is activated electrically, it produces a magnetic field in the metal core. The armature is then attracted to the core, which in turn produces a mechanical motion. When the coil is de-energized, the armature is returned to its original position by spring action.

The armature of a relay is generally designed so that electrical contact points respond to its movement. Activation of the relay coil will cause the contact points to "make" or "break" according to the design of the relay. A relay could be described as an electromagnetic switching mechanism. There are an almost endless number of special-purpose relays and switch combinations used for electrical power control. Figure 11-14 shows a typical assortment of relays and sockets, while Figure 11-15a shows a simplified diagram of the construction of a relay that is used to control a motor.

Relays use a small amount of current to create an electromagnetic field that is strong enough to attract the armature. When the armature is attracted, it either opens or closes the contacts. The contacts, then, either turn on or turn off circuits that are using large amounts of current. The minimal current that must flow through the relay coil in order to create a magnetic field strong enough to "attract" the armature is known as the "pickup" or "make" current. The current through the relay coil that allows the magnetic field to become weak enough to release the armature is known as the "break" or "dropout" current.

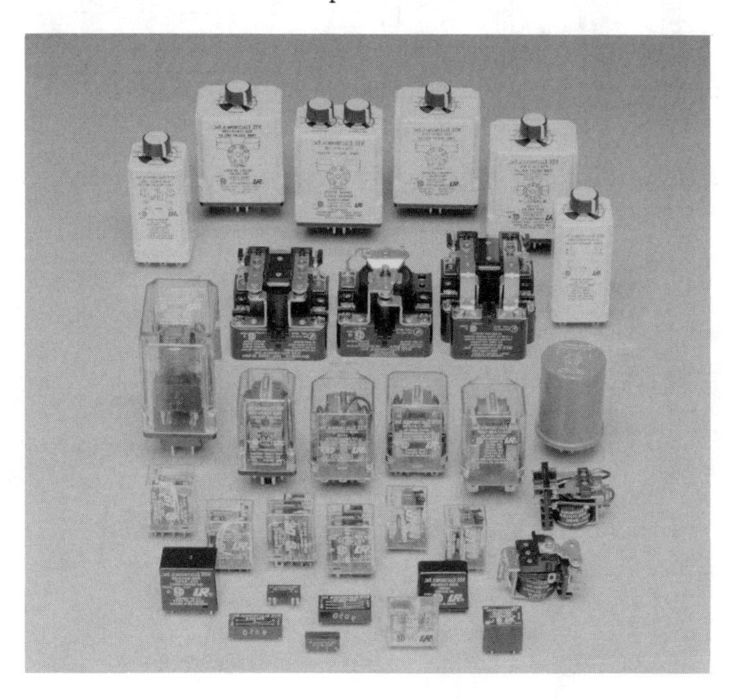

Figure 11-14. Various types of relays (Courtesy NTE Electronics).

There are two types of contacts used in conjunction with most relays—normally open (N.O.) and normally closed (N.C.). The normally open contacts remain *open* when the relay coil is de-energized and closed only when the relay is energized. The normally closed contacts remain *closed* when the relay is de-energized and open only when the coil is energized.

Solenoids

A solenoid, shown in Figure 11-15b, is an electromagnetic coil with a movable core that is constructed of a magnetic material. The core, or plunger, is sometimes attached to an external spring. This spring causes

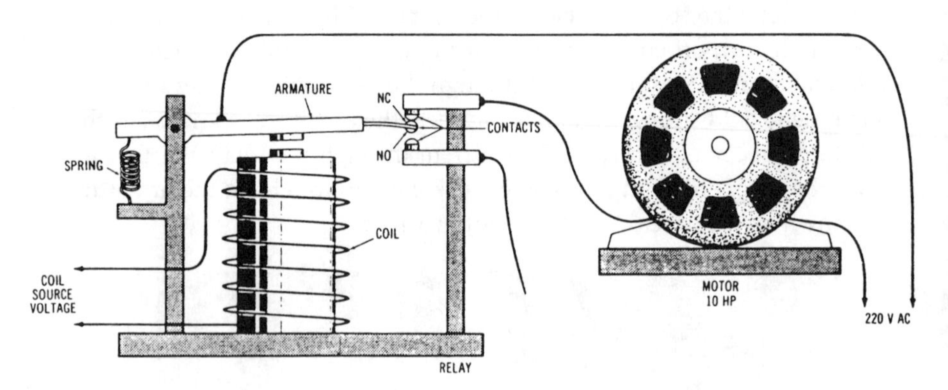

Figure 11-15a. Simplified diagram of the construction of a relay that is used to control a motor.

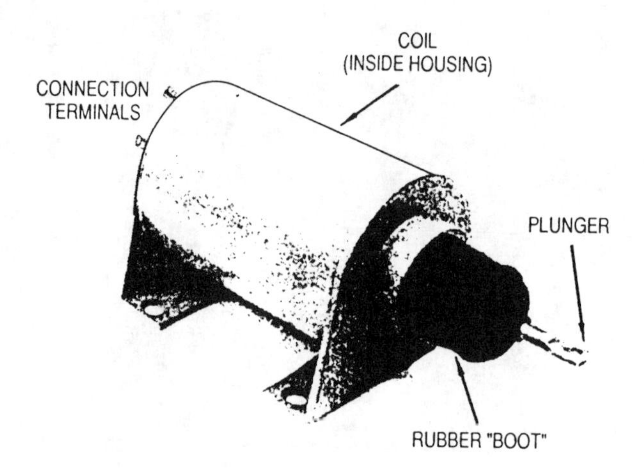

Figure 11-15b. A solenoid.

the plunger to remain in a fixed position until moved by the electromagnetic field that is created by current through the coil. This external spring also causes the core or plunger to return to its original position when the coil is de-energized.

Solenoids are used for a variety of control applications. Many gas and fuel-oil furnaces use solenoid valves to automatically turn the fuel supply on or off upon demand. Most dishwashers use one or more solenoids to control the flow of water.

Specialized Relays

There are many types of relays used for electrical power control. *General-purpose relays* are used for low-power applications. They are relatively inexpensive and small in size; Many small general-purpose relays are mounted in octal base (8-pin) plug-in sockets. *Latch*ing *relays* are another type of relay. They are almost identical to the relays discussed previously, but they have a latching mechanism which holds the contacts in position after the power has been removed from the coil. A latching relay usually has a special type of unlatching coil connected in series with a push-button stop switch. *Solid-state relays* are used where improved reliability or a rapid rate of operation are necessary. Electromagnetic relays will wear out after prolonged use and have to be replaced periodically. Solid-state relays, as do other solid-state devices, have a long life expectancy. They are not sensitive to shock, vibration, dust, moisture, or corrosion. *Timing relays* are used to turn a load device on or off after a specific period of time. One popular type is a pneumatic timing-relay such as the one shown in Figure 11-16. The operation of a pneumatic timing relay is dependent upon the movement of air within a chamber. Air movement is controlled by an adjustable orifice that controls the rate of air movement through the chamber. The air-flow rate determines the rate of movement of a diaphragm or piston assembly. This assembly is connected to the contacts of the relay. Therefore, the orifice adjustment controls the air-flow rate, which determines the time from the activation of the relay until a load connected to it is turned on or off. There are other types of timing relays, such as solid-state, thermal, oil-filled, dashpot, and motor-driven timers. Timing relays the useful for sequencing operations where a time delay is required between operations. A typical application would be as follows: (1) a "start" push button is pressed; (2) a timing relay is activated; (3) after a 5-second time delay, a motor is turned on.

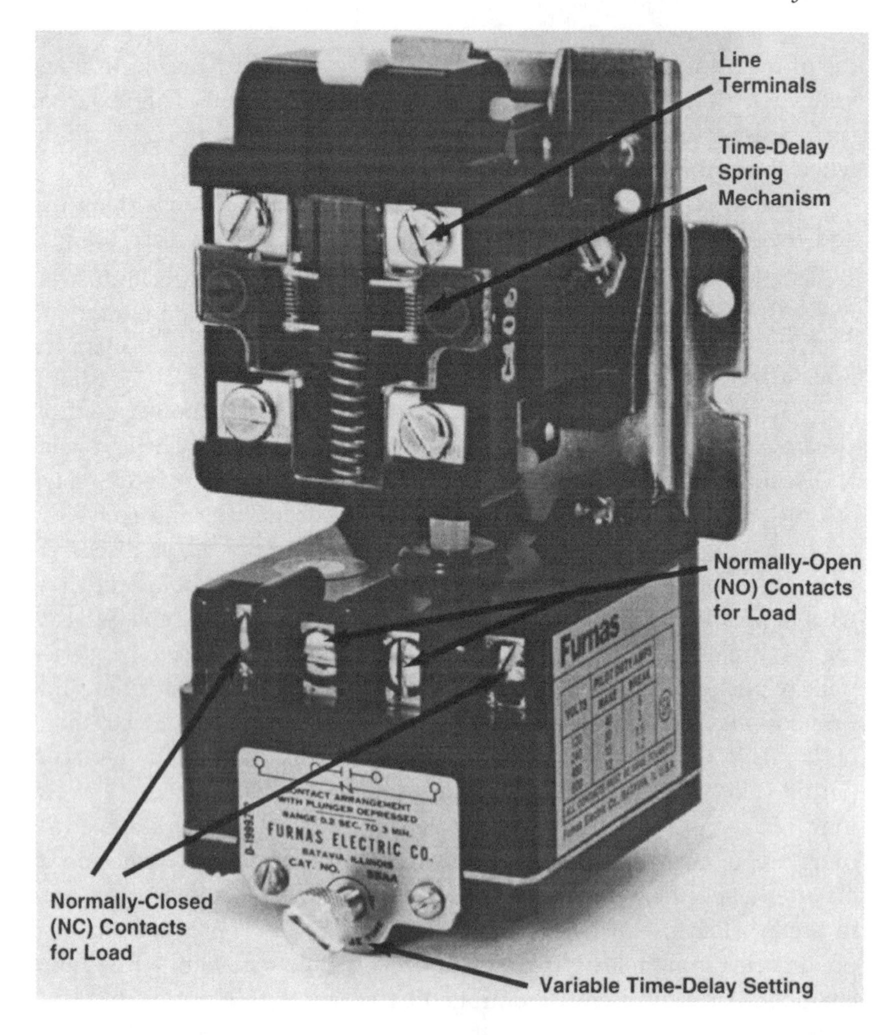

Figure 11-16. Pneumatic timing relay (Courtesy Furnas Electric Co.).

ELECTRONIC POWER CONTROL

An example of advanced electronic control is shown in Figure 11-17. This computerized power controller is capable of replacing electromagnetic circuit breakers and relays used for electrical power control. The introduction of computerized control for power equipment has brought about a new technology of machine control. Many industrial machines, such as automated manufacturing and robotic equipment, are now controlled by computerized circuits. The advances in electronics in recent years have brought about these changes. An understanding of the basic principles of electrical control is, however, still very important.

BASIC CONTROL SYSTEMS

Electrical power control systems are used with many types of loads. The most common electrical loads are motors, so our discussion will deal

Figure 11-17. Computerized control at an electrical power plant.

mainly with electric motor control. However, many of the basic control systems are also used to control lighting and heating loads. Generally, the controls for lighting and heating loads are less complex. The circuits illustrated use magnetic contactors such as shown in Figure 11-18.

Several power control circuits are summarized in Figures 11-19 through 11-27. Figure 11-19 is a start-stop push-button control circuit with overload protection (OL). Notice that the "start" push button is normally open (N.O.) and the "stop" push-button is normally closed (N.C.). Single-phase lines L1 and L2 are connected across the control circuit. When the start push button is pushed, a momentary contact is made between points 2 and 3. This causes the N.O. contact (M) to close. A complete circuit between L1 and L2 causes the electromagnetic coil to be energized. When the normally closed stop push button is pressed, the circuit between L1 and L2 will open. This causes contact M to open and turn the circuit off.

Figure 11-18. Motor starter controls: (a) across-the-line starter with thermal overload protection shown (Courtesy of Eaton Corp., Cutler-Hammer Products).

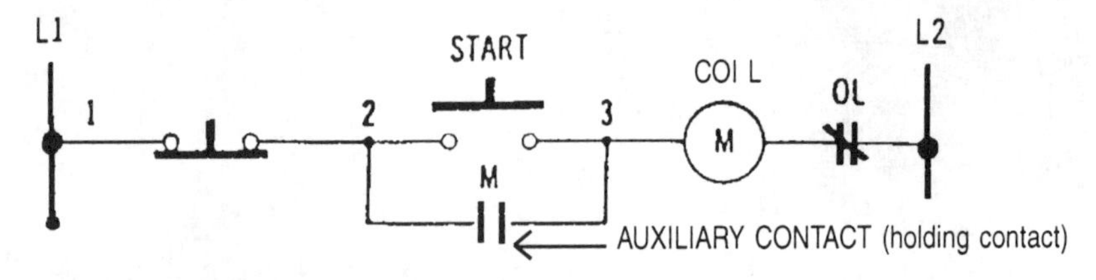

Figure 11-19. A start-stop push-button control circuit with overload protection (one-line diagram).

The circuit of Figure 11-20 is the same type of control as the circuit given in Figure 11-19. In the circuit of Figure 11-20, the start-stop control of a load can be accomplished front three separate locations. Notice that the start push buttons are connected in parallel and the stop push buttons the connected in series. The control of one load from as many locations as is desired can be accomplished with this type of control circuit.

The next circuit (Figure 11-21) is the same as the circuit in Figure 11-19, except that a "safe-run" switch is provided. The "safe" position assures that the start push button will not activate the load. A "start-safe" switch circuit often contains a key which the machine operator uses to turn the control circuit on or off.

Figure 11-22 is also like the circuit of Figure 11-19, but with a "jog-run" switch added in series with the normally open contact (M). In the "run" position, the circuit would operate just like the circuit of Figure

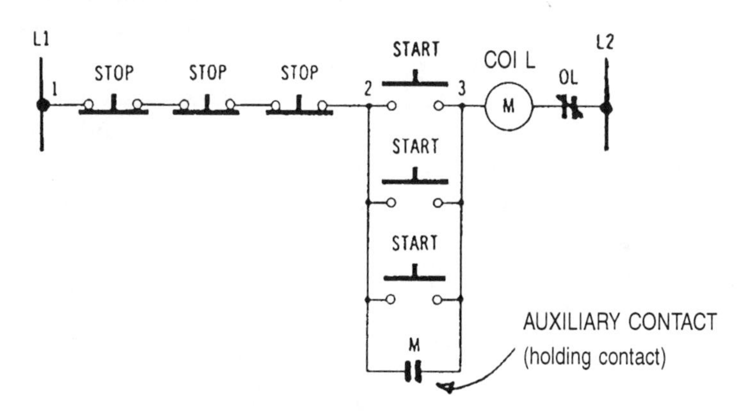

Figure 11-20. A start-stop control circuit with low-voltage protection and control from three locations.

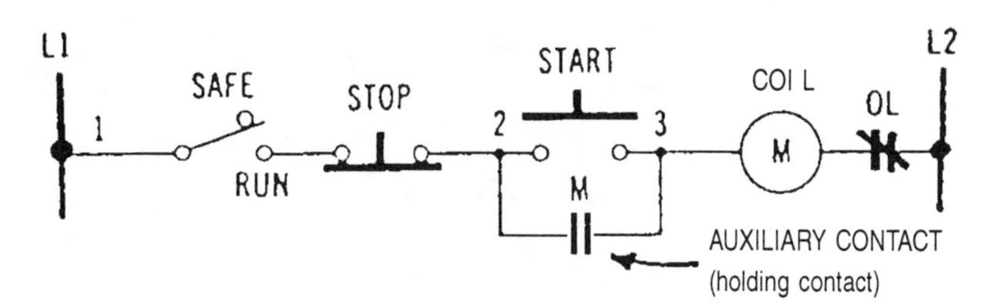

Figure 11-21. A start-stop control circuit with a safe-run selector switch.

11-19. The "jog" position is used so that a complete circuit between L1 and L2 will be achieved and sustained only while the start push button is pressed. With the selector switch in the "jog" position, a motor can be rotated a small amount, at a time, for positioning purposes. *Jogging* or *inching* is defined as the momentary operation of a motor to provide small movements of its shaft.

Figure 11-23 shows a circuit that is another method of motor-jogging control. This circuit has a separate push button for jogging which relies upon a normally open contact (CR) to operate. Two control relays are used with this circuit.

The circuit in Figure 11-24 a forward-reverse push-button control circuit with both forward and reverse limit switches (normally closed switches). When the "forward" push button is pressed, the load will operate until the "forward" limit switch is actuated. The load would, then, be turned off since the circuit from L1 to L2 would be opened. The reverse circuit operates in a similar manner. Two control relays are needed for forward-reverse operation.

The circuit of Figure 11-25 is the same as the circuit in Figure 11-24 except for the push-button arrangement. The forward and reverse push buttons are arranged in sets. Pressing the "forward" push button automatically opens the reverse circuit and pressing the "reverse" push but-

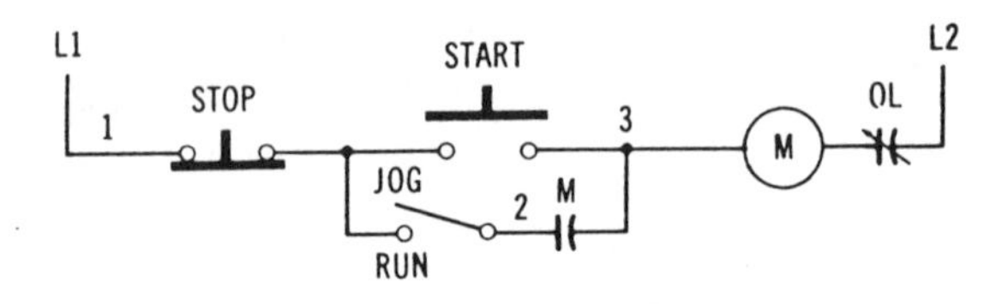

Figure 11-22.

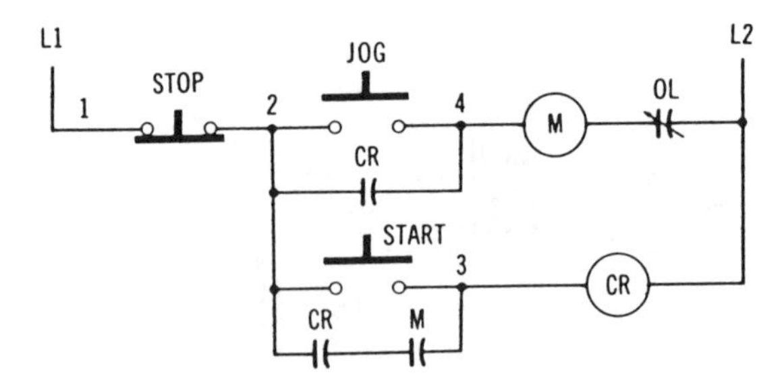

Figure 11-23.

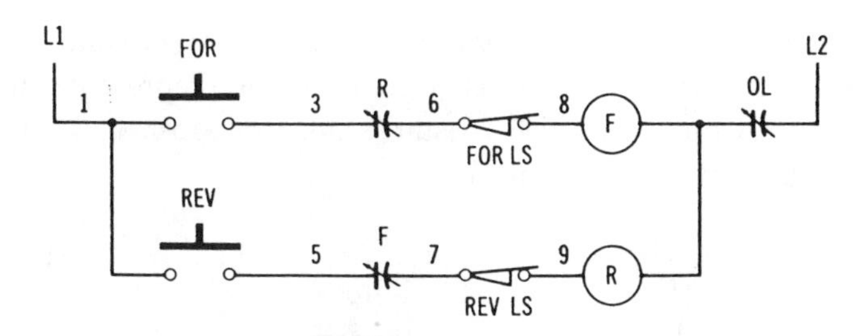

Figure 11-24

ton automatically opens the forward circuit. Limit switches are also used with this circuit of Figure 11-24. Their function is the same as in the circuit of Figure 11-24. In the circuit of Figure 11-25, when the "forward" push button is pressed, the top push button will momentarily close and the lower push button will momentarily open. When points 2 and 3 are connected, current will flow from L1 to L2 through coil (F). When coil (F) is energized, normally open contact F will close and the normally closed contact F will open. The "forward" coil will, then, remain energized. The reverse push buttons cause a similar action of the reversing circuit. Two control relays are also required.

The circuit of Figure 11-26 is similar in function to that of the circuit in Figure 11-25. The push-button arrangement of this circuit is simpler. When the normally open forward push button is pressed, current will flow through coil (F). When the forward control relay is energized, normally open contact F will close and normally closed contact F will open. This action will cause a motor to operate in the forward direction. When the normally closed stop push button is pressed, the current through coil (F) is interrupted. When the normally open reverse push button is pressed, current will flow through coil (R). When the reverse coil is energized, normally open contact F will close and normally closed contact F will open. This action will cause a motor to operate in the reverse direction, until the stop push button is pressed again.

Figure 11-27 shows a circuit that is another method of forward-reverse-stop control. This control circuit has the added feature of a high- and low-speed selector switch for either the forward or reverse direction. The selector switch is placed in series with the windings of the motor. When the selector is changed from the HIGH position to the LOW position, a modification in the windings of the motor can be made.

There are many other push-button combinations which can he

used with control relays to accomplish motor control or control of other types of loads. The circuits discussed in this section represent some basic power control functions such as start-stop, forward-reverse-stop, jogging, and multiple speed control.

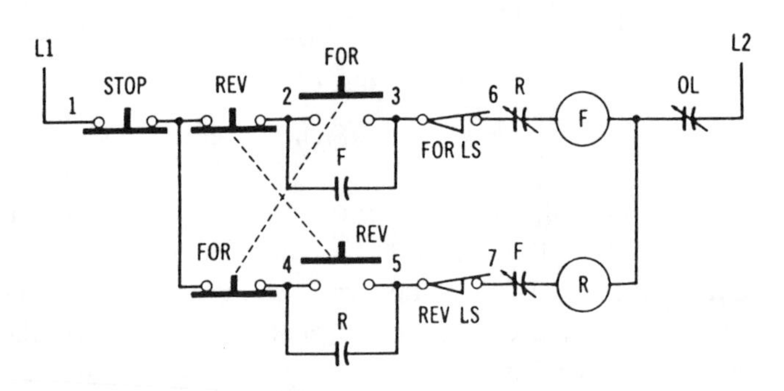

Figure 11-25. An instant forward-reverse-stop push button control circuit.

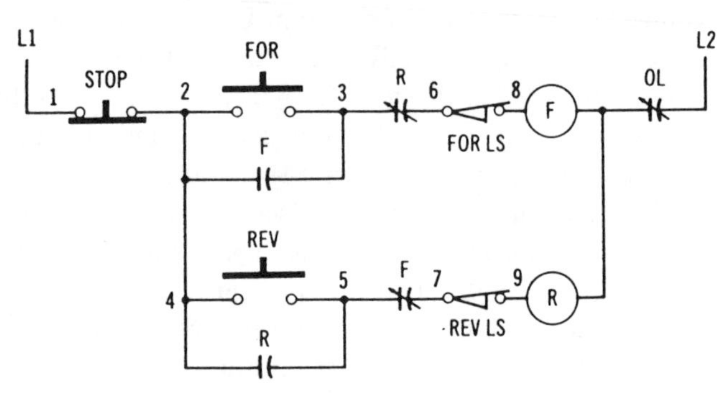

Figure 11-26. A forward-reverse-stop push button control circuit.

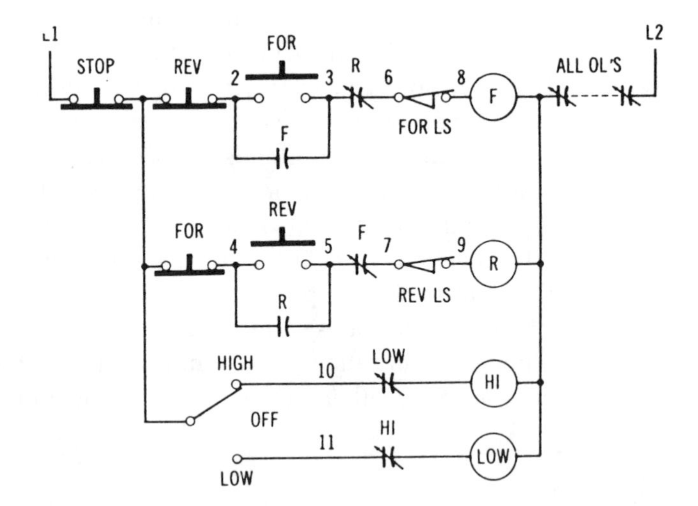

Figure 11-27. Push-button control circuit with a high- and low-speed selector switch.

MOTOR-STARTING SYSTEMS

Most motor-starting systems utilize one or more magnetic contactors. A schematic diagram of a magnetic contactor circuit used for controlling a single-phase motor is shown in Figure 11-28. Note that a magnetic contactor relies upon an electromagnetic coil which energizes when current passes through it. The activated coil performs the function of closing a set of normally open contacts. These contacts are connected in series with the power input to the motor that is being controlled.

In Figure 11-28, the "START" push-button switch is a normally open switch. When the start switch is pressed, current will flow through the coil of the magnetic contactor. This action energizes the coil and the solenoid of the contactor is drawn inward to close the contacts which are in series with the power lines. Once these contacts are closed, current will continue to flow through the electromagnetic coil through the holding contacts (No. 3 in the diagram). Current will continue to flow until the "STOP" push-button switch is pressed. The stop switch is a normally closed switch. When it is pressed, the circuit to the electromagnetic coil is broken. At this time, current no longer flows through the coil. The contacts now release and cause the power lines to be interrupted. Thus, the motor will be turned off. Magnetic contactor circuits are sometimes referred to as across-the-line starters. The relay principle is utilized since a small current through the coil controls a larger current through a motor.

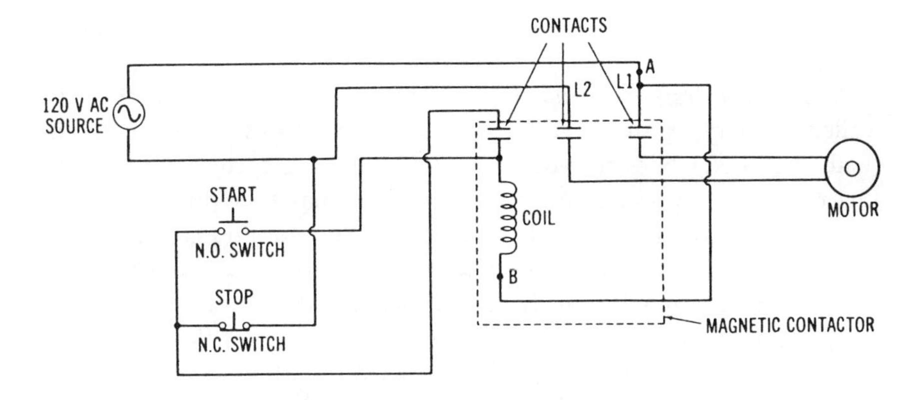

Figure 11-28. Magnetic contactor motor-control circuit.

Types of Starting Systems

Motor starting, particularly for large motors, can play an important role in the efficient operation of an electrical power system. There are several systems used to start electric motors. The motor-starting equipment that is used is placed between the electrical power source and the motor. Electric motors draw a larger current from the power lines during starting than during normal operation. Motor-starting equipment should attempt to reduce starting currents to a level that can be handled by the electrical power system where they are being used.

Full-Voltage Starting—One method of starting electric motors is called full-voltage starting. This method is the least expensive and the simplest to install. Since full power-supply voltage is applied to the motor initially, maximum starting torque and minimum acceleration time result. However, the power system must be able to handle the starting current drawn by the motor.

Full-voltage starting is illustrated by the diagram of Figure 11-29. In this power control circuit, a start-stop push-button station is used to control a three-phase motor. When the normally open start push button is pressed, current will flow through the relay coil (M) causing the normally open contacts to close. The line contacts allow full voltage to be applied to the motor when they are closed. When the start push button is released, the relay coil remains energized due to the holding contact. This contact provides a current path from L1 through the normally closed stop push button, through the holding contact, through the coil (M), through a thermal overload relay, and back to L2. When the stop push button is pressed, this circuit is opened causing the coil to be de-energized.

Primary-Resistance Starting—Another motor starting method is called primary-resistance starting. This method uses resistors in series with the power lines to reduce the motor-starting current. Ordinarily, the resistance connected into the power lines may be reduced in steps until full voltage is applied to the motor. Thus, starting current is reduced according to the value of series resistance in the power lines. Since starting torque is directly proportional to the current flow, starting torque is reduced according to the magnitude of current flow.

Figure 11-30 shows the primary-resistance starting method used to control a three-phase motor. When the start push button is pressed, coils (S) and (TR) are energized. Initially, the start contacts (S) will close, applying voltage through the primary resistors to the motor. These resis-

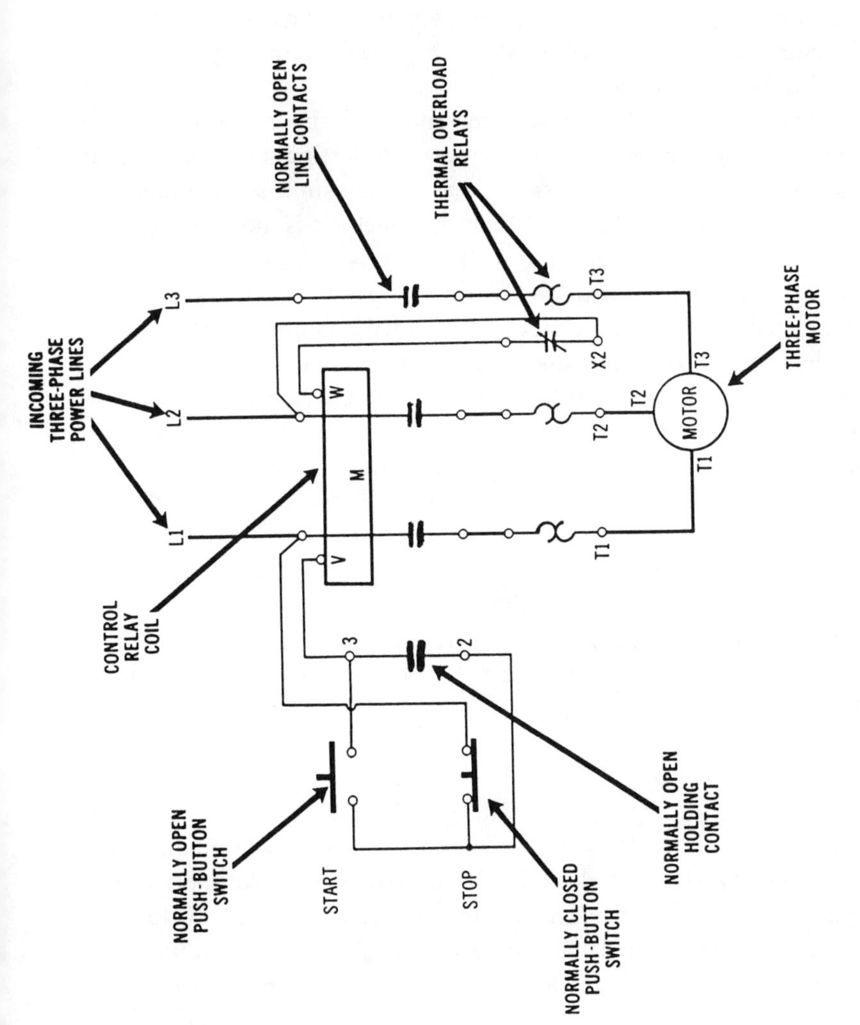

Figure 11-29. Full-voltage starting circuit for a three-phase motor.

tors reduce the value of starting current. Once the time-delay period of timing relay TR has elapsed, contact TR will close. The run contacts (R) will then close and apply full voltage to the motor. Notice that a step-down transformer is used for applying voltage to the control portion of the circuit. This is a commonly used technique for reducing the voltage applied to the relay coils.

Primary-Reactor Starting—Another method, similar to primary-resistance starting is called the primary-reactor starting method. Reactors (coils) are used in place of resistors since they consume smaller amounts of power from the AC source. Usually, this method is more appropriate for large motors that are rated at over 600 volts.

Autotransformer Starting—Autotransformer starting is another method used to start electric motors. This method employs one or more autotransformers to control the voltage that is applied to a motor. The autotransformers used are ordinarily tapped to provide a range of starting-current control. When the motor has accelerated to near its normal operating speed, the autotransformer windings are removed from the circuit. A major disadvantage of this method is the expense of the autotransformers.

An autotransformer starting circuit is shown in Figure 11-41. This is an expensive type of control that uses three autotransformers and four relays. When the start push button is pressed, current will flow through coils (1S), (2S), and (TR): The 1S and 2S contacts will then close. Voltage will be applied through the autotransformer windings to the three-phase motor. One normally closed and one normally open contact are controlled by timing relay TR. When the specified time period has elapsed, the normally closed TR contact will open and the normally open TR contact will close. Coil (R) will then energize, causing the normally open R contacts to close and apply full voltage to the motor. Normally closed R contacts are connected in series with coils (1S), (2S), and (TR) to open their circuits when coil (R) is energized. When the stop push button is pressed, the current to coil (R) will be interrupted, thus opening the power line connections to the motor.

Notice that the 65% taps of the autotransformer are used in Figure 11-31. There are also taps for 50%, 80%, and 100% to provide more flexibility in reducing the motor-starting current.

Wye-Delta Starting—It is possible to start three-phase motors more economically by using the wye-delta-starting method. Since, in a wye configuration, line current is equal to the phase current divided by 1.73

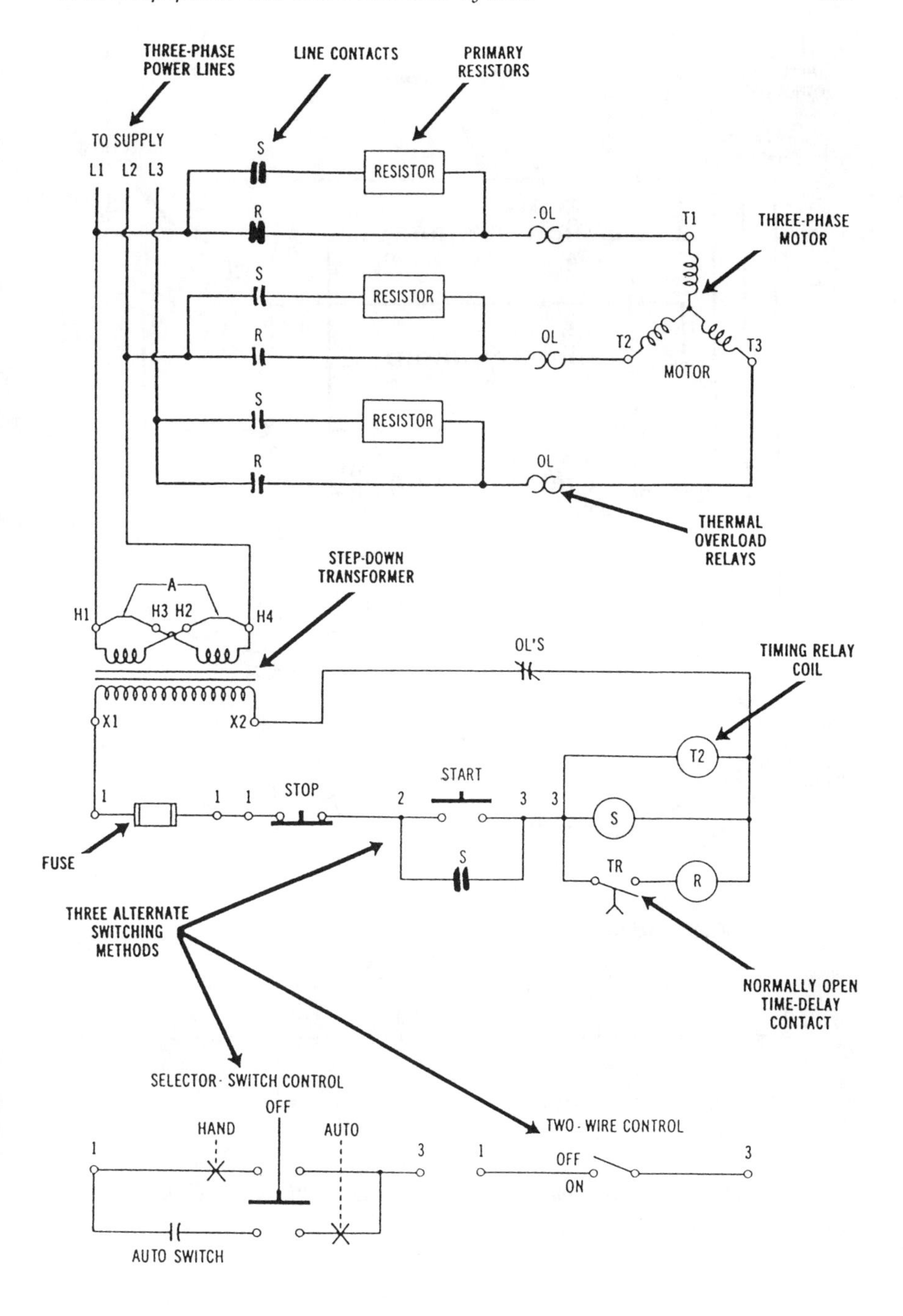

Figure 11-30. Primary-resistance starter circuit (Courtesy Furnas Electric Co.).

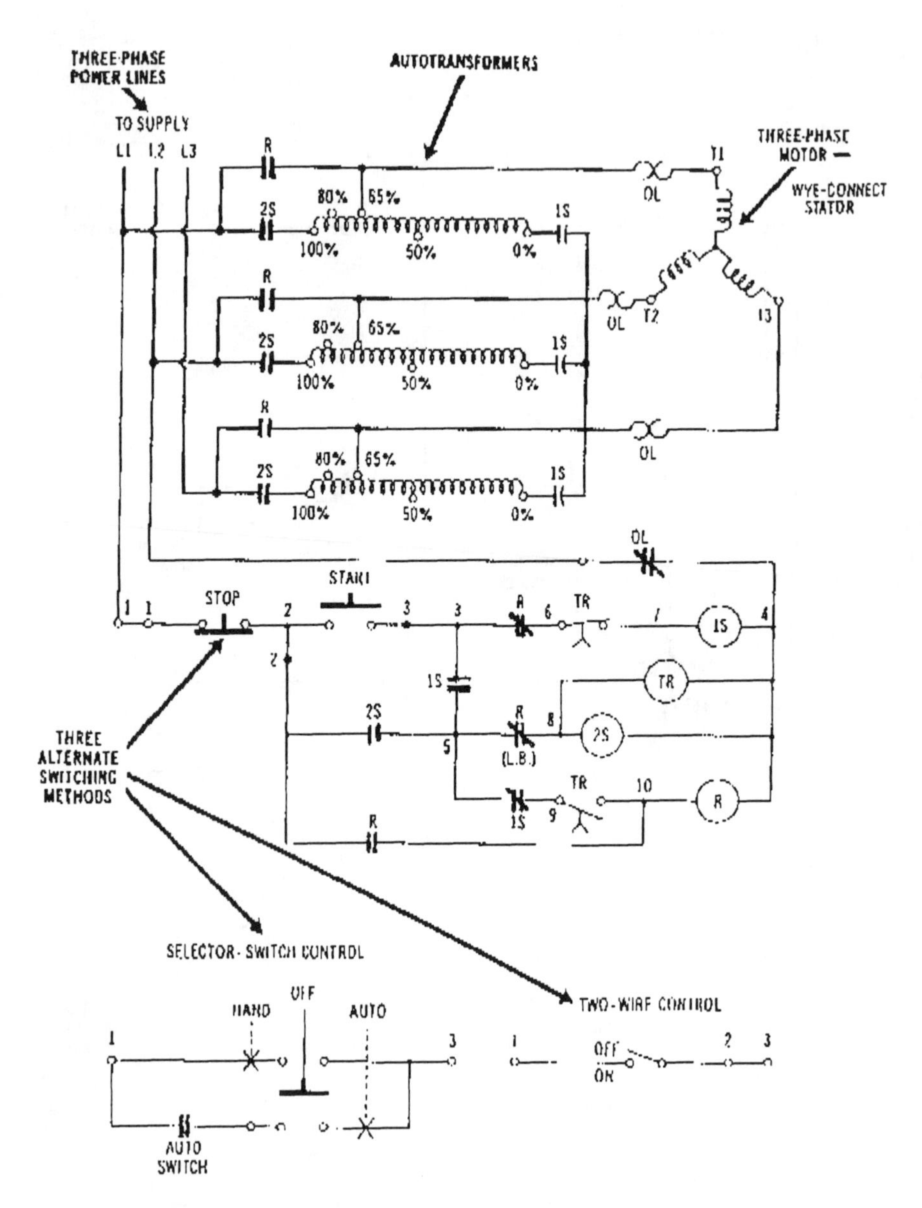

Figure 11-31. Autotransformer starter circuit used with a three-phase motor (Courtesy Furnas Electric Co.).

(or $\sqrt{3}$), it is possible to reduce the starting current by using a wye connection rather than a delta connection. This method, shown in Figure 11-32 employs a switching arrangement which places the motor windings in a wye configuration during starting and a delta arrangement for running. In this way, starting current is reduced. Although starting torque is reduced, running torque is still high since full voltage appears across each winding when the motor is connected in a delta configuration.

In Figure 11-32, when the start push button is pressed, coil (S) is energized. The normally open S contacts will close. This action connects the motor windings in a wye (or star) configuration and also activates timing relay (TR) and coil (1M). The normally open 1M contacts then close to apply voltage to the wye-connected motor windings. After the time-delay period has elapsed, the TR contacts will change state. Coil (S) will de-energize and coil (2M) will energize. The S contacts which hold the motor windings in a wye arrangement will then open. The 2M contacts will close and cause the motor windings to be connected in a delta configuration. The motor will, then, continue to run with the motor connected in a delta arrangement.

Part-Winding Starting—Figure 11-33 shows the part-winding-starting-method, which is usually more simple and less expensive than other starting methods. However, motors must be specifically designed to operate in this manner. During starting, the power-line voltage is applied across only part of the parallel-connected motor windings, thus reducing starting current. Once the motor has started, the line voltage is placed across all of the motor windings. This method is undesirable for many heavy-load applications due to the reduction of starting torque.

In Figure 11-33, when the start push button is pressed, current will flow through coil (M1) of the time-delay relay. This will cause the normally open contacts of M1 to close and a three-phase voltage will be applied to windings T1, T2, and T3. After the time-delay period has elapsed, the normally open contact located below coil (M1) in Figure 11-33 will close. This action energizes coil (M2) and causes its normally open contacts to close. The M2 contacts then connect the T7, T8, and T9 windings in parallel with the T1, T2, and T3 windings. When the stop push button is pressed, coils (M1) and (M2) will be de-energized.

Direct-Current Starting Systems—Since DC motors have no counter-electromotive force (cemf) when they are not rotating they have tremendously high starting currents. Therefore, they must use some type of control system to reduce the initial starting current. Ordinarily, a series

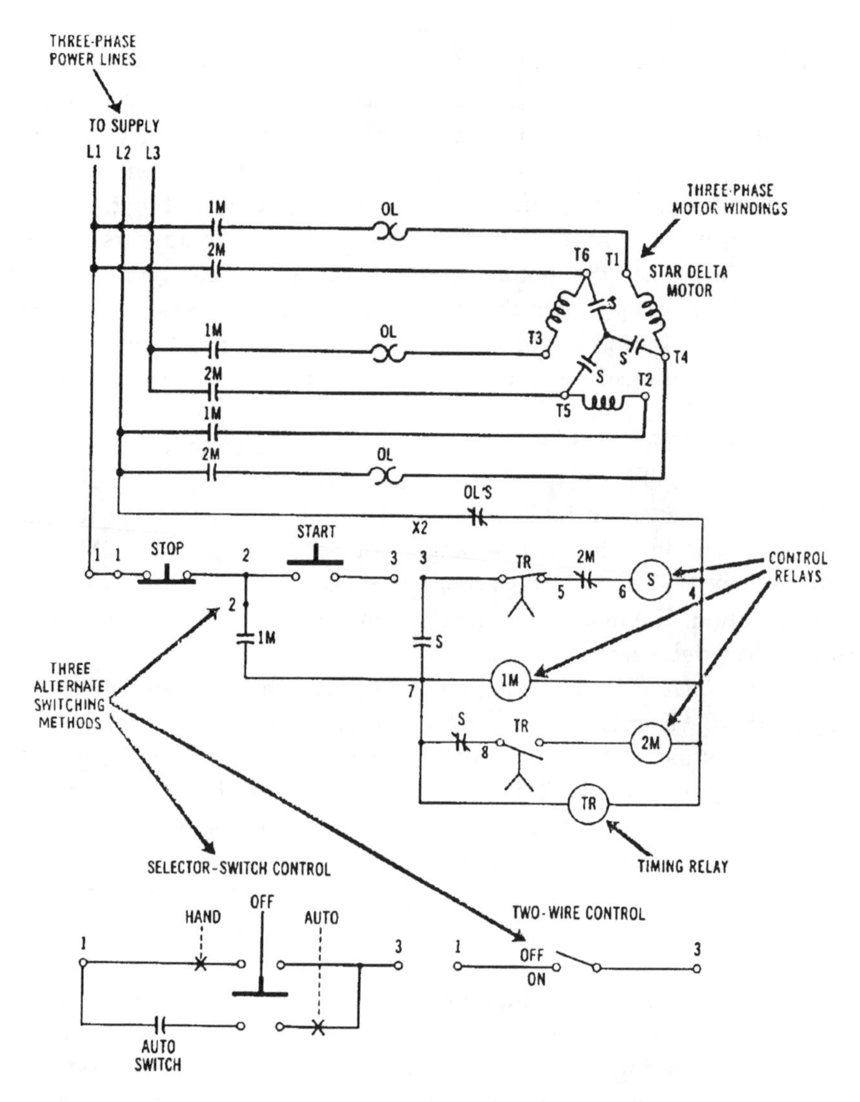

Figure 11-32. Wye-to-delta three-phase starter circuit.

resistance is used. This resistance can be manually or automatically reduced until a full voltage is applied. The four types of control systems commonly used with DC motors are (1) current limit, (2) definite time, (3) counter-emf, and (4) variable voltage. The current-limit method allows the starting current to be reduced to a specified level and, then, advanced to the next resistance step. The definite-time method causes

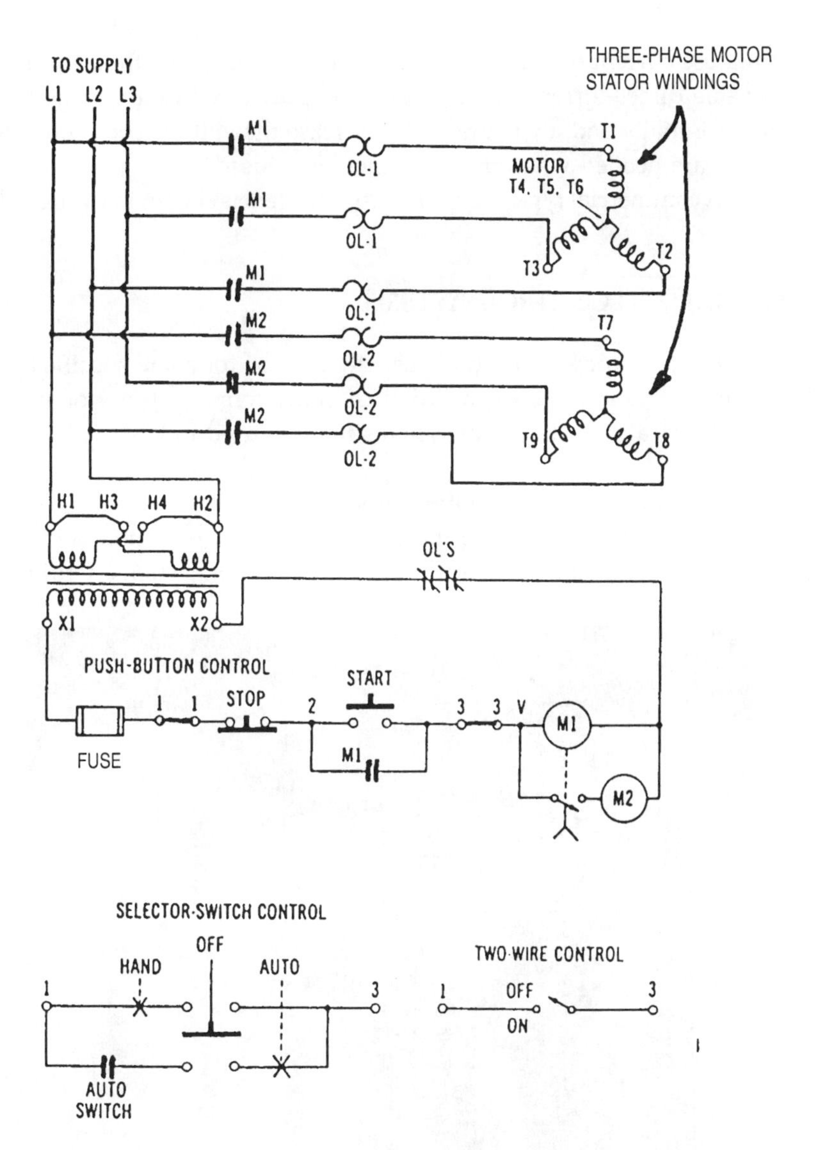

Figure 11-33. Part-winding circuit for three-phase motor starting.

the motor to increase speed in timed intervals with no regard to the amount of armature current or to the speed of the motor. The counter-emf method samples the amount of cemf generated by the armature of the motor to reduce the series resistance accordingly. This method can be used effectively since cemf is proportional to both the speed and the armature current of a DC motor. The variable-voltage method employs a variable direct-current power source to apply a reduced voltage to the motor initially and, then, gradually increase the voltage. No series resistances are needed when the latter method is used.

A commercial type of motor starting system is shown in Figure 11-34.

SPECIALIZED CONTROL SYSTEMS

Electrical power control is usually desired for some specific application. In this section, we will discuss some common types of specialized electrical power control systems that are used today.

Forward and Reverse Operation of Motors

Most types of electrical motors can be made to rotate in either direction by some simple modifications of their connections. Ordinarily,

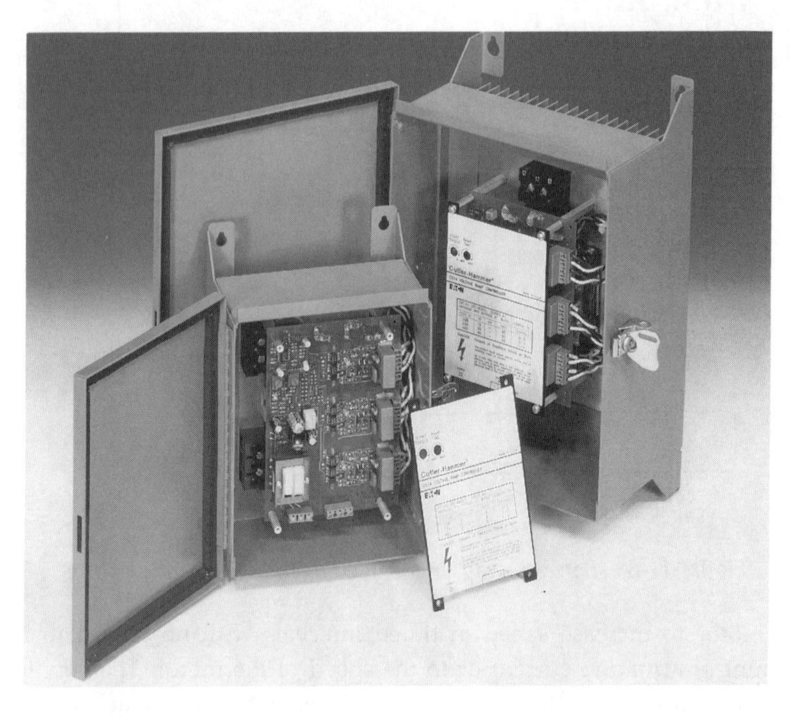

Figure 11-34. Reduced voltage starter (Courtesy Eaton Corp., Cutler-Hammer Products).

motors require two magnetic contactors, such as those shown in Figure 11-35 to accomplish forward and reverse operation. These contactors are used in conjunction with a set of three push button switches—FORWARD, REVERSE, and STOP. When the FORWARD push-button switch is depressed, the forward contactor will be energized. It is deactivated when the STOP push-button switch is depressed.

A similar procedure takes place during reverse operation.

Direct-Current Motor Reversing

Direct-current motors can have their direction of rotation reversed by changing either the armature connections or the field connections to the power source. In Figure 11-36, a DC shunt motor-control circuit is shown. When the forward push button is pressed, the F coil will be energized, causing the F contacts to close. The armature circuit is then completed from L1 through the lower F contact, up through the armature, through the upper F contact, and back to L2. Pressing the stop push button will DC-energize the F coil.

The direction of rotation of the motor is reversed when the reverse push button is pressed. This is due to the change of the current direction through the armature. Pressing the reverse push button energizes the R coil and closes the R contacts. The armature current path is then from L1 through the upper R contact, *down* through the armature, through the lower R contact, and back to L2. Pressing the stop button will deenergize the R coil.

Single-Phase Induction-Motor Reversing

Single-phase AC induction motors that have start and run windings can have their direction of rotation re versed by using the circuit in Figure 11-36. If we modify the diagram by replacing the shunt field coils with the run windings and the armature with the start windings, directional reversal of a single-phase AC motor can be accomplished. Single-phase induction motors the reversed by changing the connections of either the start windings or the run windings but not both at the same time.

Three-Phase Induction-Motor Reversing

Three-phase motors can have their direction of rotation reversed by simply changing the connections of any two power input lines. This changes the phase sequence applied to the motor. A control circuit for three-phase induction-motor reversing is shown in Figure 11-37.

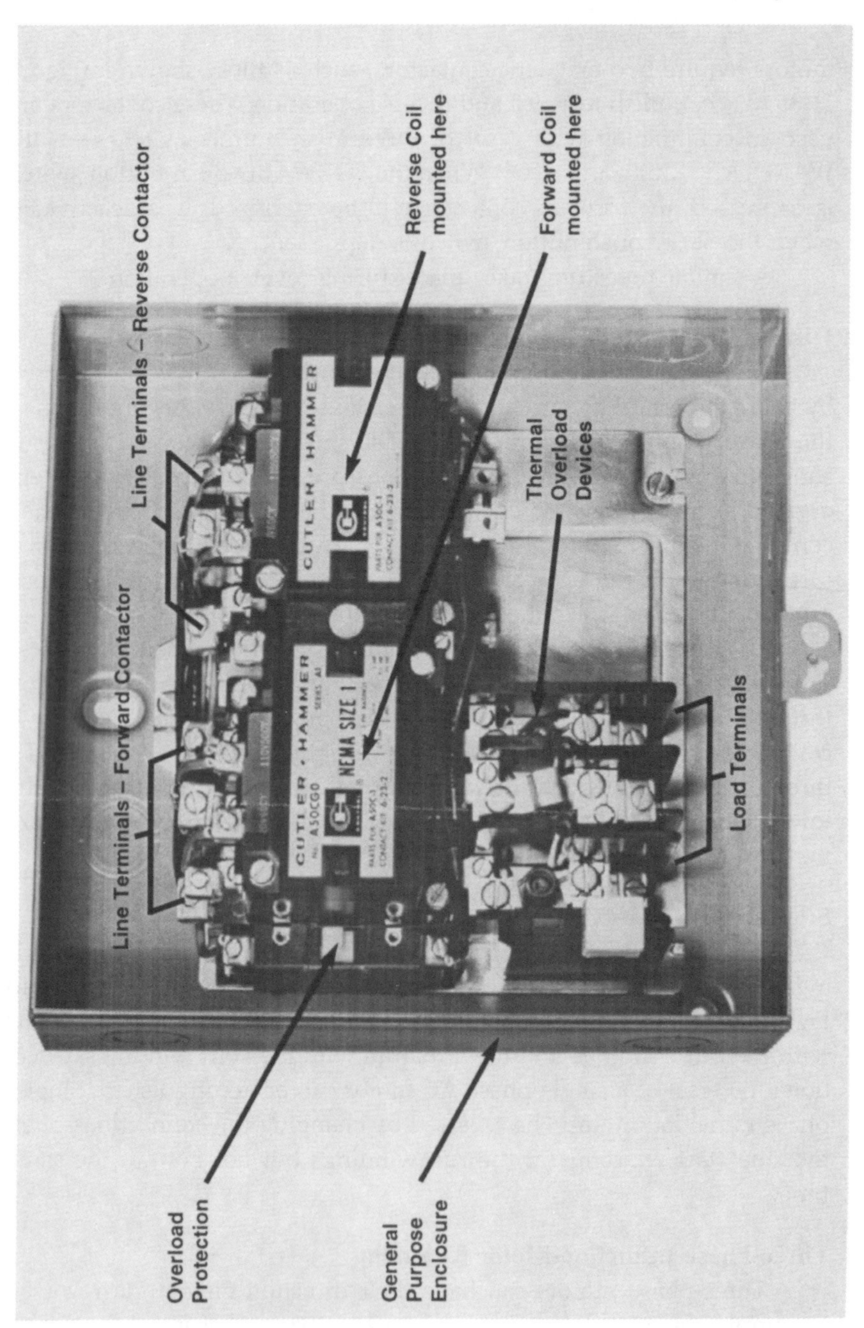

Figure 11-35. Forward and reverse motor contactors (Courtesy Eaton Corp., Cutler-Hammer Products).

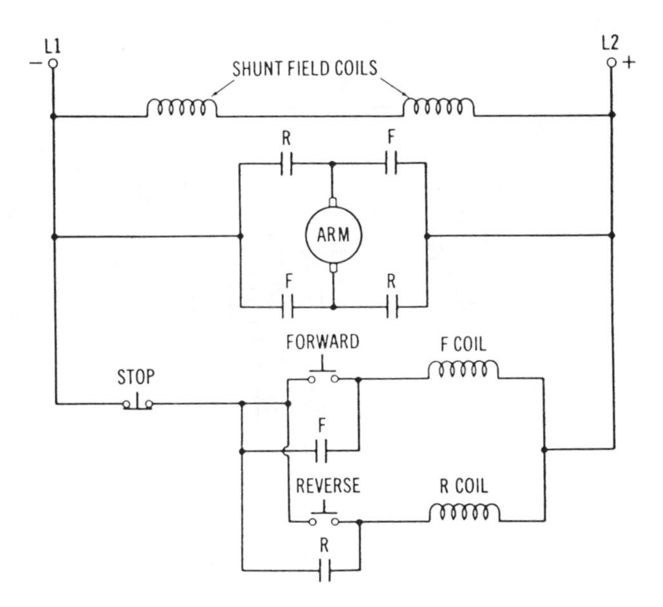

Figure 11-36. Control circuit for the forward and reverse operation of a DC shunt motor.

When the forward push button is pressed, the forward coil will energize and close the F contacts. The three-phase voltage is applied from L1 to T1, L2 to T2, and L3 to T3 to cause the motor to operate. The stop push button de-energizes the forward coil. When the reverse push button is pressed, the reverse coil is energized and the R contacts will close. The voltage is then applied from L1 to T3, L2 to T2, and L3 to T1. This action reverses the L1 and L3 connections to the motor and causes the motor to rotate in the reverse direction.

Starting Protection

Overload protection was discussed previously. Protection of expensive electric motors is necessary to extend the lifetime of these machines. The cost of overload protection is small compared to the cost of large electric motors. Motors present unique problems for protection since their starting currents are much higher than their running currents during normal operation. Solid-state overload protection and microprocessor-based monitoring systems such as shown in Figure 11-38 are now available to provide motor protection.

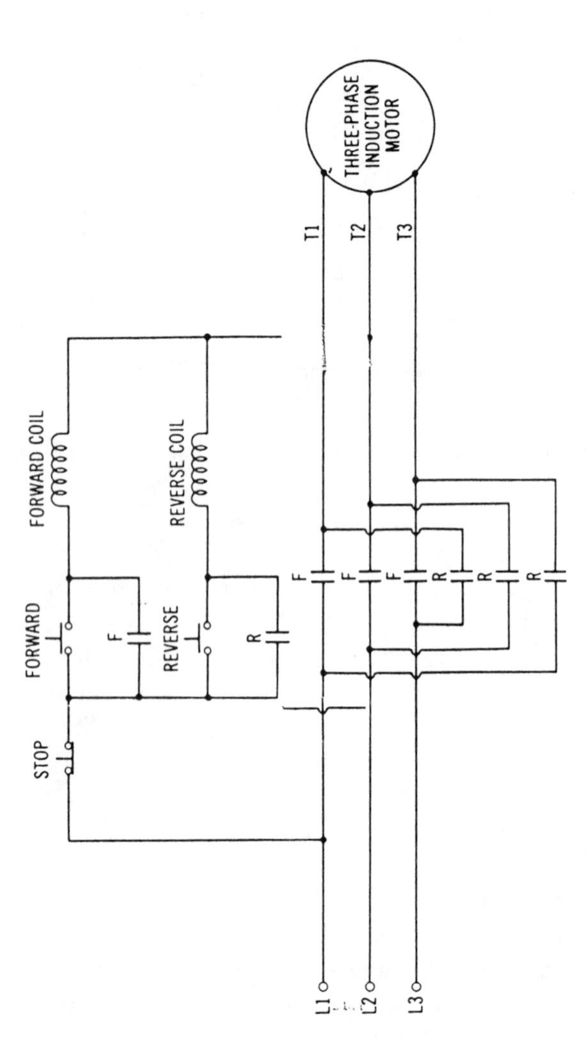

Figure 11-37. Control cirtuit for the forward and reverse operation of a three-phase induction motor.

Dynamic Braking

When a motor is turned off, its shaft will continue to rotate for a short period of time. This continued rotation is undesirable for many applications. Dynamic braking is a method used to bring a motor to a quick stop whenever power is turned off. Motors with wound armatures utilize a resistance connected across the armature as a dynamic braking method. When power is turned off, the resistance is connected across the armature. This causes the armature to act as a loaded generator, making the motor slow down immediately. This dynamic braking method is shown in Figure 11-39.

Alternating-current induction motors can be slowed down rapidly by placing a DC voltage across the winding of the motor. This DC voltage sets up a constant magnetic field which causes the rotor to slow down rapidly. A circuit for the dynamic braking of a single-phase AC induction motor is shown in Figure 11-40.

FREQUENCY-CONVERSION SYSTEM S

The power system frequency used in the United States is 60 Hertz or 60 cycles per second. However, there the specific applications that require other frequencies in order to operate properly. Mechanical frequency converters may be used to change an incoming frequency into some other frequency. Frequency converters are motor-generator sets that are connected together or solid-state variable frequency drives.

For example, a frequency of 60 Hz could be applied a synchronous motor which rotates at a specific speed. A generator connected to its shaft could have the necessary number of poles to cause it to produce a frequency of 25 Hz. Recall that frequency is determined by the following relationship:

$$\text{Freq (Hz)} = \frac{\text{Speed of Rotation (rpm)} \times \text{No. of Poles}}{120}$$

A frequency-conversion system is shown in Figure 11-41. Synchronous units such as the one shown are used wherever precise frequency control is required. It is also possible to design units which are driven by induction motors if some frequency variation can be tolerated.

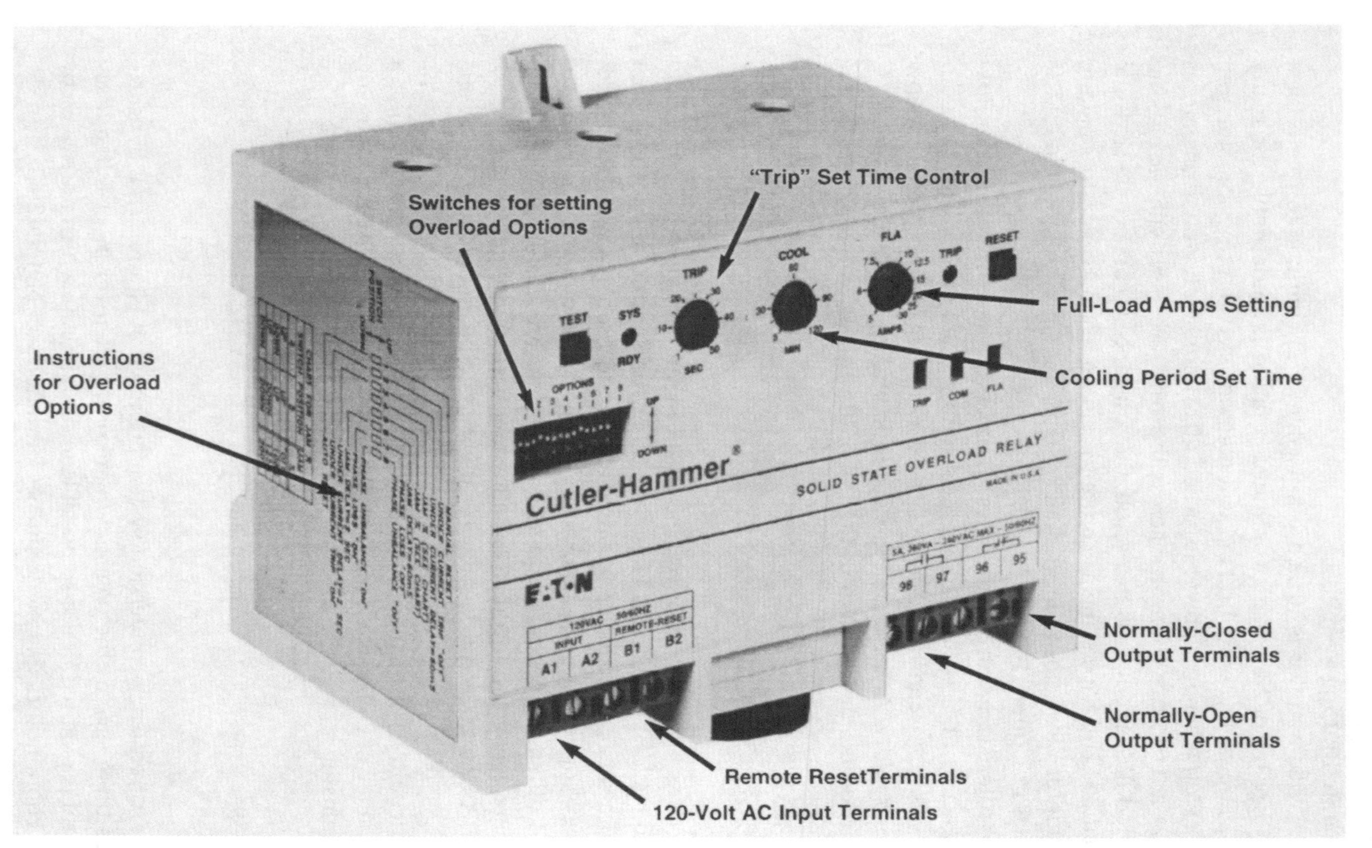

Figure 11-38. Electronic overload relay (Courtesy Eaton Corp., Cutler-Hammer Products).

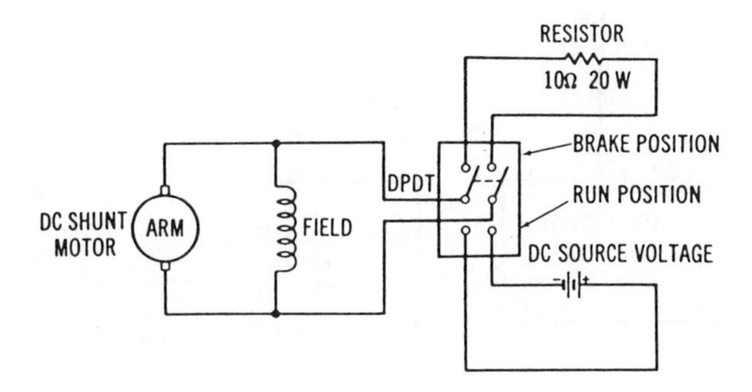

Figure 11-39. Dynamic braking circuit for a DC shunt motor.

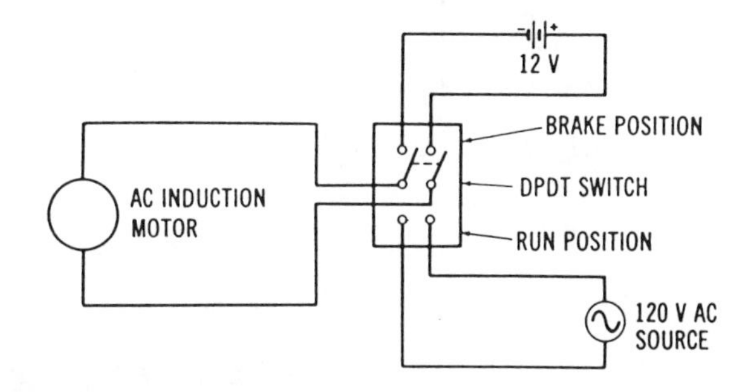

Figure 11-40. Dynamic braking circuit for a single-phase AC induction motor.

Variable frequency drives, such as those shown in Figure 11-42, are now being used to control many types of industrial machinery. They are used because they are not as expensive as variable speed controllers have been in the past. Solid state variable frequency drives are used to control overhead cranes, hoists and many other types of industrial equipment which operates from AC power lines. Variable speed can be accomplished cheaper, in most cases, than with DC motors or electromechanical variable frequency drives. Solid state drives can be used with AC induction motors to change their speed by varying the input frequency to the motor. There are fewer maintenance problems with AC induction motors than with DC motors. In addition, fewer electromechanical parts are involved in the control operation, thus equipment has a longer life expectancy.

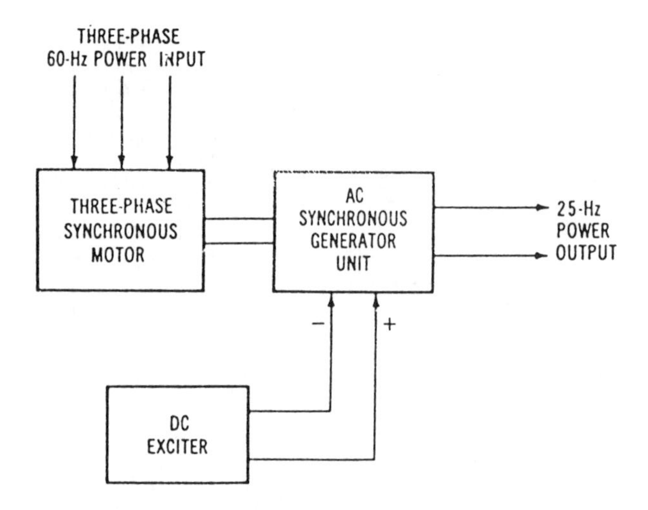

Figure 11-41. A 60-Hz to 25-Hz frequency-conversion system.

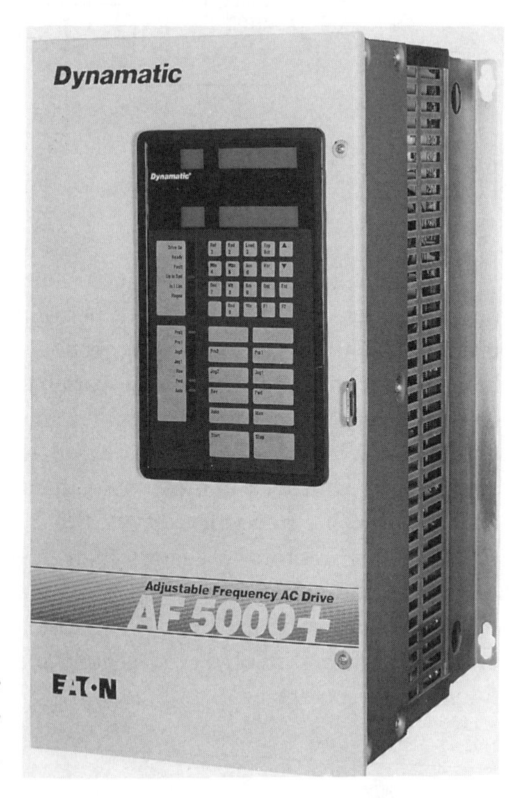

Figure 11-42a. Variable frequency drives (Courtesy Eaton Corp.).

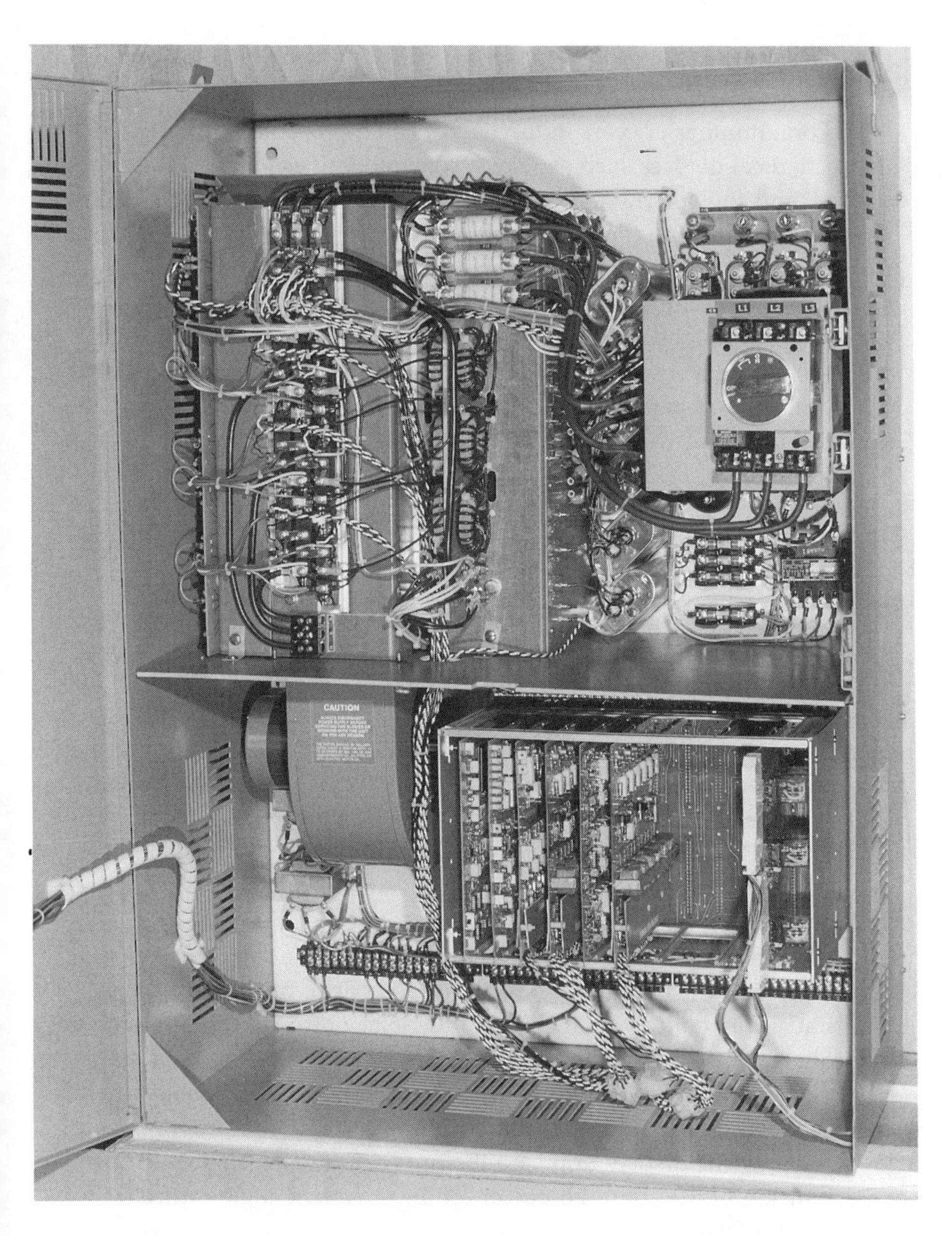

Figure 11-42b. 50 HP variable frequency drive (Courtesy Robicon Corp.).

PROGRAMMABLE LOGIC CONTROLLERS (PLCs)

For a number of years industrial control has been achieved by electromechanical devices such as relays, solenoid valves, motors, linear actuators, and timers. These devices are. used to control large production machines where only switching operations were necessary. Most controllers were used to simply turn the load device on or off In addition, some basic logic functions could also be achieved. Production line sequencing operations were achieved by motor-driven drum controllers with timers. As a rule, nearly all electromagnetic controllers were hardwired into the system and responded as permanent fixtures. Modification of the system was rather difficult to accomplish and somewhat expensive. In industries where production changes were frequent, this type of control was rather costly. It was, however, the best way and in many cases the only way that control could be effectively achieved with any degree of success.

In the late 1960s, solid-state devices and digital electronics began to appear in controllers. These innovations were primarily aimed at replacing the older electromechanical control devices. The transition to solid-state electronics has, however, been much more significant than expected. Solid-state electronic devices, digital logic ICs, and microprocessors have led to the development of programmable logic controllers (PLCs). These devices have capabilities that far exceed the older relay controllers. Programmable logic controllers are extremely flexible, have reduced downtime when making changeovers, occupy very little space, and have improved operational efficiency.

A block diagram of a simplified programmable logic control system is shown in Figure 11-43. Note that the control function of the system is achieved entirely by the programmable controller. Programmable controllers, in general, are more complex than indicated by a single block. A programmable controller has several parts. Functionally, these are described as the input/output or I/O structure, the processor, the memory unit, the display unit, and the programmer. Figure 11-44 shows an expansion of the control block of a programmable logic controller to include these parts.

The block diagram of a programmable logic controller is very similar to that of a small computer. In fact, most programmable logic controllers are classified as dedicated computers. This type of unit is usually designed to perform a number of specific control functions in the operation of a machine or industrial process.

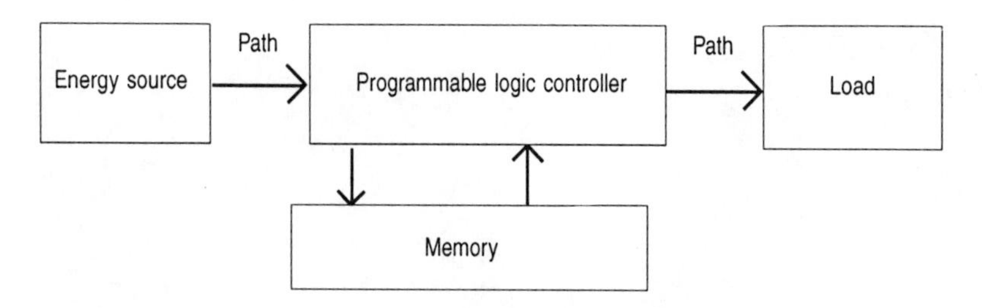

Figure 11-43. Programmable logic controller system

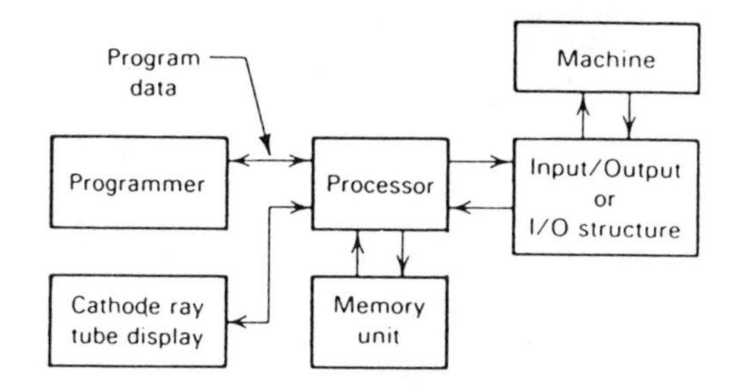

Figure 11-44. Programmable logic controller parts.

Chapter 12

Power Measurement Equipment

Chapter 12 provides an overview of types of equipment used to measure electrical power quantities. Specific applications of measurement equipment are discussed. There are many different types of equipment used to measure quantities associated with electrical power.

MEASUREMENT SYSTEMS

All measurement systems have certain basic characteristics. Usually a specific quantity is monitored either periodically or continuously. Therefore, some type of visual indication must be available of the quantity being monitored. Several types of instruments for measuring electrical and physical quantities are available. The basic types of measurement systems can be classified as: (1) analog instruments (2) oscilloscopes, (3) numerical-read-out instruments, and (4) chart-recording instruments.

Analog Instruments

Instruments which rely upon the motion of a hand or pointer are referred to as instruments. The volt-ohm-milliammeter (vom) is one type of instrument. The vom is a multifunction, multirange meter (Figure 12-1). Single function analog meters can also be used to measure electrical or physical quantities (see Figure 12-2). The basic part of an electrical analog meter is called the *meter movement* and is shown in Figure 12-3. The movement of the pointer along a calibrated scale is used to indicate an electrical or physical quantity. Physical quantities such as air flow or fluid pressure can also be monitored by analog meters.

Many meters employ an analog movement called the D'Arsonval, or moving-coil, type. The basic operational principle of this type of movement is shown in Figure 12-4. The pointer or needle of the movement remains stationary on the left portion of the calibrated scale until a current flows through the electromagnetic coil that is centrally located within a permanent magnetic field. When current flows through the coil, a reaction between the electromagnetic field of the coil and the stationary permanent magnetic field is developed. This reaction causes the hand (pointer) to deflect toward the right portion of the scale. This basic moving-coil meter movement operates due to the same principle as an electric motor. It may be used for either single function meters, which measure only one quantity, or for multifunction meters (see Figures 12-1 and 12-2). The basic meter movement may be modified so that it will measure almost any electrical or physical quantity.

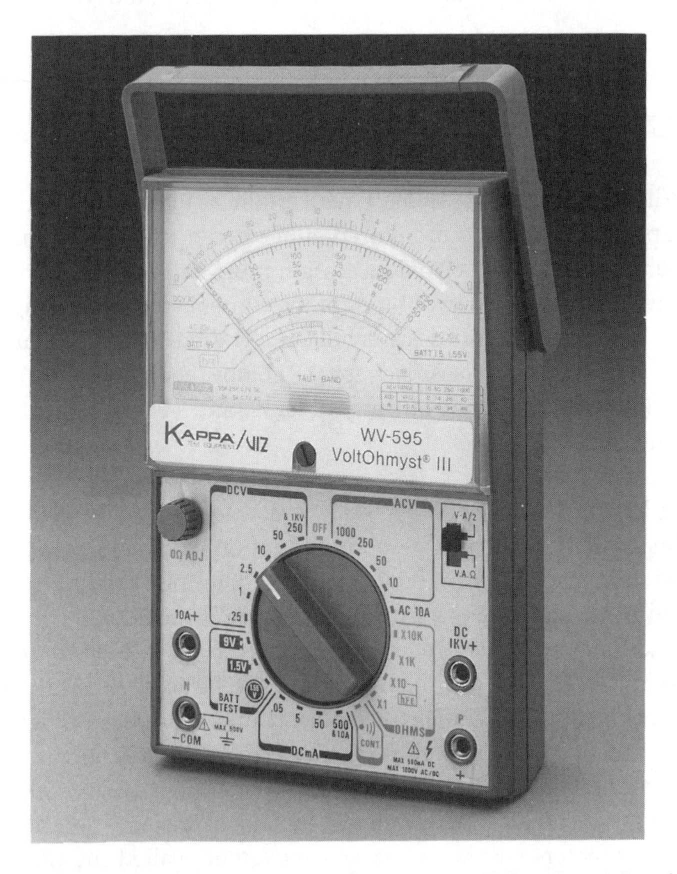

Figure 12-1. Analog multimeter (VOM) measures DC volts, AC volts, DC milliamps, and resistance (Courtesy Vector Group Inc., Instrument Division).

Figure 12-2. Single function analog meter—measures 0 to 50 DC microamperes (Courtesy Triplett Corp.).

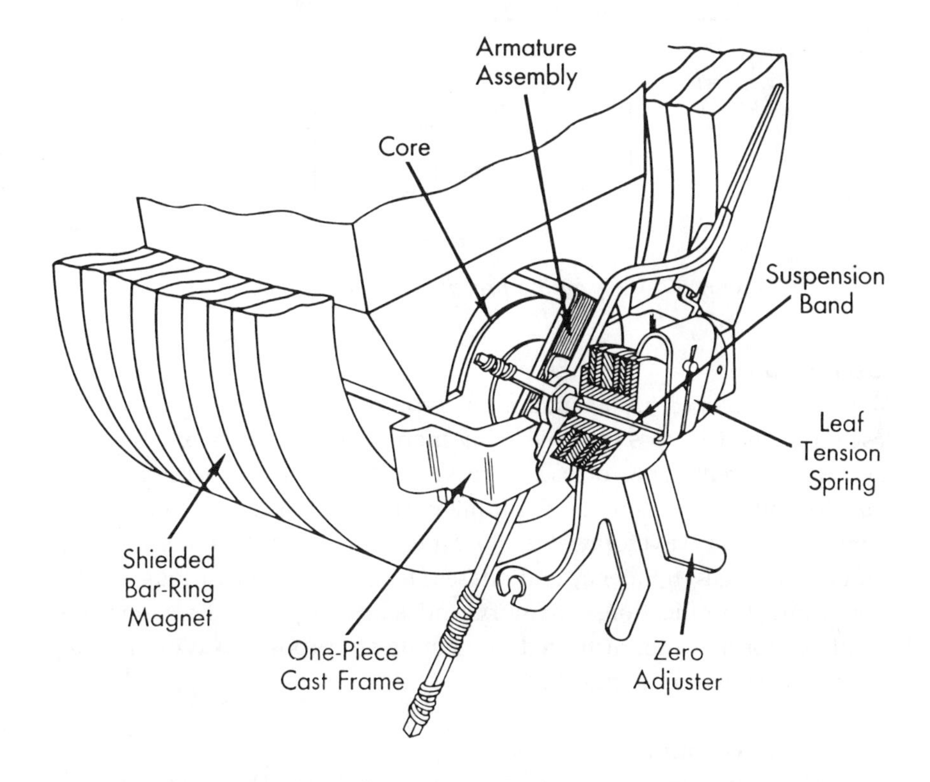

Figure 12-3. Moving-coil meter movement for analog meters (Courtesy Triplett Corp.).

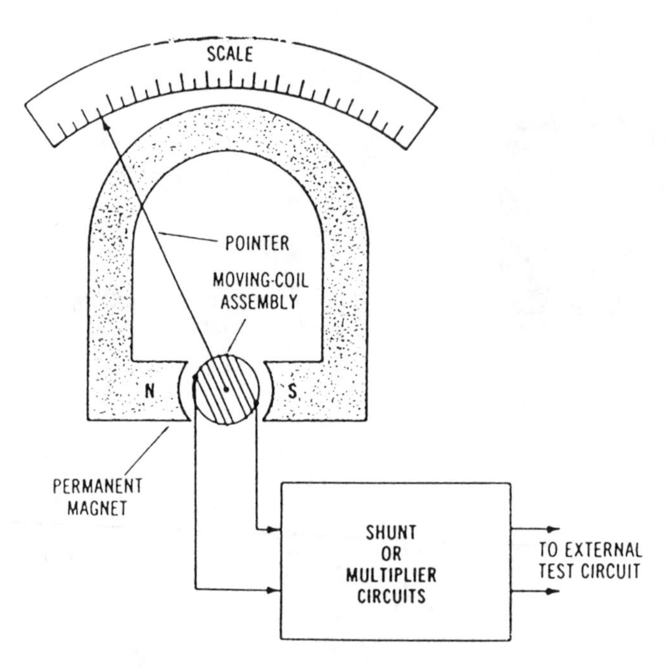

Figure 12-4. Operating principle of a meter movement.

Oscilloscopes

Oscilloscopes are an important type of instrument. By using an oscilloscope, it is possible to visually monitor the voltages of a system. Several different types of oscilloscopes are available. General-purpose oscilloscopes are used for equipment servicing and for displaying simple types of waveforms. Oscilloscopes may be used to measure AC and DC voltages, frequency and phase relationships, and various timing and control applications. Memory and storage-type instruments are available for more sophisticated measurement purposes. A typical oscilloscope is shown in Figure 12-5.

Numerical-readout Instruments

Many instruments employ numerical readouts. These simplify the measurement processes and permit making more accurate measurements. Numerical-readout instruments, such as those shown in Figure 12-6 rely upon the operation of digital circuitry in order to produce a numerical display of the measured quantity. They are very popular measuring instruments.

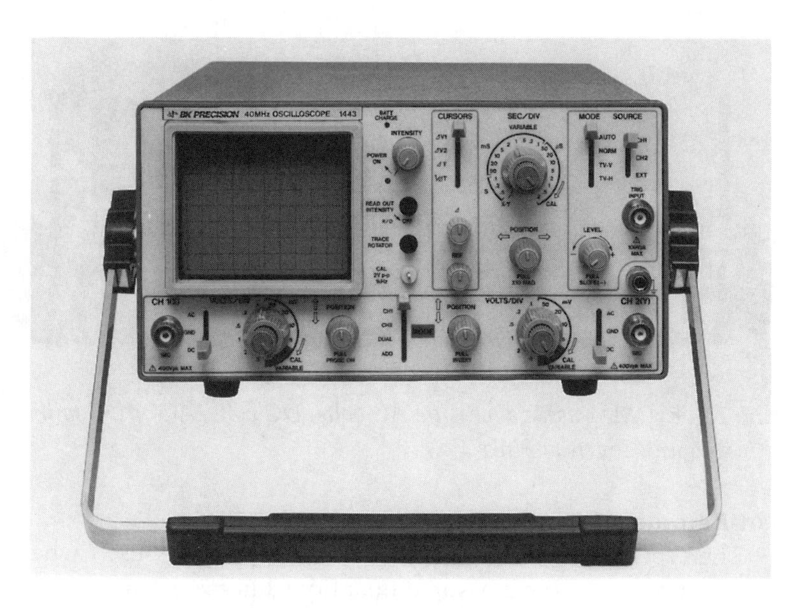

Figure 12-5. Oscilloscope (Model 1443, 40 MHz Oscilloscope, w/Cursors & Readouts) (Courtesy Maxtec International Corp.).

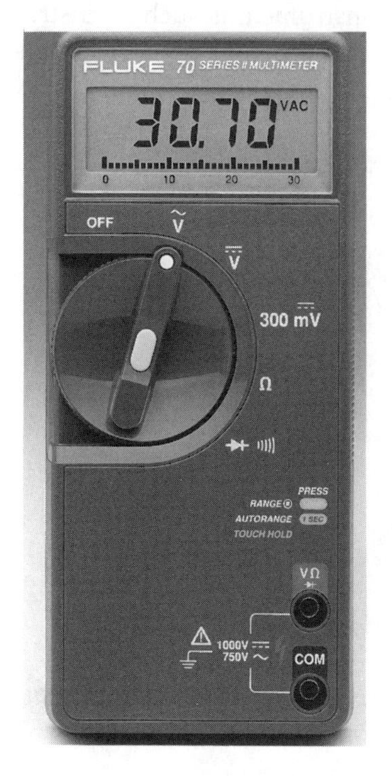

Figure 12-6a. Digital multimeters (DMMs)—autoranging meter for AC and DC volts and resistance measurement.

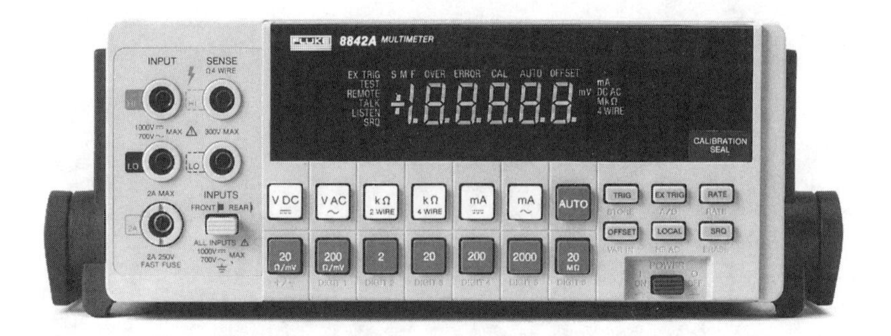

Figure 12-6b. DMM used to measure AC volts, DC volts, AC or DC milliamps and resistance (Courtesy Fluke Co.).

Chart-Recording Instruments

The types of instruments discussed previously are used when no permanent record of the measured quantity values is needed. However, instruments can be employed that provide a permanent record of the measured values. Also, values that are measured over a specific time period can be recorded. A chart-recording instrument is such an instrument. The types of chart recorders include both pen and ink recorders and inkless recorders (see Figure 12-7).

Figure 12-7a. Chart recorders: pen and ink recorder with three color pens and digital display (Courtesy ABB-Kent Taylor Co.).

Figure 12-7b. Chart recorders: inkless recorders used to monitor steam turbine functions at a power plant.

Pen and ink recorders have a pen attached to the instrument. This pen is caused by either electrical or mechanical means to touch a paper chart and leave a permanent record of the measured quantity on the chart. The charts utilized may be either roll charts (which revolve on rollers under the pen mechanism) or circular charts (which revolve on an axis under the pen). Chart recorders may use more than one pen to record several quantities simultaneously. In this case, each pen mechanism would be connected to measure a specific quantity. The pen of a chart recorder is a capillary tube device that is actually an extension of the basic meter movement. The pen must be connected to a constant source of ink. The pen is moved by the torque that is exerted by the meter movement just as the pointer of a hand-deflection type of meter is moved. The chart used for recording the measured quantity usually contains lines that correspond to the radius of the pen movement. Increments on the chart are marked according to time intervals. The chart must be moved under the pen at a constant speed. Either a spring-drive mechanism, a synchronous AC motor, or a DC servomotor can be used to drive the chart. Recorders are also available which use a single pen to make permanent records of measured quantities on a single chart. In this case, either coded lines or different colored ink could be used to record the quantities.

Inkless recorders may use a voltage applied to the pen point to produce an impression on a sensitive paper chart. In another process, the pen is heated to cause a trace to be melted along the chart paper. The obvious advantage of inkless recorders is that ink is not required.

Chart-recording instruments are commercially available for measuring almost any electrical or physical quantity. For many applications, the recording system may be located a great distance from the device being measured. For accurate system monitoring, a central instrumentation system might be used. Power plants, for instance, ordinarily use chart recorders at a centralized location to monitor the various electrical and physical quantities involved in the power-plant operation (see Figure 12-7b).

The operation of a typical roll-chart recording instrument involves several basic principles. The user should assure that the chart roll has enough paper to last throughout the duration of the time that it is to be used. The ink supply should be checked. If ink is needed, the well should be filled to the proper level. Also, the pen should be checked for proper pressure on the roll chart and for accurate adjustment along the incremental scale of the roll chart. The user should also assure that the meter is properly connected to the external circuit

MEASURING ELECTRICAL POWER

Electrical power is measured with a wattmeter. An analog meter movement called a *dynamometer movement,* shown in Figure 12-8, is used for analog wattmeters. Note that this movement has two electromagnetic coils. One coil, called the *current* coil, is connected in series with the load to be measured. The other coil, called the *potential* coil, is connected in parallel with the load. Thus, the strength of each electromagnetic field affects the movement of the meter pointer. The operating principle of this movement is similar to that of the moving-coil type, except that there is a fixed electromagnetic field rather than a permanent-magnet field.

When measuring DC power, the total power is the product of voltage times current (P = V × I). However, when measuring AC power, we must consider the power factor of the load, since P = V × I × pf. The true power of an AC circuit may be read directly with a wattmeter. When a load is either inductive or capacitive, the true power will be less than the apparent power (V × I).

MEASURING ELECTRICAL ENERGY

The amount of electrical energy used over a certain period of time may be measured by using a watt-hour meter. A watt-hour meter, illustrated in Figure 12-9 relies upon the operation of a small motor mounted

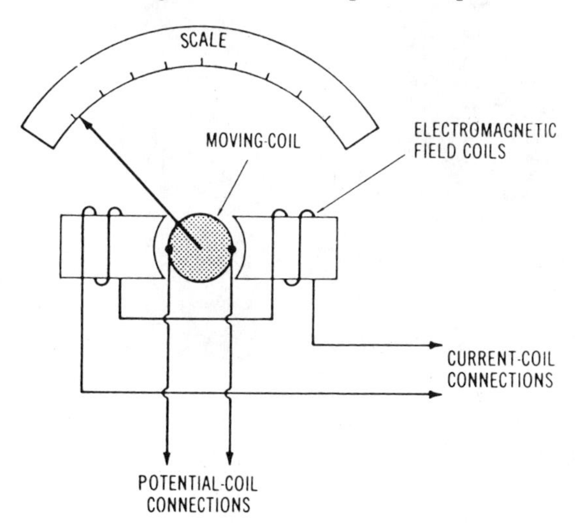

Figure 12-8. Measuring electrical power. (a) Basic dynamometer movement showing potential coil and current coil connection.

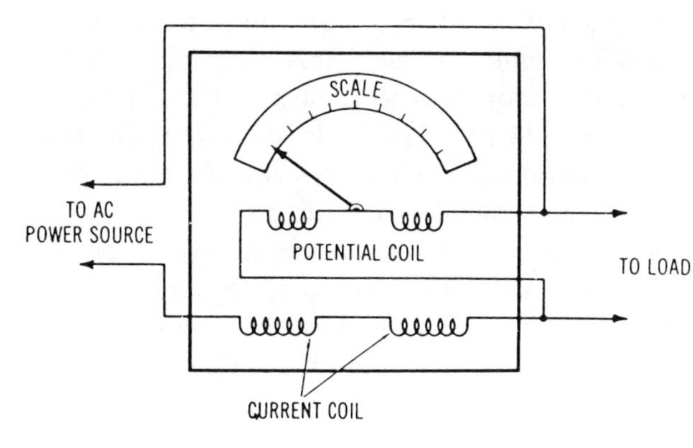

Figure 12-8. (b) Wattmeter movement using dynamometer movement showing source and load correction.

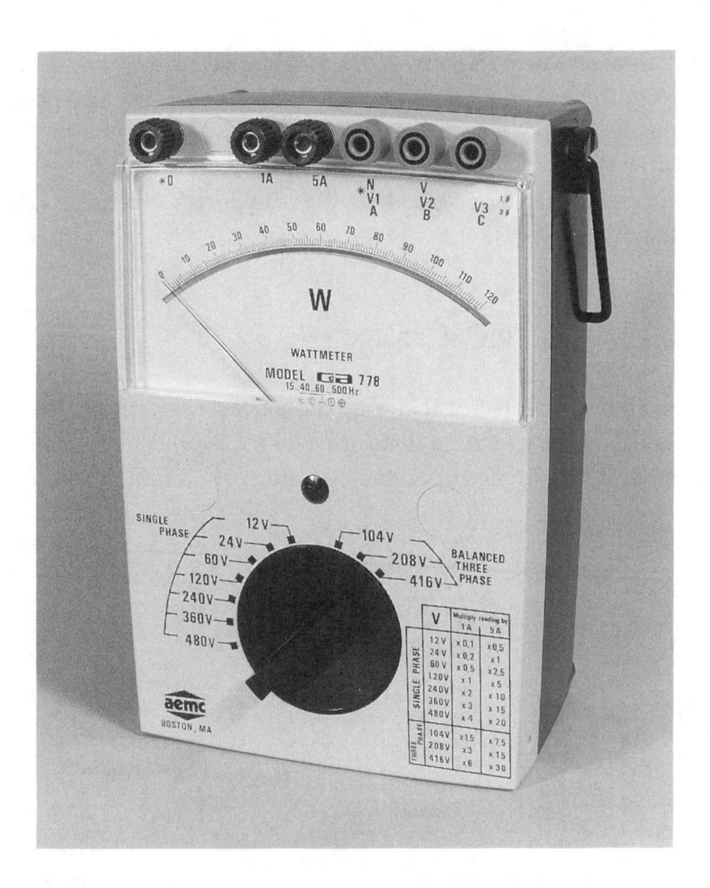

Figure 12-8. (c) Analog wattmeter with electrodynamometer. Meter movement used to measure single-phase or three-phase power (Courtesy AEMC Corp., Boston, MA).

inside its enclosure. The speed of the motor is proportional to the power applied to it. The rotor is an aluminum disk that is connected to a numerical register, which usually indicates the number of kilowatt-hours of electrical energy used. Figure 12-10 shows the dial-type face plate that is frequently used on watt-hour meters. Other types of watt-hour meters have direct numerical readout of the kilowatt-hours used.

The watt-hour meter is connected between the in coming power lines and the branch circuits of an electrical power system. In this way, all electrical energy that is used must pass through a kilowatt-hour meter. The same type of system is used for home, industrial, or commercial service entrances.

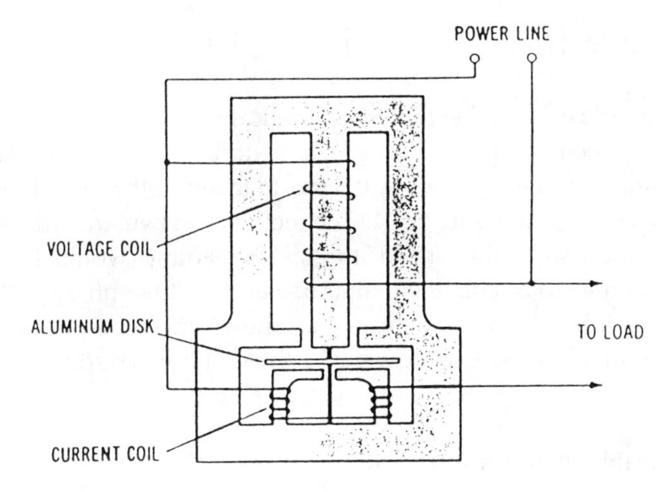

Figure 12-9. Construction diagram of a watt-hour meter.

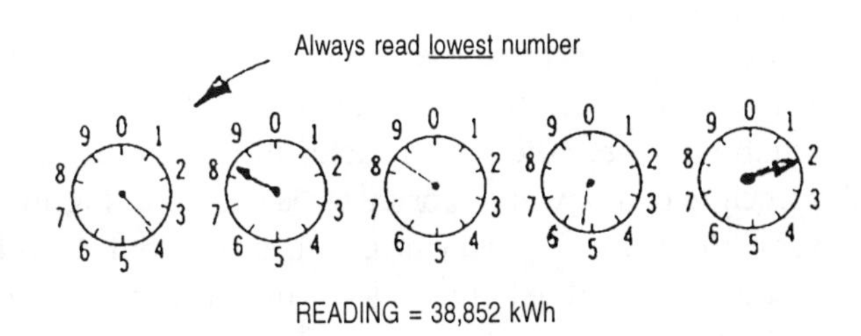

Figure 12-10. Registering dials of a watt-hour meter.

The operation of a watt-hour meter is in many ways similar to the conventional wattmeter. A potential coil is connected across the incoming power lines to monitor voltage while a current meter is placed in series with the line to measure current. Both meter sections are contained within the watt-hour meter enclosure. The voltage and current of the power system affect the movement of an aluminum-disk rotor which is part of the watt-hour meter assembly. The operation of the watt-hour meter may be considered as similar to an AC induction motor. The stator is an electromagnet that has two sets of windings—the voltage windings and the current windings. The field developed in the voltage windings causes a current to be induced into the aluminum disk. The torque produced is proportional to the voltage and the *in-phase* current of the system. Therefore, the watt-hour meter will monitor the true power converted in a system.

MEASURING THREE-PHASE ELECTRICAL ENERGY

For industrial and commercial applications, it is usually necessary to monitor the three-phase energy that is utilized. It is possible to use a combination of single-phase wattmeters to measure the total three-phase power, as shown in Figure 12-11. The methods shown are ordinarily not very practical since the sum of the meter readings would have to be found in order to calculate the total power of a three-phase system.

Three-phase power analyzers, as shown in Figure 12-12, are designed to monitor the true power of a three-phase system.

MEASURING POWER FACTOR

Power factor is the ratio of the true power of a system to the apparent power (volts × amperes). To determine power factor, we could use a wattmeter, voltmeter, and an ammeter and use the relationship of pf = W/VA. However, it would be more convenient to use a power factor meter in situations where the power factor must be monitored.

The principle of a power factor meter is shown in Figure 12-13. The power factor meter is similar to a wattmeter, except that it has two armature coils that will rotate due to their electromagnetic field strengths. The armature coils are mounted on the same shaft so that their alignment is about 90° apart. One coil is connected across the AC line (in series with a resistance) while the other coil is connected across the line through an inductance. The resistive path through the coil reacts to produce a flux proportional to the *in-phase* component of the power. The inductive path reacts in proportion to the *out-of-phase* component of the power.

If a unity (1.0) power factor load is connected to the meter, the current in the resistive path through coil A would develop full torque. Since there is no out-of-phase component, no torque would be developed through the inductive path. The meter movement would now indicate full-scale or unity power factor. As the power factor decreases below 1.0, the torque developed by the inductive path through coil B would become greater. This torque would be in opposition to the torque developed by the resistive path. Therefore, a power factor of less than 1.0 would be indicated. The scale must be calibrated to measure power factor ranges from zero to unity. Figure 12-14 shows a type of power factor meter.

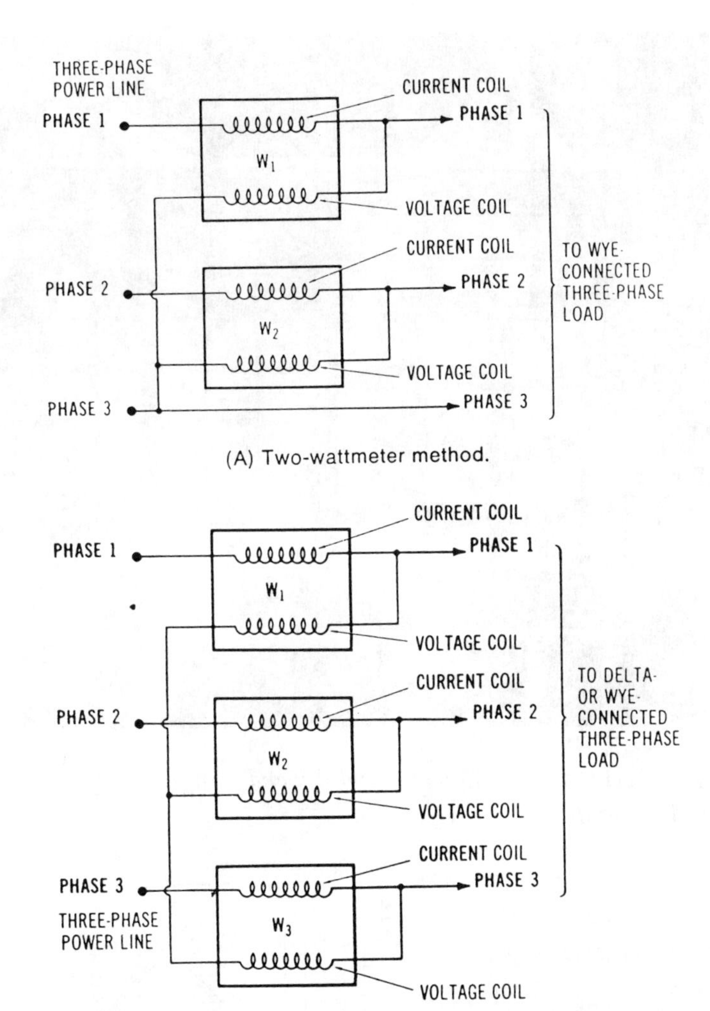

Figure 12-11. Using single-phase wattmeters to measure three-phase power.

POWER-DEMAND METERS

Figure 12-15 shows a power-demand meter. These instruments perform, an important industrial function. Power demand is expressed as:

$$\text{power demand} = \frac{\text{peak power used (kW)}}{\text{average power used (kW)}}.$$

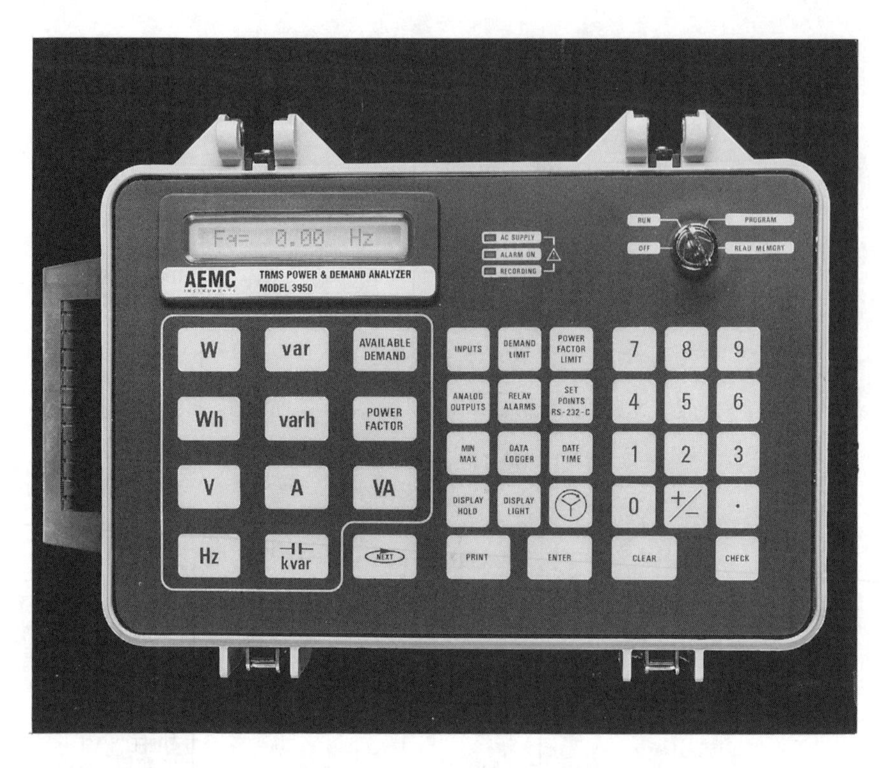

Figure 12-12. True RMS power and demand analyzer, digital type (Courtesy AEMC Corp.).

Figure 12-13. Schematic diagram of a power factor meter circuit.

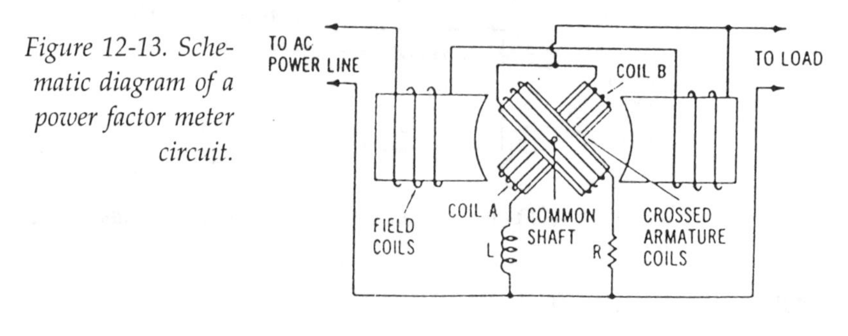

Sample Problem:

Given: An industry uses 5000 kW peak power and 3880 kW average power over a 24-hour period.

Find: The industry's demand factor over the 24-hour period.

Figure 12-14. Power factor meters, analog type (Courtesy Westinghouse Electric Corp.).

Solution:

$$\text{Demand} = \frac{\text{Peak Power}}{\text{Avg. power}} = \frac{5000\,\text{kW}}{3800\,\text{kW}}$$

Demand = 1.31

This ratio is important since it indicates the amount above the average power consumption that a utility company must supply an industry. Power demand is usually calculated over a 15-, 30-, or 60-minute interval and, then, converted into longer periods of time.

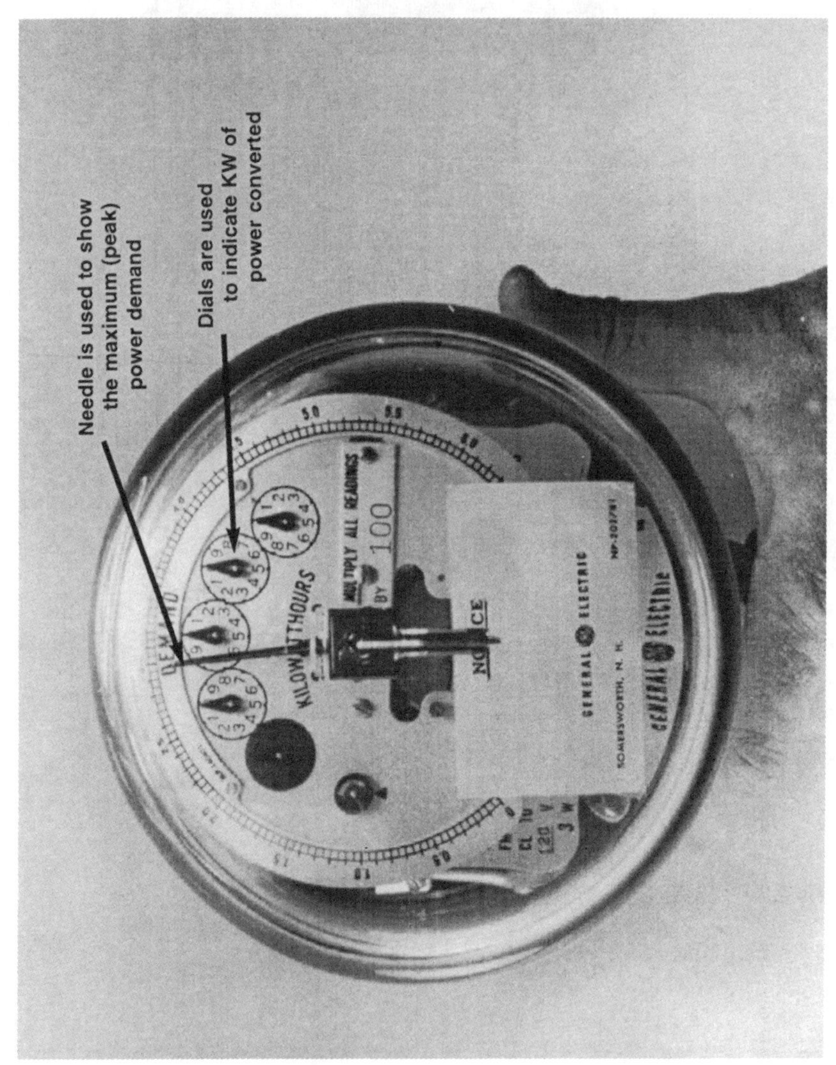

Figure 12-15. Power demand meter.

Industries may be penalized by a utility company if their peak power demand far exceeds their average demand. Power-demand monitors help industries to better utilize their electrical power. A high peak demand means that the equipment for an industrial power-distribution system must be rated higher. The closer that the peak demand approaches the value of the average demand, the more efficient the industrial power system is in terms of power utilization.

FREQUENCY MEASUREMENT

Another power measurement that is very important is the frequency. The frequency of the power source must remain stable or the operation of many types of equipment can be affected. Frequency refers to the number of cycles of voltage or current which occur in a given period of time. The unit of measurement for frequency is the hertz (Hz), which means cycles per second. A table of frequency bands is shown in Figure 12-16. The standard power frequency in the United States is 60 hertz. Some other countries use 50 hertz.

Frequency can be measured with several different types of meters. An electronic counter is one type of frequency indicator. Vibrating-reed instruments are also commonly used for measuring power frequencies. An oscilloscope can also be used to measure frequency. Graphic recording instruments may be used to provide a visual display of frequency over a period of time. The electrical power industry commonly uses this method to monitor the frequency output of its alternators.

GROUND-FAULT INDICATORS

Ground-fault indicators are used to locate faulty system or equipment grounding conditions. The equipment of electrical power systems must be properly grounded. Proper grounding procedures, principles, and ground-fault circuit interrupters are discussed in Chapter 9.

A ground-fault indicator may be used to check for faulty grounding at various points in an electrical power system. Several conditions can exist that might be hazardous. These faulty wiring methods include: (1) hot and neutral wires reversed, (2) open equipment ground wire, (3) open neutral wire, (4) open hot wire, (5) hot and equipment ground

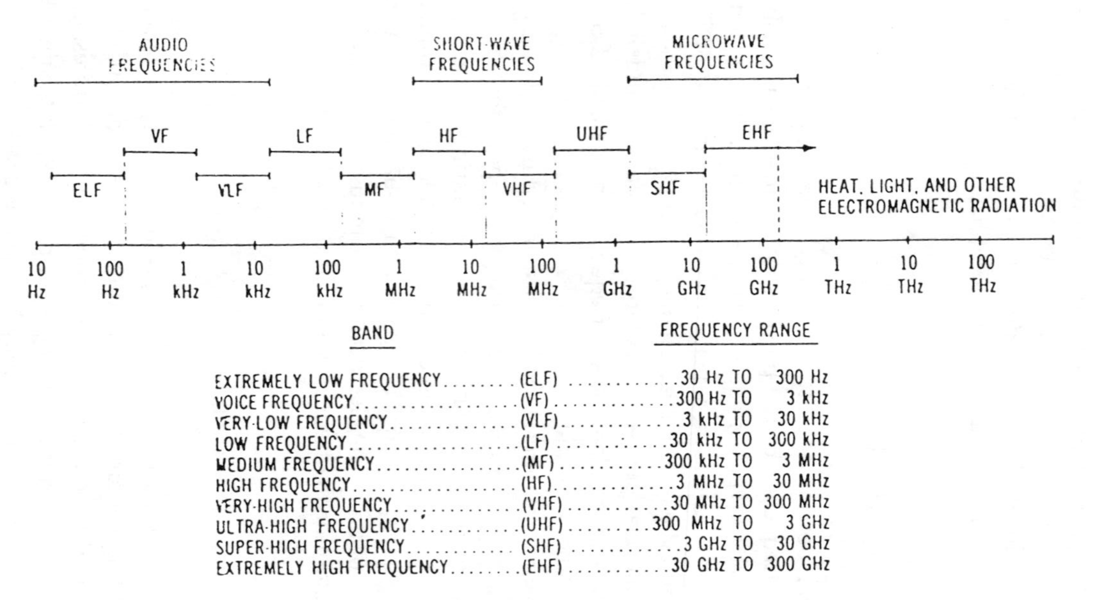

Figure 12-16. Classification of frequency bands.

wires reversed, or (6) hot wire on the neutral terminal and the neutral is not connected. Each of these conditions would present a serious problem in the electrical power system. Although proper wiring eliminates most of these problems, a periodic check with a ground-fault indicator would assure that the electrical wiring is safe and efficient.

Figure 12-17 shows equipment which may be used to test for faults in electrical systems.

MEGOHMMETERS

Megohmmeters, such as shown in Figure 12-18, are used to measure high resistances which are beyond the range of a typical ohmmeter. These indicators are used primarily for checking the quality of insulation on electrical power equipment (mainly motors). The quality of equipment insulation varies with age, moisture content, and the applied voltage. The megohmmeter is similar to a typical ohmmeter except it uses a high voltage DC source rather than a battery. Figure 12-19 shows a diagram of one type of megohmmeter circuit. This circuit is essentially the same as an ohmmeter with the exception that a DC generator or other high-voltage DC source is used.

Periodic insulation tests should be made on all large power equipment. As insulation breaks down with age, the equipment starts to malfunction. A good method is to develop a periodic schedule for checking and recording insulation resistance. Then, it can be predicted when a piece of equipment needs to be replaced or repaired. A resistance-versus-

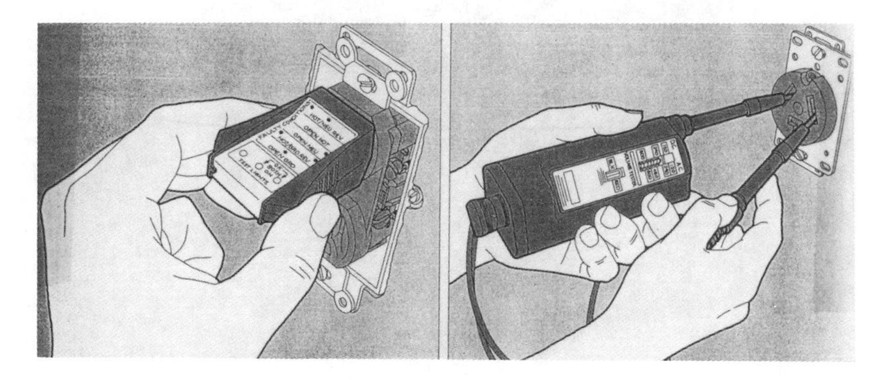

Figure 12-17. Instruments used to check for electrical system faults.

time graph Figure 12-20) can be made and the trend shown on the graph can be noted. A downward trend (a decrease in insulation resistance) over a specific time period indicates that an insulation problem exists.

CLAMP-ON METERS

Clamp-on meters, such as shown in Figure 12-21, are popular for measuring current in power lines. This indicator may be used for periodic checks of the current by clamping it around a power line. It is an easy-to-use and convenient maintenance and testing instrument.
The simplified circuit of a clamp-on analog current meter is shown in Figure 12-22. Current flow through a conductor creates a magnetic field around the conductor. The varying magnetic field induces a current into the iron core of the clamp portion of the meter. The meter scale is calibrated so that when a specific value of current flows in a power line, it will be indicated on the scale. Of course, the current flow in the power line is proportional to the current induced into the iron core of the clamp-on meter. The clamp-on meter may also have a voltage and resistance function which utilizes external test leads. Thus, the meter can be used to measure other quantities.

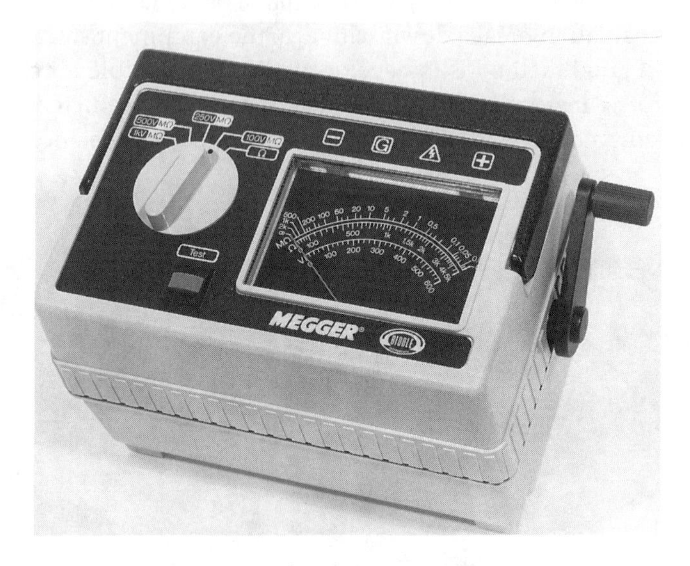

Figure 12-18. Megohmmeter (Courtesy Biddle Instruments).

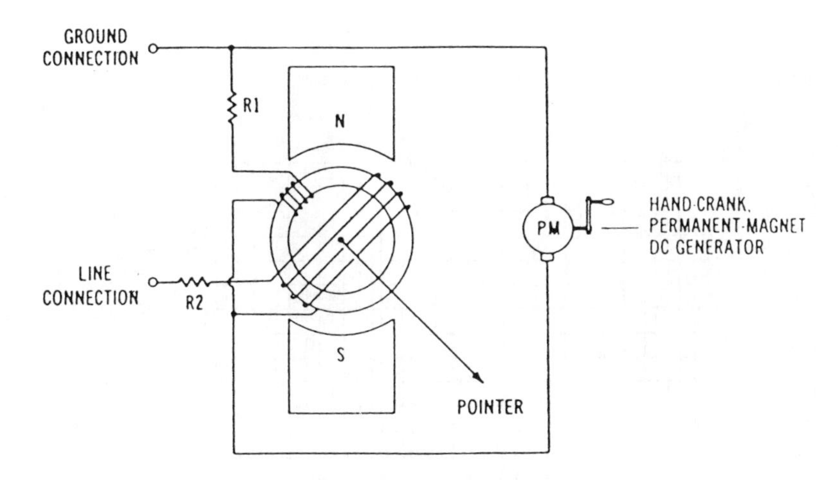

Figure 12-19. Circuit diagram of a megohmmeter, analog type.

TELEMETERING SYSTEMS

When a quantity being measured is indicated at a location some distance from its transducer or sensing element, the measurement process is referred to as *telemetering*. Many types of metering systems fit this definition. However, telemetering systems are usually used for long-distance measurement or for centralized measurement systems. For instance, many industries group their indicating systems together to facilitate process control. Another example of telemetering is the centralized monitoring (on a regional basis) of electrical power by utility companies. These systems are similar to other measuring systems except that a transmitter/receiver communication system is usually involved.

Many types of electrical and physical quantities can be monitored by using telemetering systems. The most common transmission media for telemetering systems are: (1) wire—such as telephone lines, (2) superimposed signals—which are 30- to 200-kHz signals carried on electrical power distribution lines, and (3) radio-frequency signals—from am, fm, and phase-modulation transmitters. The block diagram of one type of telemetering system is shown in Figure 12-23. In this type of system, a DC voltage from the transducer is used to modulate an am or fm transmitter. The radio-frequency (rf) signal is then received at another location and converted back into a DC voltage to activate some end device. The end device, which may be located a considerable dis-

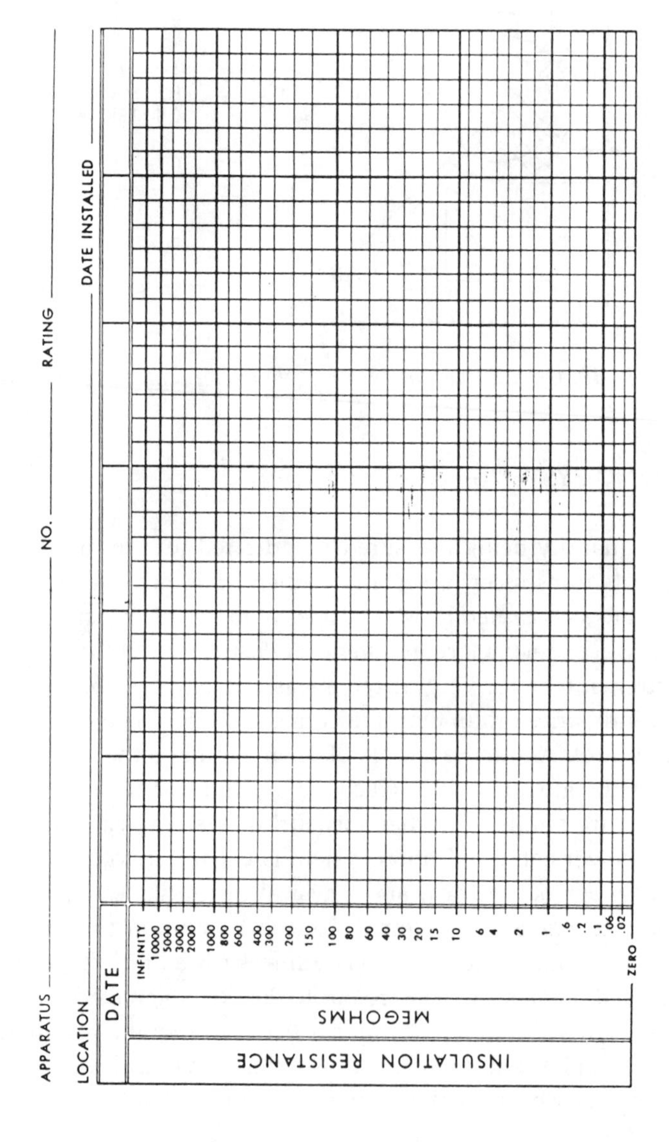

Figure 12-20. Resistance versus time chart to be used with a megohmmeter (Courtesy Biddle Instruments).

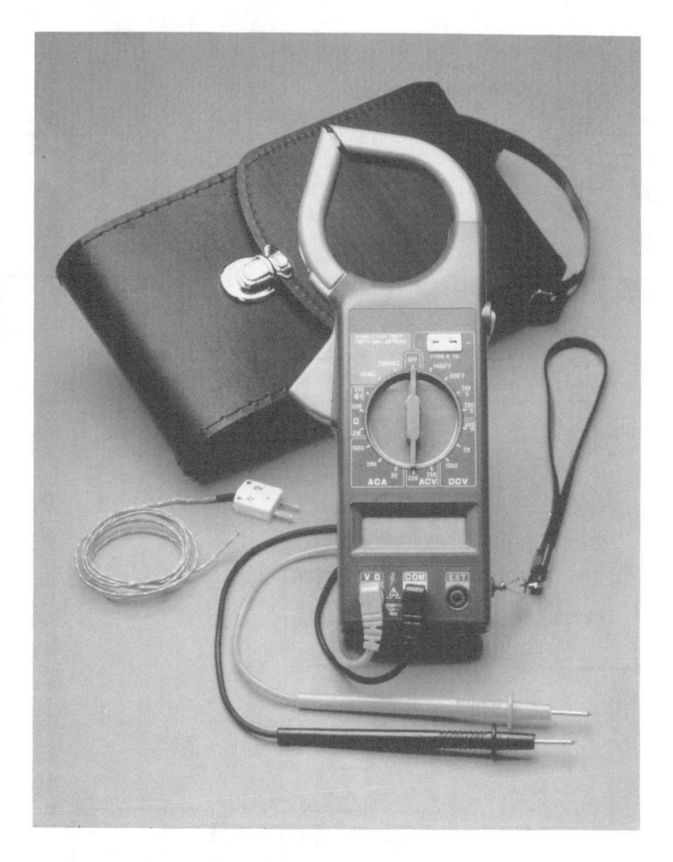

Figure 12-21. Clamp-on meter-digital type (Courtesy Omega Corp.).

tance from the transducer, might be a chart recorder, a hand-deflection meter, or possibly, a process controller. Digital telemetering is also used since binary signals are well suited for data transmission. In this system, the transducer output is converted to a binary code for transmission.

Telemetering is the measurement of some quantity at an area that is distant from its origin. For instance, it is possible, by using telemetering systems, to monitor on one meter the power used at several different locations. Almost any quantity value, either electrical or non-electrical, can be transmitted by using some type of telemetering system. A basic telemetering system has: (1) a transmitting unit, (2) a receiving unit, and (3) an interconnection method. Electrical power systems frequently utilize telemetering systems for the monitoring of power.

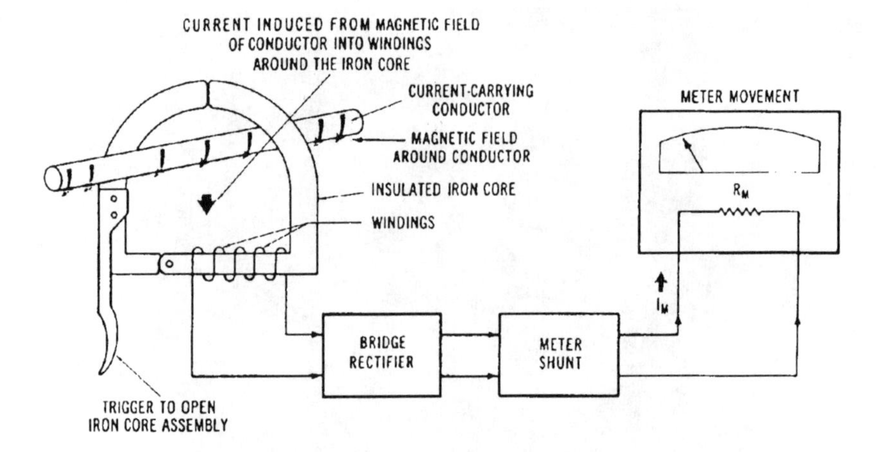

Figure 12-22. Circuit diagram of a clamp-on current meter.

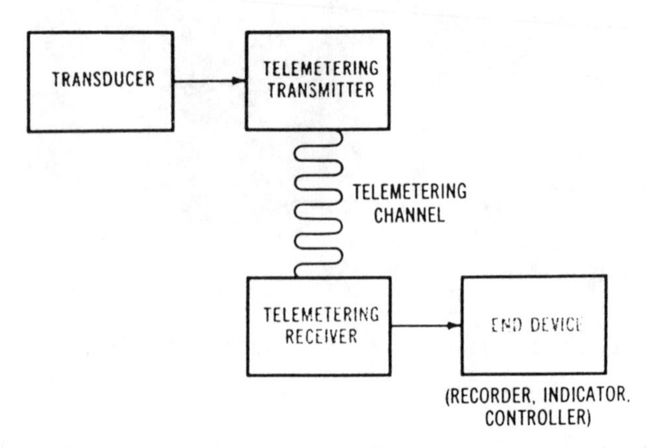

Figure 12-23. Block diagram of a basic telemetering system.

Chapter 13

Quality Considerations for Electrical Distribution Systems

Electrical power is a most useful form of energy. Almost every home, industry and commercial building in our country is served by our electrical distribution system. This distribution system furnishes power to provide for our heating, lighting, and equipment needs. Our electrical requirements are typically met so well that we take this high quality service for granted.

Several factors of electrical power distribution systems affect the quality of operation. On the National level, we must be concerned with the quality of the environment and conservation of natural resources used to produce electrical power. System reliability is an important quality indicator. Careful plans for load demand control must be made to assure the availability of electrical power. Many customers, such as hospitals, can not withstand power outages. Emergency generation systems and uninterruptible power systems must be used in these facilities, even though the likelihood of a power outage is very improbable.

Power quality considerations include planning for transients and harmonics in power lines. Surge suppression and shielding to avoid electromagnetic interference (EMI) provide additional power quality assurance.

Economic factors are also important quality considerations for electrical distribution systems. Economic measures are used to increase the efficiency of electrical power systems. The measures include electrical load management, computerized monitoring and control demand factor and load factor planning.

Many of these important quality considerations for electrical power distribution systems will be discussed in this chapter.

ENERGY CONSERVATION AND ENVIRONMENTAL CONSIDERATIONS

Energy conservation and environmental quality must always be considered by the electrical power industry. About 95% of the energy we use comes from fossil fuels -coal oil and natural gas. Since the earth has only a limited supply of the natural resources, sources of electrical energy must be studied from both a conservation and environmental perspective. The primary sources of energy throughout the world, in order of importance, are fossil fuels, water power, and nuclear fission energy. Also, solar energy, wind energy, tidal energy, and geothermal power provide small amounts of energy. Other systems (discussed in Chapter 4) in the experimental stages include nuclear fusion, MHD, fuel cells, biomass, and other potential sources.

Petroleum supplies almost one-half of the energy used in the U.S. and in the world, including most of the energy used for transportation and residential heating. Coal provides about one-third of the energy used in the world, but less than 25% in the U.S. The electrical power system is a major user of coal in the U.S., as discussed in Chapter 4. Natural gas and fuel oil also play a significant role in the production of electrical power in our country.

We must keep in mind that electrical power itself is not a source of energy. Electrical power plants typically burn coal, natural gas or fuel oil to produce high temperature steam. The steam furnishes the energy to rotate steam turbines that produce the mechanical energy to rotate three phase AC generators which provide our requirement for electrical energy. This is an **indirect energy conversion** process, as compared to a solar cell for example. A solar cell provides **direct energy conversion** by converting solar energy directly to electrical energy in the form of direct current (DC). We must continue to be conservation conscious in order to assure that electrical power will be available. The experimental research and development efforts of electrical power companies must continue. We have become accustomed to low cost, "unlimited" supply of electrical power.

The environmental concerns with electrical power systems are rather obvious. Air pollution and water pollution must be considered during the production of electrical power. When dealing with distribution systems, the construction of large steel towers, wooden poles and outside enclosures for equipment are considered by some to be unsightly. However, the rather inexpensive supply of electrical power to almost everyone has been aided by outdoor construction methods. Underground distribution, such as used by large cities, is much more expensive. The shapes, sizes and colors of poles have been made more aesthetic. In addition, environmental planning has been evident in the placement of power lines in areas less likely to be seen by large numbers of people. Also, power, communication, and other utility lines are placed on common poles to reduce clutter. Proper planning in some cases has influenced the design of underground utilities at a greater cost to improve appearance.

ELECTRICAL POWER QUALITY

Electronic equipment has become an important part of today's industrial and commercial everyday activity. Uninterrupted, high quality power is essential is essential to the operation of many facilities. This continues to be more important as industries become more automated and sophisticated electronic equipment becomes more commonplace. The proper performance of electronic and computer controlled equipment has made power quality an important consideration for industrial and commercial consumers.

Changes in electrical power usage which have brought about an increased awareness of power quality include the increased industrial and commercial use of sensitive equipment such as personal computers and main frame computers, electronic process controllers, and timing equipment. The effects of certain types of loads such as electronic power converters, electric furnaces, arc-discharge vapor lighting, and large motors produce problems associated with power quality.

A power quality problem is any voltage, current or frequency deviation which causes electronic equipment to fail or malfunction. Years ago, the types of loads connected to electrical distribution systems were not sensitive to power quality. Motors, for example, were not adversely affected by lowered voltage levels. With today's equipment, power quality problems can cause expensive process shutdowns and equipment malfunctions.

Clean, high quality power must have a perfect sine wave shape of constant frequency and voltage value. Voltage must be constantly maintained at 60 Hertz frequency and at the proper level. The waveshape must be free of harmonics, noise, and transients. Electrical power systems have extreme difficulty maintaining such power.

Power Quality Problems

Several factors which contribute to power quality problems are discussed in the following sections.

Voltage variations—caused by poorly regulated distribution systems which experience significant load changes over time. They can occur when a power system switches systems for interconnecting to a particular customer.

Overvoltages (Swells) and undervoltages (Sags)—due to deliberate changes by the power company, sudden load changes, or incorrect transformer tap settings. These may cause shortened equipment life. These conditions can last for more than a half cycle and should not be confused with surges which are only milliseconds in duration. A swell is an increase in voltage of less than 2 second duration outside the normal tolerance of electronic equipment, caused by a sudden load decrease. A sag is a decrease in voltage of less than 2 second duration outside the normal tolerance of equipment caused by starting of heavy loads or power system faults.

Transients (Surges)—caused by lightning, power line switching, turning large motors on or off, and short circuits. They may also be caused common office equipment such as copiers or printers. They may cause the gradual degradation of equipment life due to the continued effects of surges. Transients fall into two categories: impulsive transients and oscillatory transients. Impulsive transients may be caused by lightning strikes or power system load switching. Oscillatory transients are usually caused by power factor corrective capacitor switching by the power company in anticipation of higher power demand in early morning hours. Transients are commonly caused by inductive or capacitive devices.

Transient voltage surge suppression (TVSS) devices are a cost effective way to protect computer controlled equipment. Transient voltages may be produced either inside a facility or externally. TVSS devices are designed to suppress transients regardless of their origin. They form a low impedance path which bypasses the transient away from the protected equipment. An important rating of TVSS devices is **clamping level.** This term is used to describe a TVSS's ability to suppress incoming transients from their original voltage to some lower level.

Noise—results from lighting switches, tool operation, and other types of equipment. Also electromagnetic interference (EMI) and improper grounding can cause noise problems. Noise can take place due to electromagnetic induction or be distributed in a building's wiring. Noise can cause problems with data transmission and possibly to computer hardware.

Harmonics—are multiples of the power line frequency (60 Hz) which develop as a result of non-linear loads, rectifiers, electronic lighting ballasts, and switching power supplies used with data processing equipment. Harmonics can cause overheating of conductors and transformers, insulation breakdown, and excessive heat buildup in electronic equipment.

The presence of third harmonic (180 Hz) voltage across neutrals and grounds seems to be a persistent problem. For example, the neutral current in a three phase, four-wire (wye) system, neutral current can exceed phase current levels if harmonic problems exist. The use of personal computers and peripherals such as laser printers, adjustable speed machine drives, solid state heater controls and the increased quantity of fluorescent lights have increased the awareness of harmonic problems in facilities.

Signs that a facility might have harmonic problems include unexplained conductor or transformer overheating, power factor correction capacitor failure, unexplained blown fuses or tripped circuit breakers, or faulty adjustable speed machine drive system operation. Measurements must be made to determine if harmonic problems exist. Because harmonics likely are not present throughout the entire system, several locations must be monitored to determine the extent of the problem.

Neutral power lines in branch circuits are probably more susceptible to overcurrent damage due to harmonics. Due to their multiple frequencies, harmonic currents have different characteristics than 60 Hertz currents. Harmonic currents can flow to ground at unexpected locations. One example is at power factor correction capacitors in buildings. The capacitors form a circuit with the building wiring system whose impedance varies with frequency. At higher frequencies, capacitors are a low impedance path to ground. This situation causes current flow greater than the capacitor's protective equipment. Then, the capacitor's overload protection will be interrupted, causing the capacitors to fail due.

Electromagnetic Radiation—Electromagnetic fields (EMFs) may not be a power distribution problem; however, they exist all around us and should be considered. They come from computers, video display terminals, fluorescent lights, transformers and almost all electrical equipment. Magnetic fields have been found to cause disturbances in computers and medical imaging equipment for example. Recent publicity has caused liability concerns in regard to EMFs.

EMFs result from the flow of electrical current in conductors and exists wherever electrical equipment is plugged in. Power lines, electrical wiring in buildings, and electrical equipment have EMFs present. Fluorescent lamps are a major source of EMFs. A common way to reduce the effects of EMFs is by **shielding** of equipment where EMFs may be a problem. Shielding equipment should be installed during construction or through careful renovation of facilities. Shielding material is usually either ferromagnetic or conductive material which is placed around the source of EMFs.

Electromagnetic Interference (EMI) can occur visibly on computer monitors, audio-visual equipment or in data transmission. EMI is caused by EMFs which are not properly suppressed. Buildings near high voltage transmission lines, transformers, or subways can experience high levels of EMFs. Communications equipment can also cause EMI problems. EMFs may be monitored by using a gaussmeter to measure magnetic flux density which is transferred from electrical equipment. Exposure level can be monitored over a period of time and at different locations. EMF exposure diminishes quickly over distance.

Brownouts—are undervoltages which occur during heavy demand periods on electrical power systems and when heavy industrial or commercial loads are started.

Blackouts (outages)—are caused by power system failure or extreme system overload. They can cause loss of unsaved computer data and potential hardware and software problems.

Solving Power Quality Problems

Each of these problems discussed produces different effects on the electrical power system. They must typically be dealt with on an individual basis at industrial and commercial facilities. Three ways of dealing with power quality problems include surge protection equipment, power conditioners, and back-up power supplies. Monitoring power quality is necessary in order to find solutions to power quality problems and improve system reliability.

Surge Protection—These devices divert surges and transients away from sensitive equipment. They range from plug-in strips which can be used in the home to sophisticated units used in industrial and commercial settings.

Power Conditioners—Power conditioners, such as shown in Figure 13-1, provide clean, stable, isolated power by suppressing surges, noise, overvoltage, undervoltage, and harmonics. They do not protect against power outages or blackouts.

Figure 13-1. Power Conditioners—"Z-Line" power distribution and control system (Courtesy Pulizzi Engineering, Ins.).

Back-Up Power Supplies—This broad category includes **uninterruptible power supplies (UPS) and emergency (stand-by) power supplies.** They can be used in an industrial or commercial facility to guard against spikes, transients, voltage variations, and blackouts.

An **uninterruptible power supply (UPS)** is intended to provide constant power to equipment which is protected from power outages. They remain connected to the building's power system and require no transfer time during power interruptions. The purpose of these power protection devices is to provide dependable, continuous, clean, regulated power to critical equipment (loads) under all load conditions. UPS systems are available in three types: on-line, off-line and line interactive. During normal conditions, an **on-line system** converts AC power to DC to keep batteries charged and converts the DC battery power back to AC. The connected load is continuously supplied with the constant battery power, regardless of the utility power line conditions. During normal conditions, an off-line UPS supplies the load from the utility power line. When an outage occurs, the batteries of the UPS deliver power to an **inverter** which converts the DC power back to AC power for the load. The major problem of this system is that the time required to transfer from AC to DC can be a problem for some computerized systems and electronic equipment. A **line interactive UPS,** under normal conditions, filters the line current to the load and keeps the batteries fully charged. When a power failure is detected, power is provided from the battery to the load. Since the battery and the inverter are constantly connected to the load, no power interruption occurs.

Emergency power supplies provide power to selected equipment only in the event of a power failure. They do so with a very short delay called transfer time, which is in the range of a few milliseconds. Automatic transfer switches, such as shown in Figure 13-2, are used to make transfer time unnoticed by the equipment connected to the power supply.

Automatic Transfer Switch

Automatic transfer switches (ATSs), allow switching of multiple power sources to supply an electrical load. ATSs are required for emergency and stand-by electrical power distribution systems. ATSs contain voltage monitors set to sense when the primary distribution drops out and when it returns. When electrical power drops below the preset level, the ATS transfers to the alternate power source, such as an emergency

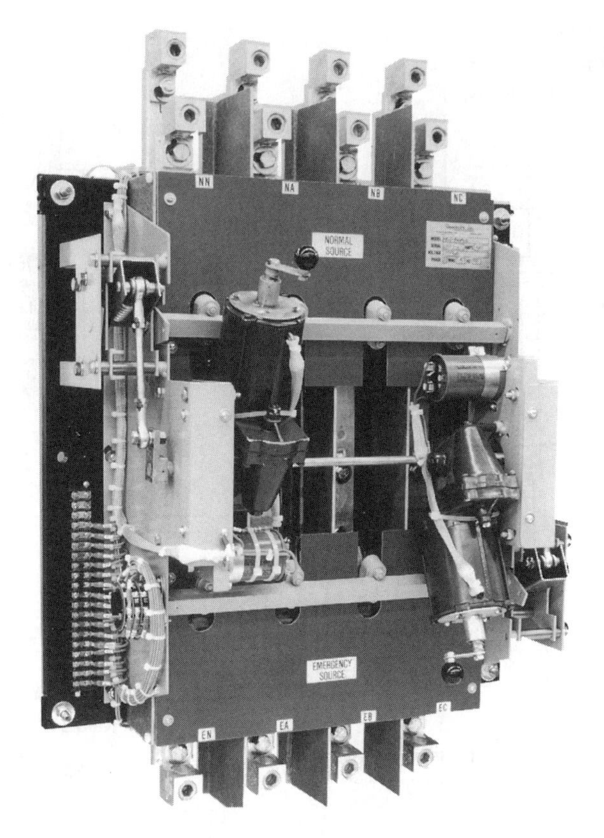

Figure 13-2. Automatic transfer switch.

generator in the facility. Retransfer back to normal power then can take place automatically at a preset delay time after the primary distribution has been restored. A typical facility diagram with emergency power and automatic transfer switching provided is shown in Figure 13- 3.

Another similar switch, called a closed transition transfer switch (CTTS), is used enable a facility to momentarily parallel its stand-by generating system with the utility's power distribution system to avoid disturbances while transferring the load to the stand-by system.

Monitoring Power Quality

A key to improving power quality is improved monitoring with power analysis equipment. The need for UPS and on-site stand-by power supplies is growing as equipment and processes become more

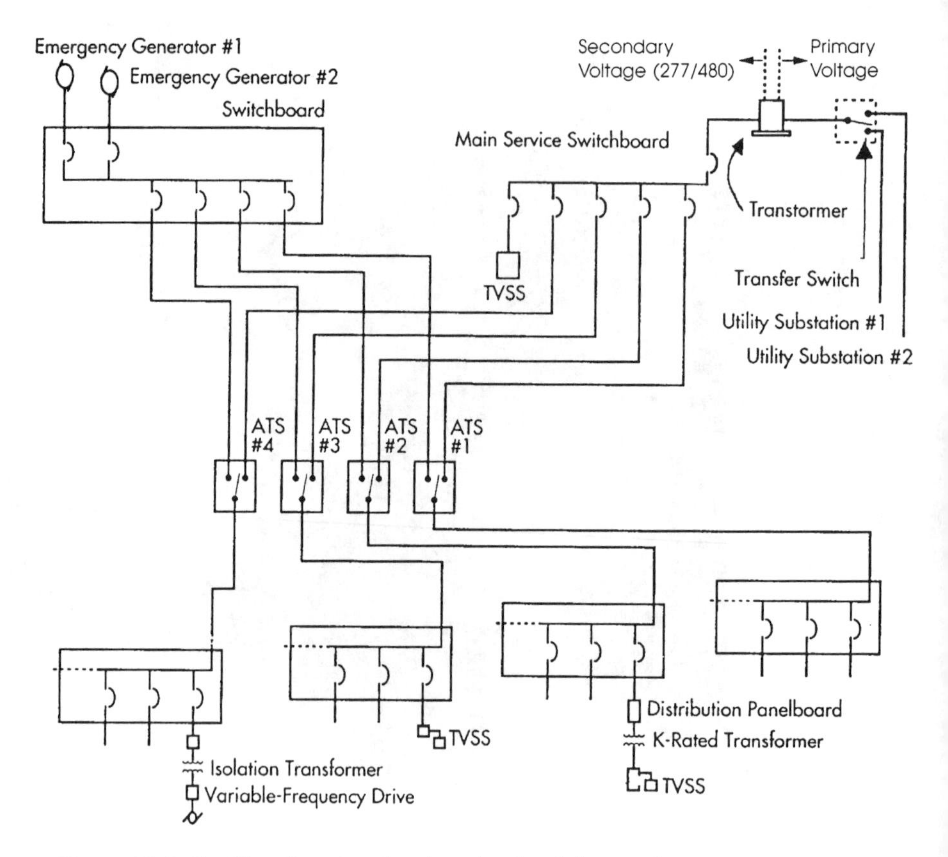

Figure 13-3. Facility diagram showing the circuit layout for emergency generators and automatic transfer switches (ATSs).

complex and downtime costs increase. A problem in power quality assessment is determining if the problem is on-site or caused by the electrical utility supplying power to the facility. Power monitors are connected to a facility's power lines for extended periods of time. They typically measure and record brownout conditions, overvoltage, harmonic distortion, voltage spikes and noise, line phase shift and frequency changes. They also provide information on current levels and power consumption, both actual and reactive (VARs). Some analyzers, such as shown in Figure 13-4 are programmable or equipped with a personal computer interface to simplify analysis and recording of data. Monitoring for power quality improvement requires the proper power analysis equipment.

Figure 13-4. Portable notebook computer with interface and software used to monitor electrical systems—notice the display on the right of an AC sine wave (Courtesy of IO Tech, Inc.).

System Reliability

The reliability of an electrical power system usually refers to the lack of interruption of service. Not all customers are concerned about system reliability to the same extent. However, some facilities, such as hospitals, military bases, hotels, and apartment complexes, where life and safety are concerned must have uninterrupted electrical service. Most of these facilities have emergency stand-by power systems to supplement the electrical power distribution system. Utilities companies utilize data of system reliability such as number of interruptions per customer served, hours of interruption per customer served, and average duration of interruption per customer to provide an index of system reliability. These statistics can be evaluated and used to improve system efficiency.

ECONOMIC CONSIDERATIONS

Many organizations now consider electrical power as not just a monthly expense but a critical resource. If this resource is monitored and controlled effectively, productivity is improved, scrap is eliminated, costs are reduced and quality is improved. Companies must evaluate their power delivery costs, power interruption costs and power quality costs. Power delivery costs are a huge investment for most industrial and commercial companies. Problems associated with electrical power, including equipment failures and decreased efficiency should be considered. Poor power management can cost lost production and decreased productivity.

Power management

Power management should include surveying electrical equipment in a facility in terms of age and condition to identify potential weaknesses. Periodic, on-going preventive maintenance of all equipment should be scheduled. Software is now available to perform maintenance scheduling. Power monitoring and control equipment can be used to effectively study the performance of electrical equipment. Proper monitoring can result in energy cost savings, productivity improvements, and increased system reliability.

Load Economic Considerations

The planning which takes place to predict electrical power system load requirement at any given time is very critical to the efficient system operation. The **maximum demand** of electrical power is the greatest Kilowatt-hour usage during a specified period of time. The actual load demand varies from hour to hour and time of day. For example, maximum demand might take place at a residence at 6:30 p.m. when maximum lighting, heating or cooling, and appliance usage takes place. An industry might experience maximum demand at 7:30 a.m. when machines are being started and daily production is beginning.

Demand factor. The ratio of maximum demand to total connected load is called **demand factor.** For example, a large industry might have a connected load of 20 Megawatts, but if only 75% of its electrical equipment is operating, the demand factor would be only 75% of maximum. Typical demand factor estimating is important in determining the size of equipment to be installed for consumers. These estimates are important for planning new systems and for improving the efficiency of existing systems.

Load factor. The ratio of average electrical power usage for a period of time to the maximum demand during that period is called **load factor.** For example, a customer might have a maximum demand of 20 kW during a 24 hour period. However, the average demand during that period might be 10 kW. The load factor would equal 10 kW divided by 20 kW or 50%. Load factor estimates for various types of loads provide additional information for improving the efficiency of electrical power systems.

Diversity Factor. Variation occurs in the time of maximum usage for different customers. Maximum demands of combined loads must be considered in electrical distribution system planning. **Diversity factor** is the ratio of the combined maximum demands of individual loads to the maximum demand of the entire load connected. For example, if three loads each have a maximum demand of 10 kW and the maximum demand of the system is 15 kW, the diversity factor would equal 30 kW (10 kW + 10 kW + 10 kW) divided by 15 kW or 200%. Diversity factor is always greater than 100% while demand factor is always less that 100%. Each of these factors is important in the design of electrical power systems.

Factors that affect electrical power bills

1. Energy charge—is the number of kWh used during a billing period or the number of kVA used in the case of an industrial or commercial user.

2. Demand charge—provides an incentive for industrial or commercial customers to reduce maximum load demand. This charge during a billing period compensates the utility company for costs incurred to serve the facility's maximum load. Demand charges may be a large portion of an electrical bill. These charges may be reduced by lowering energy usage peaks, reducing kVA and improving power factor. The basis of the billing reduction is to cause maximum demand to be closer to average demand over the billing

period, thus reducing the demand factor of the facility.

3. Power factor charge—is a penalty assessed to encourage the industrial customer to improve power factor. Utilities companies impose a power factor penalty when PF drops below 90% typically or sometimes as high as 95%. This charge can be reduced by installing power factor correction capacitors or synchronous capacitors. Figure 13-3 can be used to determine the amount of capacitance needed for power factor correction.

Appendix A

Important Terms

This glossary is an alphabetical list by topic of electrical terms. Some of the terms are discussed in the text, while others are covered in other courses. This glossary will provide you with brief discussions of terms when you need them.

Ac: abbreviation for alternating current.

Alternating current: current produced when electrons move first in one direction and then in the opposite direction; two types include single-phase the three-phase.

Alnico: an alloy of aluminum, nickel, iron, and cobalt used to make permanent magnets.

Alternator: a rotating machine that generates AC voltage.

Ammeter: a meter used to measure current (amperes).

Ampere: the electrical charge movement which is the basic unit of measurement for current flow in an electrical circuit.

Air-core inductor: a coil wound on an insulated core or a coil of wire that does not have a metal core.

Ampere-turn: the unit of measurement of magnetic field strength; amperes of current times the number of turns of wire.

Amplitude: the vertical height of an AC waveform.

Angle of lead or lag: the angle between applied voltage and current flow in an AC circuit, in degrees; in an inductive (L) circuit, voltage (E)

leads current (*I*)—*ELI*; in a capacitive (*C*) circuit, current (*I*) leads voltage (*E*)—*ICE*.

Apparent-power (volt-amperes): the total power *delivered* to an AC circuit; applied voltage times current.

Armature: the movable part of a relay, rotating coils of a motor or the part of a generator into which current is induced.

Attenuation: a reduction in value.

Average voltage: the value of an AC sinewave voltage which is found by this formula: $V_{avg} = V_{peak} \times 0.636$.

Ballast: an inductive coil placed in a fluorescent light circuit to cause the light to operate. Battery: an electrical energy source that is two or more cells connected together.

Block diagram: a diagram used to show how the parts of a system fit together.

Branch: a path of a parallel circuit.

Branch Circuit: That portion of a wiring system extending beyond the final **load** protective device of the circuit.

Branch current: the current through a parallel branch of a circuit.

Branch resistance: the total resistance of a parallel branch of a circuit.

Branch voltage: the voltage across a parallel branch of a circuit.

Brush: a sliding contact made of carbon and graphite which touches the commutator of a generator or motor.

BX Cable: Flexible metal conduit with conductors already in place. Used to wire buildings where local codes permit.

Cable: A general term applied to larger sizes of wire, either solid or stranded, using singly or in combination. A cable is usually heavily insulated.

Capacitance: the property of a device to oppose changes in voltage due to energy stored in its electrostatic field.

Captive heating: a method of heating nonconductive materials by placing them between two metal plates and applying, a high-frequency AC voltage.

Capacitive reactance: the opposition to the flow of AC current caused by a capacitive device (measured in ohms).

Capacitor: a device that has capacitance and is usually made of two metal plate materials separated by a dielectric material (insulator).

Cathode ray tube (CRT): a vacuum tube in which electrons are emitted from the cathode in the shape of a narrow beam and speeded up so that they will strike the screen and produce light.

Center-tap: a terminal connection made to the center of a transformer winding.

Choke coil: an inductor coil used to block the flow of AC current and pass DC current.

Circuit: a path through which electrical current flows in a circuit.

Circuit Breaker: Switch which is operated either by hand or automatically for opening a circuit when the current becomes too high.

Closed circuit: a circuit which forms a complete path so that electrical current can flow through it.

Coefficient of coupling: a decimal value that indicates the amount of magnetic coupling between coils.

Color Code: Any system of colors used to identify or specify items.

Combination circuit: a circuit that has a portion connected in series with the voltage source and another part connected in parallel.

Component: An electrical device that is used on a circuit.

Commutation: the process of applying direct current of the proper polarity at the proper time to the rotor windings of a DC motor through split rings and brushes; or the process of changing AC induced into the rotor of a DC generator into DC applied to the load circuit.

Commutator: an assembly of copper segments that provide a method of connecting rotating coils to the brushes of a DC generator.

Conductance: the ability of a resistance of a circuit to conduct current, measured in siemens or mhos; the inverse of resistance $G = 1/R$.

Conductor: Material which provides a path for electric current between two points. Thus, a material which offers little resistance to the continuous flow of electric current.

Conduit: A metal pipe through which electric wires run.

Conduit Bender: A special tool for bending conduit; often called a MICKEY

Continuity: The property of having a continuous or complete path for current.

Continuity check: a test to see if a circuit is an open or closed patch.

Control: the part of an electrical system which affects what the system does; a switch to turn a light on and off is a type of control.

Copper losses: heat losses in electrical machines due to the resistance of the copper wire used for windings; also called 1^2R losses.

Core: laminated iron or steel that is used in the internal construction of the magnetic circuit of electrical machines or transformers.

Counter electromotive force (CEMF): "generator action" that takes place in motors when voltage (EMF) is induced into the rotor conductors which opposes the source voltage (EMF).

Coupling: the amount of mutual inductance between coils.

Current: The movement of electrons through a conductor. Current is measured in amperes.

Cycle: a sequence of events that causes one complete pattern of alternating current from a zero reference, in a positive direction, then back to zero, then in a negative direction and back to zero.

D'Arsonval meter movement: a stationary magnet which has a moving electromagnetic coil inside that causes a meter needle to deflect.

DC: abbreviation for direct current.

DC generator: a rotating machine that produces a form of direct-current (DC) voltage.

Delta connection: a method of connecting the stator windings of three phase machines in which the beginning of one winding is connected to the end of the next winding; power input lines come into the junctions of the windings.

Dielectric: an insulating material placed between the metal plates of a capacitor.

Direct current: the flow of electrons in one direction from negative to positive.

Difference in potential: the voltage across two points of a circuit.

Direct current (DC): tile type of electrical power that is produced by batteries and power supplies; used mostly for portable and specialized power applications; the flow of electrons in one direction from negative (–) to positive (+).

E: an abbreviation sometimes used for voltage; *V* is used in this book.

Eddy currents: induced current in the metal parts of rotating machines which causes heat losses.

Effective voltage: a value of an AC sinewave voltage that has the same effect as an equal value of DC voltage:

$$V_{eff} = V_{peak} \times 0.707$$

Efficiency: the ratio of output power and input power of a machine or system.

$$\% \text{ Efficiency} = \frac{P_{out}}{P_{in}} \times 100$$

Electricity: A general term used when referring to the energy associated with electrons at rest or in motion.

Electromagnet: a coil of wire wound on an iron core so that as current flows through the coil it becomes magnetized.

Electromotive force (EMF): the "pressure," or force, that causes electrical current to flow.

Electrolytic capacitor: a capacitor that has a positive plate made of aluminum and a dry paste or liquid used to form the negative plate.

Electron: a small particle which is part of an atom that is said to have a negative (–) electrical charge; electrons cause the transfer of electrical energy from one place to another.

Electron current flow: current flow that is assumed to be in the direction of electron movement from a negative (–) potential to a positive (+) potential.

Electrostatic field: the field that is developed around a material due to the energy of an electrical charge.

ELI: a term used to help remember that voltage (E) leads current (I) in an inductive (L) circuit.

Energy: the capacity to do work.

Farad: the unit of measurement of capacitance that is produced when a charge of 1 C causes a potential of 1 V to be developed.

Field coils: electromagnetic coils that develop the magnetic fields of electric machines.

Field poles: the laminated metal which serves as the core material for field coils.

Fluorescent lighting: a lighting method that uses lamps which are tubes filled with mercury vapor to produce a bright white light.

Flux (symbol is Φ): invisible lines of force that extend around a magnetic material.

Flux density: the number of lines of force per unit area of a magnetic material or circuit.

Fossil fuel system: a power system that produces electrical energy due to the conversion of heat from coal, oil, or natural gas.

Frequency: the number of AC cycles per second, measured in hertz (Hz).

Fuse: an electrical overcurrent device that opens a circuit when it melts due to excess current flow through it.

Gauss: a unit of measurement of magnetic flux density.

Generator: a rotating electrical machine that converts mechanical energy into electrical energy.

Gilbert: a unit of measurement of magnetomotive force (MMF).

Ground: of two types: *system grounds*, current-carrying conductors used for electrical power distribution; and *safety grounds,* not intended to carry electrical current but to protect individuals from electrical shock hazards.

Ground fault: an accidental connection to a grounded conductor.

Ground-fault circuit interrupter (GFCI): a device used in electrical wiring for hazardous locations; it detects fault conditions and responds rapidly to open a circuit before shock occurs to an individual or equipment is damaged.

Henry: the unit of measurement of inductance which is produced when a voltage of 1 volt is induced when the current through a coil is changing at a rate of 1 A per second.

Hertz: the international unit of measurement of frequency equal to 1 cycle per second.

Horsepower: 1 horsepower (hp) = 33,000 ft-lb of work per minute or 550 ft-lb per second; it is the unit of measuring mechanical energy; 1 hp = 746 W.

Hot conductor: a wire that is electrically energized or live and is not grounded.

Hydroelectric system: a power system that produces electrical energy due to the energy of flowing water.

ICE: a term used to help remember that current (*I*) leads voltage (*E*) in a capacitive (*C*) circuit.

Impedance (Z): the total opposition to current flow in an AC circuit which is a combination of resistance (R) and reactance (X) in a circuit; measured in ohms:

$$Z = \sqrt{R^2 + X^2}$$

Incandescent lighting: a method of lighting which uses bulbs that have tungsten filaments which when heated by electrical current produce light.

Indicator: the part of an electrical system that shows if it is on or off or indicates some specific quantity.

Induced current: the current that flows through a conductor due to magnetic transfer of energy.

Induced voltage: the potential that causes induced current to flow through a conductor which passes through a magnetic field.

Induction motor: an AC motor that has a solid squirrel-cage rotor and operates on the induction (transformer action) principle.

Inductance: the property of a circuit to oppose changes in current due to energy stored in a magnetic field.

Inductive circuit: a circuit that has one or more inductors or has the property of inductance such as an electric motor circuit.

Inductive heating: a method of heating conductors in which the material to be heated is placed inside a coil of wire and high-frequency AC voltage is applied.

Inductive reactance (X_L): the opposition to current flow in an AC circuit caused by an inductance (L) measured in ohms: $X_1 = 2\pi f L$.

Inductor: a coil of wire that has the property of inductance and is used in a circuit for that purpose.

In phase: two waveforms of the same frequency which pass through their minimum and maximum values at the same time and polarity.

Insulator: a material that offers a high resistance to electrical current flow.

Isolation transformer: a transformer with a 1:1 turns ratio used to isolate an AC power line from equipment with a chassis ground.

Junction box: a metal or plastic box where conductors are joined together in wiring.

Kilowatt-hour (kWh): 1000 W per hour; a unit of measurement for electrical energy,

Kinetic energy: energy that exists because of movement.

Knockout: A disc fitted into a hole in outlet boxes which is pressed or knocked out when it is desired to run wires in" that particular point.

Laminations: thin pieces of sheet metal used to construct the metal parts of motors and generators.

Lagging phase angle: the angle by which current *lags* voltage (or voltage *leads* current) in an inductive circuit.

Laws of magnetism: (1) like magnetic poles repel; (2) unlike magnetic poles attract.

Leading phase angle: the angle by which current *leads* voltage (or voltage *lags* current) in a capacitive circuit.

Load: the part of an electrical system that converts electrical energy into another form of energy, such as an electric motor which converts electrical energy into mechanical energy.

Magnet: a metallic material, usually iron, nickel, or cobalt, which has magnetic properties.

Magnetic circuit: a complete path for magnetic lines of force from a north to a south polarity.

Magnetic field: magnetic lines of force that extend from a north polarity and enter a south polarity to form a closed loop around the outside of a magnetic material.

Magnetic materials: metallic materials such as iron, nickel, and cobalt that exhibit magnetic properties.

Magnetic poles: areas of concentrated lines of force on a magnet that produce north and south polarities.

Magnetomotive force (MMF): a force that produces magnetic flux around a magnetic device.

Maximum power transfer: a condition that exists when the resistance or impedance of a load (R_L) equals that of the source which supplies it (R_S).

MCM: a unit of measure for large circular conductors, equal to 1000 circular mils.

Megohmeter: a meter used to measure very high resistances.

Mho: ohm spelled backward; a unit of measurement sometimes used for conductance, susceptance, and admittance; being replaced by the *siemens.*

Motor: a rotating machine that converts electrical energy into mechanical energy.

Multifunction meter (multimeter): a meter that measures two or more electrical quantities, such as a VOM.

Multirange meter: a meter that has two or more ranges to measure an electrical quantity.

Mutual inductance: when two coils are located close together so that the magnetic flux of the coil affect one another in terms of their inductance properties.

N.E.C.: National Electrical Code—A set of rules governing construction and installation of electrical equipment and wiring design.

NEMA (National Electrical Manufacturers' Association): an organization that establishes standards for electrical equipment.

Neutral: a grounded conductor of an electrical circuit which carries current and has white insulation.

Nuclear fission system: a power system that produces electrical energy due to the heat developed by the splitting of atoms in a nuclear reactor.

Ohm: (symbol is Ω): the unit of measurement of electrical resistance.

Ohmmeter: a meter used to measure resistance (ohms).

Ohm's Law: the law that explains the relationship of voltage, current, and resistance in electrical circuits.

Open circuit: a circuit that has a broken path so that no electrical current can flow through it.

Oscilloscope: an instrument that has a cathode ray tube to allow a visual display of voltages.

Overcurrent device: a device such as a fuse or circuit breaker which is used to open a circuit when an excess current flows in the circuit.

Overload: a condition that results when more current flows in a circuit than it is designed to carry.

Parallel circuit: a circuit that has two or more current paths.

Path: the part of an electrical system through which the energy travels from a source to a load, such as the electrical wiring used in a building.

Peak-to-peak voltage: the value of AC sine-wave voltage from positive peak to negative peak.

Peak voltage (V_{peak}): the maximum positive or negative value of AC sine-wave voltage: $V_{peak} = V_{eff} \times 1.41$.

Period (time): the time required to complete one AC cycle: time = 1/frequency.

Permanent magnet: bars or other shapes of materials which retain their magnetic properties.

Phase angle: the angular displacement between applied voltage and current flow in an AC circuit.

Polarity: the direction of an electrical potential (- or +) or a magnetic charge (north or south).

Primary winding: the coil of a transformer to which AC source voltage is applied.

Prime mover: a system that supplies the mechanical energy to rotate an electrical generator.

Potential energy: energy that exists due to position.

Potentiometer: (Rheostat) a variable-resistance component used as a control device in electrical circuits.

Power (P): the rate of doing work in electrical circuits found by $P = I \times V$.

Power factor (pf): the ratio of power converted (true power) in an AC circuit and the power delivered (apparent power):

$$PF = \frac{\text{true power (watts}}{\text{apparent power (volt–amperes)}}$$

Pulsating DC: a DC voltage that is not in "straight line" or "pure" DC form, such as the DC voltage produced by a battery.

Reactance (X): the opposition to AC current flow due to inductance (X,) or capacitance (X_c).

Reactive circuit: an AC circuit that has the property of inductance or capacitance.

Reactive power (var): the quantity of "unused" power developed by reactive components (inductive or capacitive) in an AC circuit or system; this unused power is delivered back to the power source since it is not converted to another form of energy; the unit of measurement is the volt-ampere reactive (var).

Regulation: a measure of the amount of voltage change which occurs in the output of a generator due to changes in load.

Relay: an electromagnetic switch that uses a low current to control a higher current circuit.

Resistive circuit: whose only opposition to current flow is resistance; a nonreactive circuit.

Resistive heating: a method heating which relies on the heat produced when electrical current moves through a conductor.

Resistor: a component used to control the amount of current flow in a circuit.

Root-mean-square (RMS) voltage: same as *Effective voltage*.

Rotating-armature method: the method used when a generator has DC voltage applied to produce a field to the stationary part (stator) of the machine and voltage is induced into the rotating part (rotor).

Rotating-field method: the method used when a generator has DC voltage applied to produce a field to the rotor of the machine and voltage is induced into the stator coils.

Rotor: the rotating part of an electrical generator or motor.

Schematic diagram: a diagram used to show how the components of an electronic system fit together.

Self-starting: the ability of a motor to begin rotation when electrical power is applied to it.

Semiconductor: a material that has electrical resistance somewhere between a conductor and an insulator.

Series circuit: a circuit that has one path for current flow.

Shaded-pole motor: a single-phase AC motor that uses copper shading coils around its poles to produce rotation.

Short circuit: a fault condition that results when direct contact (zero resistance) is made between two conductors of an electrical system (usually by accident).

Siemens: same as mho.

Signal: an electrical waveform of varying value which is applied to a circuit.

Sine wave: a waveform of one cycle of AC voltage.

Single phase: the type of electrical power that is supplied to homes and is in the form of a sine wave when it is produced by a generator.

Single-phase AC generator: a generator that produces single-phase AC voltage in the form of a sine wave.

Single-phase motor: any motor that operates with single-phase AC voltage applied to it.

Slip rings: solid metal rings mounted on the end of a rotor shaft and connected to the brushes and the rotor windings; used with some types of AC motors.

Solenoid: an electromagnetic coil with a metal core which moves when current passes through the coil.

Source: the part of an electrical-system that supplies energy to other parts of the system such as a battery that supplies energy for a flashlight.

Speed regulation: the ability of a motor to maintain a steady speed with changes in load.

Split-phase motor: a single-phase AC motor which has start windings and run windings and operates on the induction principle.

Split rings: a group of copper bars mounted on the end of a rotor shaft of a DC motor and connected to the brushes and the rotor windings.

Squirrel-cage rotor: a solid rotor used in AC induction motors.

Starter: a resistive network used to limit starting current in motors.

Stator: the stationary part of a motor.

Steam turbine: a machine that uses the pressure of steam to cause rotation which is used to turn an electrical generator.

Step-down transformer: a transformer that has a secondary voltage lower than its primary voltage.

Step-up transformer: a transformer that has a secondary voltage higher than its primary voltage.

Susceptance: the ability of an inductance (B_L) or a capacitance (B_C) to pass AC current; measured in siemens or mhos: $B_L = 1/X_L$ and $B_C = 1/X_C$.

Switch: a control device used to turn a circuit on or off.

Symbol: used as a simple way to represent a component.

Synchronous motor: an AC motor that operates at a constant speed regardless of the load applied.

Theta (θ): a Greek letter used to represent the phase angle of an AC circuit.

Three-phase AC: the type of electrical power that is generated at power plants and transmitted over long distances.

Three-phase AC generator: a generator that produces three AC voltages.

Torque: mechanical energy in the form of rotary motion.

Total current: the current that flows from the voltage source of a circuit.

Total resistance: the total opposition to current flow of a circuit, which may be found by removing the voltage source and connecting an ohmmeter across the points where the source was connected.

Total voltage: the voltage supplied by a source.

Transformer: an AC power control device that transfers energy from its primary winding to its secondary winding by mutual inductance and is ordinarily used to increase or decrease voltage.

True power (watts): the power actually *converted* to another form by an AC circuit; true power is measured with a wattmeter.

Turns ratio: the ratio of the number of turns of the primary winding *(N_p)* of a transformer to the number of turns of the secondary winding *(N_S)*.

Universal motor: a motor that operates with either AC or DC voltage applied.

Vapor lighting: a method of lighting that uses lamps which are filled with certain gases that produce light when electrical current is applied.

VAR (volt-amperes reactive): the unit of measurement of reactive power.

Vector: a straight line which indicates a quantity that has magnitude and direction.

Voltage: electrical force, or "pressure," that causes current to flow in a circuit.

Voltage drop: the electrical potential (voltage) that exists across two points of an electrical circuit; found by $V = I \times R$.

Voltage drop: the reduction in voltage caused by the resistance of conductors of an electrical distribution system; it causes a voltage less than the supply voltage at points near the end of an electrical circuit that is farthest from the source.

Volt-ampere (VA): the unit of measurement of apparent power.

Voltmeter: a meter used to measure voltage.

Volt-ohm-milliammeter (VOM): a multifunction, multirange meter which is usually designed to measure voltage, current, and resistance; also called a multimeter.

Watt: the basic unit of electrical power; the amount of power converted when 1 A of current flows under a pressure of 1 V.

Watt-hour: a unit of energy measurement equal to one watt per hour.

Watt-hour meter: a meter that monitors electrical energy used over a period of time. Wattmeter: a meter used to measure the actual electrical energy that is converted in a circuit or system.

Waveform: the pattern of an AC frequency derived by looking at instantaneous voltage values that occur over a period of time; on a graph a waveform is plotted with instantaneous voltages.

Work: the transforming or transferring of energy.

Wye connection: a method of connecting the windings of three-phase equipment in which the three beginnings or ends of the windings are joined together to form a common (neutral) point and the other wires connect to the power lines.

Appendix B

Electrical Symbols

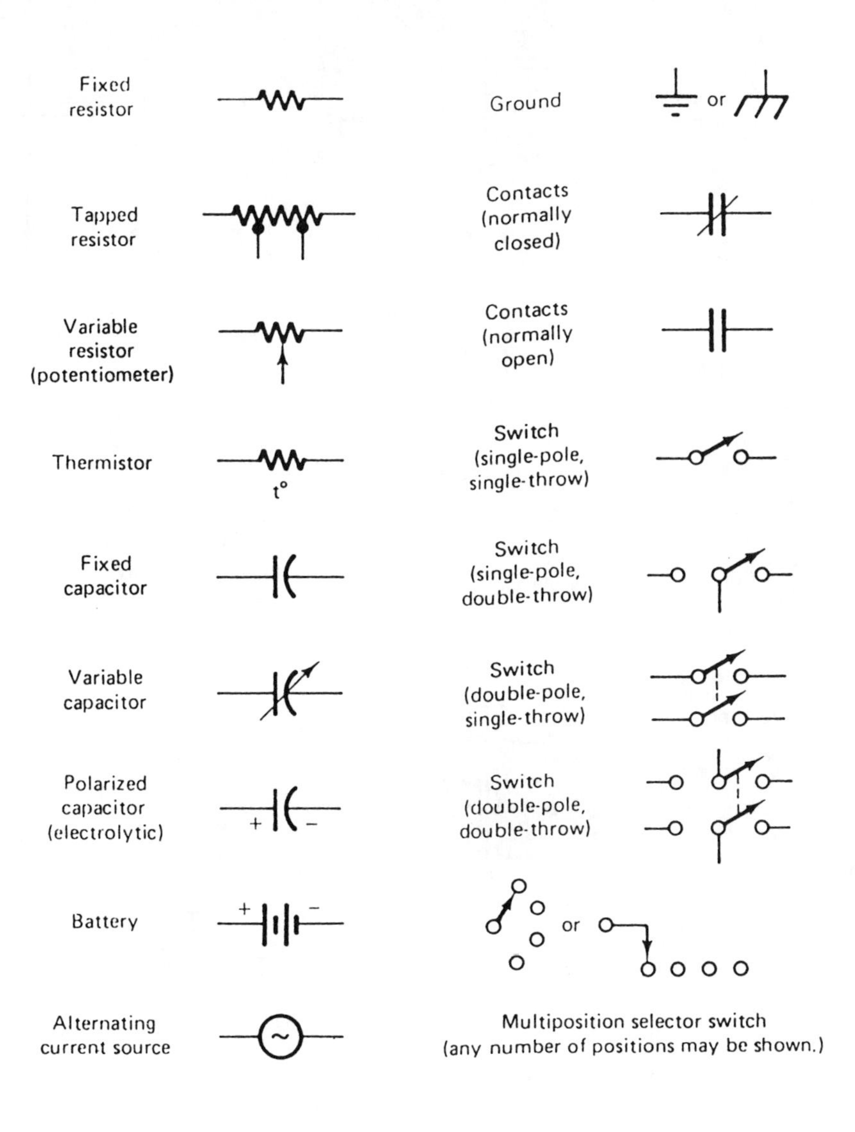

Fuse		Thermocouple	
Circuit breaker (single pole)		Thermal cutout device	
Circuit breaker (three pole)	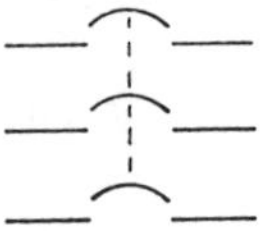	Wires crossing; not connected	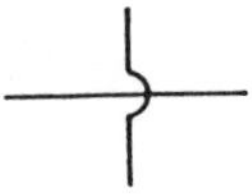
Coil (air core)		Wires connected	
Coil (iron core)		Female connector	
Coil tapped (air core)		Male connector	
Coil adjustable (air core)	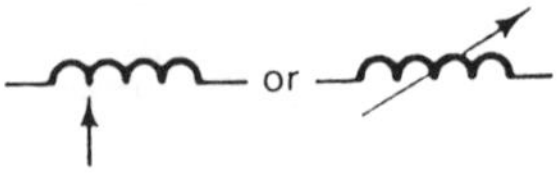	Joined connectors	
Transformer (air core)	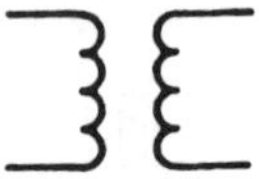	Jack (2-conductor)	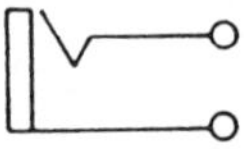
Transformer (iron core)	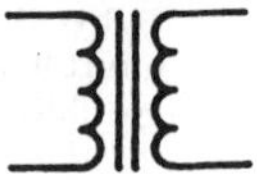	Plug (2-conductor)	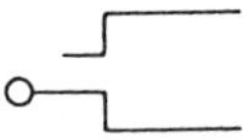

Pushbutton switch (normally open)

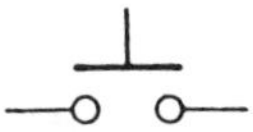

Autotransformer

Pushbutton switch (normally closed)

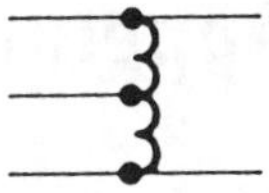

Generator or motor field coil

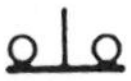

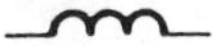

Pushbutton switch (double circuit)

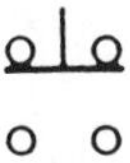

Antenna

or

Limit switch (normally open)

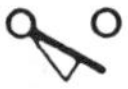

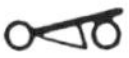

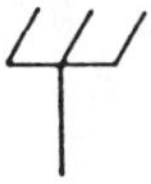

Photovoltaic cell
Solar cell

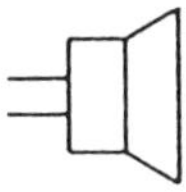

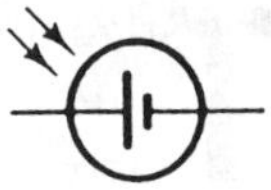

Limit switch (normally closed)

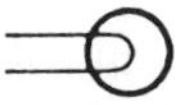

Synchro unit

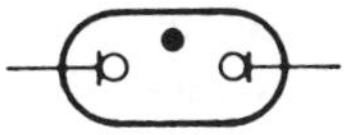

S_1 S_2 S_3 R_1 R_2

Electrical bell

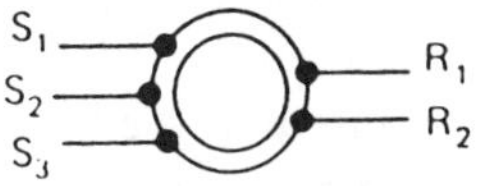

Meter

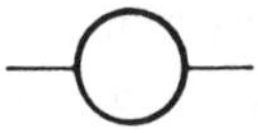

Replace with letter(s) designating type: V, A, MA, μA, W, etc.

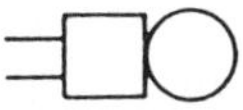

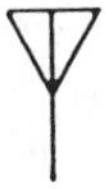

Loudspeaker

Microphone

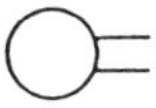

or

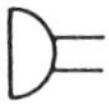

Generator

G or GEN

Incandescent lamp

Motor

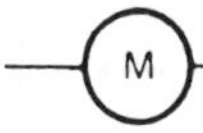

M or MOT

Fluorescent lamp

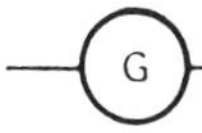

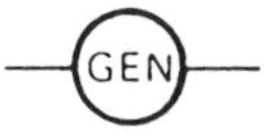

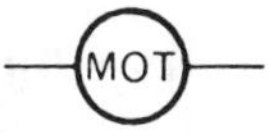

Appendix C

Trigonometry for Electrical Power Systems

Trigonometry is a very valuable mathematical tool for anyone who studies AC circuits. Trigonometry deals with angles and triangles, particularly the right triangle, which has one angle of 90°. An electronic example of a right triangle is shown in Figure C-1. This example illustrates how resistance, reactance, and impedance are related in AC circuits. We know that resistance (R) and reactance (X) are 90° apart, so their angle of intersection forms a right angle. We can use the law of right triangles, known as the *Pythagorean tizeorem,* to solve for the value of any side. This theorem states that in any right triangle, the square of the hypotenuse is equal to the sum of the squares of the other two sides. With reference to Figure C-1, we can express the Pythagorean theorem mathematically as:

$$Z^2 = R^2 + X^2$$

or

$$Z = \sqrt{R^2 + X^2}$$

By using trigonometric relationships, we can solve problems dealing with phase angles, power factor, and reactive power in AC circuits. The three most used trigonometric functions are the *sine,* the *cosine,* and the *tangent.* Figure C-2 illustrates how these functions are expressed mathematically. Their values can be found easily by using a calculator.

This process can be reversed to find the size of an angle when the ratios of the sides are known. The term *inverse function is* used to indicate this process. For example, the notation inv sin $x = \theta$ means that θ is the angle whose sine is x. Thus, inv sin 0.9455 = θ, easily solved with a calculator (θ = 71°).

Trigonometric ratios hold true for angles of any size; however,

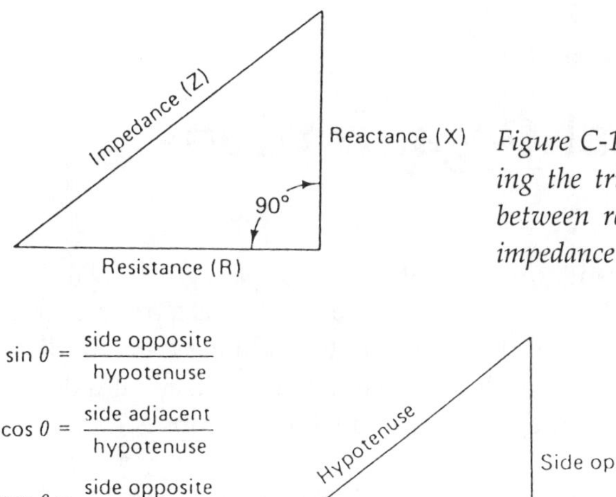

Figure C-1. Right triangle illustrating the trigonometric relationships between resistance, reactance, and impedance in AC circuits.

Figure C-2. Illustration of the trignometric relationships of the sides of a right triangle to the angel θ.

angles in the first quadrant of a standard graph (0° to 90°) are used as a reference, and in order to solve for angles greater than 90° (second-, third-, and fourth-quadrant angles), they must first be expressed as equivalent first-quadrant angles (see Figure C-3). All first-quadrant angles have positive functions, whereas angles in the second, third, and fourth quadrants have two negative functions and one positive function.

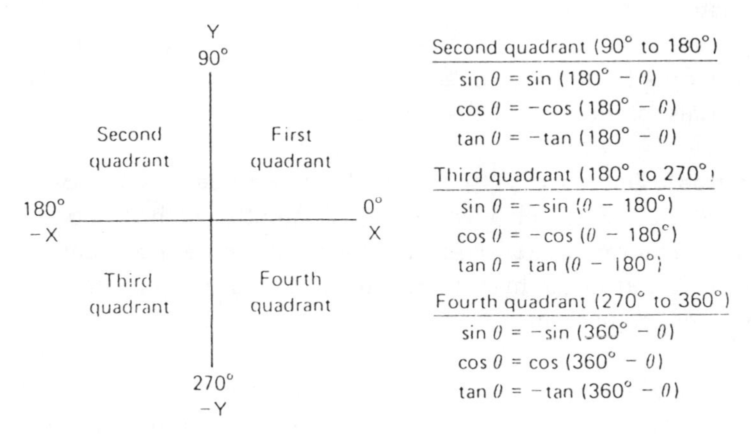

Figure C-3. First-, third- and fourth-quadrant angles.

Appendix D

Professional Organizations

American National Standards Institute (ANSI), New York, NY, 212-642-4900; Fax: 212-398-0023: Web: www.ansi.org/ANSI is a private, non-profit administrator and coordinator of the US voluntary standardization system. ANSI does not develop standards rather, it facilitates development by establishing consensus among qualified groups.

American Society for Quality (ASQ), Milwaukee, WI, 414-272-8575; Web: www.asqc.org. The ASQ facilitates continuous improvement and increased customer satisfaction by promoting the use of quality principles, concepts, and technologies.

American Society of Test Engineers (ASTE), Chariton City, MA, E-mail: mkeller@s1.drc.com; Web: www.astetest.org. The ASTE is dedicated to promoting test engineering as a profession.

American Society for Testing and Materials (ASTM), West Conshohocken, PA, 610-832-9655; Fax: 610-832-9555; E mail: service@astm.org; Web: www.astm.org. ASTM is an international society that works to develop high-quality, voluntary technical standards for materials, products, systems, and services.

Electronic Industries Association (EIA), Arlington, VA, 703-907-7500; Fax: 703-907-7794; Web: www.eia.org. Committed to the competitiveness of the American producer, EIA represents the entire spectrum of companies involved in the manufacture of electronic components, parts, systems, and equipment.

Institute of Electrical and Electronics Engineers (IEEE), Piscataway, NJ, 908-981-0060; Web: www.ieee.org/. The IEEE is a technical professional society that advances the theory and practice of electrical, electronics, and computer engineering as well as computer science.

Institute of Environmental Sciences (IES), Mount Prospect, IL, 847-255-1561: Fax: 847-255-1699; E-mail: instenvsci@aol.com; Web: instenvsci.org. The IES is an international professional society that serves members and the industries they represent through education and the development of recommended practices and standards.

International Organization for Standardization (ISO), Geneva, Switzerland, +41-22-749-01-11; Fax: +41-22-733-34-30; E-mail: central@iso.ch; Web: www.iso.ch/. The ISO promotes the development of standardization to facilitate the international exchange of goods and services and to develop cooperation in intellectual, scientific, technological, and economic activity. (The scope of ISO does not cover electrical and electronic engineering).

The International Society for Measurement & Control, Research Triangle Park, NC, 919-549-8411; Fax: 919-549-8288; E-mail: info@isa.org; Web: www.isa org/. The ISA is a society of more than 49,000 professionals involved in instrumentation measurement, and control. The ISA conducts training programs, publishes literature, and organizes an annual conference and exhibition of instrumentation and control.

Japan Electric Measuring Instruments Manufacturers' Association (JEMIMA), Tokyo, Japan, +81-3-3502-0601; Fax: +81-3-3502-0600. JEMIMA is a nonprofit industrial organization authorized by the Japanese government. JEMIMA is devoted to a variety of activities—cooperation with the government, providing statistics about the electronics industry, and sponsoring exhibitions.

National Institute of Standards and Technology (NIST), Gaithersburg, MD, 301-975-3058; Fax: 301-926-1630; E-mail: inquiries@nist.gov; Web: www.nist.gov/. As a non-regulatory agency of the Commerce Department's Technology Administration, NIST promotes U.S. economic growth by working with industry to develop and apply technology, measurements, and standards.

National Society of Professional Engineers (NSPE), Alexandria, VA, 703-684-2800; Fax: 703-836-4875; E-mail: customer.service@nspe.org; Web: www.nspe.org/. The NSPE is an interdisciplinary professional engineering society representing more than 70,000 engineers.

Society of Automotive Engineers (SAE), Warrendale, PA, 412-776-4841- Fax: 412-776-5760; E-mail: sae@sae.org; Web: www.sae.org. The SAE produces technical publications, conducts numerous meetings, seminars and educational activities, and fosters information exchange among the worldwide automotive and aerospace communities.

Underwriters Laboratories (UL), Northbrook, IL, 847-272-8800; Web: web138.bbnplanet com/. The UL is an independent, nonprofit certification organization that has evaluated products in the interest of public safety for 100 years.

Index